MAMMALIAN PROTEASES:
A Glossary and Bibliography

MAMMALIAN PROTEASES:
A Glossary and Bibliography

Volume I
Endopeptidases

A. J. BARRETT
Strangeways Research Laboratory
Worts' Causeway, Cambridge, U.K.

J. Ken McDONALD
Department of Biochemistry
Medical University of South Carolina
Charleston, South Carolina, U.S.A.

1980

ACADEMIC PRESS

A Subsidiary of Harcourt Brace Jovanovich, Publishers
London New York Toronto Sydney San Francisco

ACADEMIC PRESS INC. (LONDON) LTD.
24/28 Oval Road,
London NW1

United States Edition published by
ACADEMIC PRESS INC.
111 Fifth Avenue
New York, New York 10003

British Library Cataloguing in Publication Data
Mammalian proteases
Vol. 1: Endopeptidases
1. Proteinase – Handbooks, manuals, etc.
2. Proteinase – Bibliography
I. Barrett, A J   II. McDonald, J K
599'.01'9256   QP609.P75   79-41797
ISBN 0-12-079501-9

Printed in Great Britain by
Whitstable Litho Ltd., Whitstable, Kent

PREFACE

The field of proteolytic enzymes (proteases) is currently the scene of vigorous research in many countries. The main stimulus for this is the growing body of evidence that the enzyme systems responsible for protein breakdown are vitally important for the normal health and development of living organisms. The research is being carried out by scientists whose primary expertise ranges from chemistry through biochemistry and physiology to pharmacology and medicine.

As is inevitably the case, the most rapid accumulation of thoroughly tested and generally accepted data is occurring at the more fundamental levels in this area of knowledge. Simultaneously, the development of a relatively firm foundation of chemical and biochemical information is leading to rapidly accellerating progress at the more complex levels of research. An excellent example of this is the way in which the work of chemists and emzymologists has led to the discovery of new classes of selective inhibitors for individual proteolytic enzymes. These agents are being used to investigate the roles of the enzymes in normal physiology, and tested for usefulness as drugs in the control of disordered protein metabolism.

Knowledge of the proteases has now gone far beyond the stage at which an individual scientist can expect to have read most of the original papers that may be relevant to his work, and reviews are clearly essential. It has been our feeling, however, that review articles and books written and printed in conventional ways are increasingly inadequate to the needs of the research worker. We have therefore developed a computer-aided system for the codification and presentation of information, of which the present volume is the first product. Our object has been to provide a concise summary of the properties of each enzyme, together with a bibliography that includes titles, and is sufficiently complete to allow the scientist to find whatever further information he may need with the minimum of difficulty. The data are stored in a computer-accessible form, and we plan to keep the text under constant revision so that further editions can be produced.

Protein breakdown is generally initiated by endopeptidases (proteinases), which hydrolyze internal peptide bonds within the long polypeptide chains, and is completed by exopeptidases (peptidases), which remove amino acid residues one, two or three at a time from the ends of the fragments. For this and other reasons, it seemed appropriate that volume 1 of the Glossary and Bibliography should deal with the endopeptidases. The exopeptidases will be covered by volume 2, and there are plans for a third volume on proteinase inhibitors.

The scope of volume 1 is extremely broad: we are not aware of any previous publication which has attempted so comprehensive a review of the mammalian endopeptidases. This breadth of approach facilitates comparison of, say, a proteinase of the blood coagulation system with a cellular enzyme or an alimentary one. We hope that some of these comparisons will prove interesting and useful in both basic and applied studies concerning the mammalian proteases.

A.J.B.

J.K.McD.

ACKNOWLEDGEMENTS

We are grateful for valuable help and advice received from many
friends and colleagues, including S. Ansorge, A. Baici, U. Bretz,
A. F. Bury, E. W. Davies, W. C. Duckworth, F. Fiedler, B. Foltmann,
H. Fritz, M. Hobart, D. A. W. Grant, L. M. Greenbaum, W. Hornebeck,
T. Inagami, C. M. Jackson, A. J. Kenny, H. Kirschke, J. Langner,
N. Marks, D. B. Morton, W. Müller-Esterl, G. Murphy, H. Nagase,
M. Orlowski, E. Phillip, D. B. Rifkin, A. Ryle, G. S. Salvesen,
W. N. Schwartz, E. Shaw, E. E. Slater, P. M. Starkey, C. Szego,
W. H. Taylor, R. Walter, Z. Werb, S. Wilk and J. F. Woessner, Jr.
  Nevertheless, it must be said that we have not been able to act
on all the advice, and all errors and omissions are solely the
authors' responsibility! We shall be very pleased to have these
pointed out to us with a view to rectification in possible further
editions.
  We thank Mrs. Eveline Broad, Mrs. Ann Hall and Miss Nancy
Watkins for their skilled secretarial and library assistance, and
are most grateful to Mrs. Judy Sibley of Academic Press for her
advice on the format of the text.

A.J.B.

J.K.McD.

CONTENTS

Section 2
CYSTEINE PROTEINASES

CONTENTS OF VOLUME 2
(In preparation)

INTRODUCTION

The present volume has been compiled as part of an effort to provide a convenient source of information on the mammalian proteolytic enzymes and their inhibitors. This first volume deals with the endopeptidases, and the second volume covering the exopeptidases is in preparation. The text is being assembled in Cambridge by a computer-assisted technique, and our intention is to continue to up-date and revise the information after this first publication.

The organization of Volume 1 is simple. The endopeptidases are divided into five groups, the first four being based on catalytic mechanism. In accordance with our previous suggestion (Barrett, 1980), the terms "cysteine proteinases" and "aspartic proteinases" are used for the groups which have previously been called "thiol proteinases" and "acid" or "carboxyl proteinases", respectively. The fifth group contains proteinases that cannot yet be positively categorized.

The order of entries within each main section is somewhat arbitrary, but we have tried to group together enzymes of distinct physiological systems, such as blood coagulation or complement. When a sequence of activity in such a "cascade" as the coagulation system is recognized, we have tended to use this as a basis for the sequence of entries, but in other cases a simple alphabetical or numerical arrangement has been followed.

Each entry starts with a short summary of the properties of the enzyme under a dozen or so standard headings. The name adopted for each enzyme has been our own choice, but generally reflects current usage. In a few instances we have felt that there are strong reasons to advocate a new name. The name is followed by the "EC" number of the enzyme (International Union of Biochemistry, 1978) if one has been allocated, or a partial number in parentheses. The section "Earlier names" sometimes includes alternative names still current, and "Distribution" is seldom known with any completeness. "Source" (for purification), like most of the subsequent headings, is self-explanatory. Under "Comment" we have felt free to express opinions which may sometimes be controversial, such as the possible identity of the enzyme with others listed elsewhere. In this connection, we should emphasize that the present volume in no way represents an "official" catalogue of proteinases in the way that the IUB (1978) list (groups 3.4.21-99) does – indeed, some of the entries have been created largely to provide us with an opportunity to point out that there are grounds for serious doubt as to whether the enzyme truly exists in its own right!

The bibliography section of each entry is arranged chronologically, and alphabetically within each year. (Occasionally, the strict alphanumerical ordering, as done by the computer, may seem slightly odd, but we doubt that this will inconvenience the reader appreciably.) For enzymes that have been the subject of only a few papers we feel that the bibliographies are often nearly complete. For the well-known enzymes we have had to be selective, but we have tried to give enough references to allow the reader access to the full literature of any mammalian proteinase. Our compilation of references for this volume was completed early in 1980, but we anticipate that these references will prove useful as starting points in searches of the available citation indexes for subsequent publications, too.

References

Barrett, A. J. Introduction: The classification of proteinases. In: Protein Breakdown in Health and Disease (Ciba Foundation Symposium 75), (Evered, D. & Whelan, J. eds), pp. 1-13, Excerpta Medica, Amsterdam.

International Union of Biochemistry. Enzyme Nomenclature 1978. Academic Press, New York, 1979.

ABBREVIATIONS AND CONVENTIONS

Generally, the names of amino-acids and their derivatives have
been abbreviated according to common practice (IUPAC-IUB
Commission on Biochemical Nomenclature, 1972). Amino-acids may
be assumed to be in the L configuration unless D or DL is
specified.

Less common amino-acid residues

Aala	Aza-alanine (the analogue of alanine in which the α-carbon has been replaced by a nitrogen atom)
Ahx	6-Aminohexanoic acid
Aval	Aza-valine
Gla	γ-Carboxyglutamic acid
γ-Glu	γ(Linked)-glutamic acid
PCA	Pyrrolid-2-one-5-carboxylic acid
Pip	Pipecolic acid
$Tyr(I_2)$	2,4-Diidotyrosine

Substituents at the amino group

Ac-	Acetyl-
Boc-	t-Butyloxycarbonyl-
Bz-	Benzoyl-
Glt-	Glutaryl- (i.e. 4-carboxybutyryl-)
HCO-	Formyl-
MeOSuc-	Methoxysuccinyl- (i.e. 3-methyloxycarbonyl-propionyl-)
Pz-	Phenylazobenzyloxycarbonyl-
Suc-	Succinyl- (i.e. 3-carboxypropionyl-)
Tos-	Tosyl- (i.e. p-toluylsulfonyl-)
Z-	Benzyloxycarbonyl-

Substituents at the carboxyl group

$-CH_2Br$	-Bromomethane (note 1)
$-CH_2Cl$	-Chloromethane (note 1)
$-CHN_2$	-Diazomethane (note 1)
-H	Aldehyde
$-NH_2$	Amide

-NMec	4-Methyl-7-coumarylamide (note 2)
-2-NNap	2-Naphthylamide
-NNapOMe	4-Methoxy-2-naphthylamide
-NPhNO$_2$	4-Nitroanilide
-OEt	Ethyl ester
-OMe	Methyl ester
-OMec	4-Methylumbelliferyl ester (i.e. 4-methyl-7-coumaryl ester) (note 2)
-1-ONap	1-Naphthyl ester
-2-ONap	2-Napthyl ester
-ONapAS	Naphthol AS ester (i.e. 3-[carboxamidophenyl]-2-naphthyl ester)
-ONapAS-D	Naphthol AS-D ester (i.e. 3-[carboxamido-o-toluyl]--2-naphthyl ester)
-OPhNO$_2$	4-Nitrophenyl ester
-SBzl	Thiobenzyl ester

Notes

1. We feel that the terminology of the "chloromethylketones", "diazomethylketones" and related compounds is simplified by treating them as amino-acyl or peptidyl derivatives of chloromethane, diazomethane, etc. Accordingly, we use hyphens in their abbreviations.

2. The abbreviations we have used for the esters of β-methyl-umbelliferone (7-hydroxy-4-methylcoumarin) and the amino-derivatives of 4-methyl-7-aminocoumarin, -OMec and -NMec, are intended to emphasize the structural analogy between these groups of fluorigenic compounds. Thus, by analogy with -ONap and -NNap for the naphthol esters and naphthylamides, respectively, and -OPhNO$_2$ and -NPhNO$_2$ for the nitrophenyl esters and anilides, these compounds can be thought of as being derivatives of 4-methylcoumarin (Mec) linked through an oxygen or a nitrogen atom at the 7 position.

Other abbreviations

Dip-	Di-isopropylphosphoryl-
Dip-F	Di-isopropyl fluorophosphate
Gbz-OMec	4-Guanidinobenzoic acid 4-methylumbelliferyl ester (or MUGB, methylumbelliferone guanidino-benzoate)
Gbz-OPhNO$_2$	4-Guanidinobenzoic acid 4-nitrophenylester (or NPGB, nitrophenol guanidinobenzoate)
IAcNH$_2$	Iodoacetamide
IAcOH	Iodoacetic acid

MalNEt   N-Ethylmaleimide
Nbs$_2$ 5,5-Dithiobis(2-nitrobenzoic acid) (Ellman's
                reagent)
Pms-F   Phenylmethanesulfonyl fluoride
SDS   Sodium dodecyl sulfate

Conventions

Antipain, chymostatin, elastatinal, leupeptin, pepstatin,
phosphoramidon.
  Microbial proteinase inhibitors (Umezawa & Aoyagi, 1977).
Aprotinin.
  Bovine trypsin/kallikrein inhibitor, Kunitz pancreatic trypsin
  inhibitor, etc. marketted as Trasylol (the registered tradename
  of Bayer AG, Leverkusen, Federal Republic of Germany).
Complement component fragments.
  We have attempted to follow the convention by which the "a"
  fragment is that of lower molecular weight, and "b" the larger
  fragment, for all components, including C2 (which has often been
  treated differently).
Polyanetholesulfonate.
  "Liquoid", a synthetic heparin-like substance.
S-2302, etc., and Chromozym PK, etc.
  The S- and Chromozym designations for peptide nitroanilide
  substrates of serine proteinases are trade names used by Kabi
  Diagnostica, S-11287, Stockholm, Sweden, and Pentapharm Ltd.,
  CH-4002, Basle, Switzerland, respectively.

References

IUPAC-IUB Commission on Biochemical Nomenclature. Symbols for
  amino-acid derivatives and peptides. Recommendations 1971. J.
  Biol. Chem. 247: 977-983, 1972.
Umezawa, H. & Aoyagi, T. Activities of proteinase inhibitors of
  microbial origin. In: Proteinases in Mammalian Cells and Tissues
  (Barrett, A. J. ed.), pp. 637-662, North-Holland Publishing Co.,
  Amsterdam, 1977.

Section 1

SERINE PROTEINASES

Entry 1.01

ENTEROPEPTIDASE

Summary

EC Number: 3.4.21.9

Earlier names: Enterokinase.

Distribution: Duodenal brush border membranes and intestinal
contents of man, pig, etc.

Source: As above.

Action: Highly specific for activating trypsinogen by cleavage: Val-
(Asp)$_4$-Lys$\neq$Ile-; also cleaves peptide analogues of the
activation sequence, and may show some activity against
synthetic substrates of trypsin.

Requirements: pH optimum 8.4. Ca^{2+} (10 mM) stimulates.

Substrate, usual: Trypsinogen (activation being followed with
Z-Gly-Gly-Arg-NNap, or other substrates of trypsin).

Substrate, special: Gly-(Asp)$_4$-Lys-NNap. Gbz-OPhNO$_2$ for titration.

Inhibitors: Tos-Lys-CH$_2$Cl, Dip-F, 4-aminobenzamidine (K$_i$ 10^{-5} M).
(Not α_2-macroglobulin, α_1-proteinase inhibitor, CĪ inhibitor or
other plasma proteins, soybean trypsin inhibitor or aprotinin).

Molecular properties: Glycoprotein containing up to 57%
carbohydrate (but not bound by concanavalin A-Sepharose). Bovine
enzyme reported to be of mol.wt. 145,000, with disulfide-bonded
chains of 57,000 and 82,000. Catalytic His and Ser residues are
on the light chain of the porcine enzyme. Human enzyme may be
300,000 mol.wt., and composed of three chains, the smallest
(53,000) containing the catalytic Ser. pI 4.8.

Comment: Physiological role is initiation of activation of
pancreatic zymogens. Bovine and porcine enzymes have specificity
for the -(Asp)$_4$-Lys- sequence in the activation peptide; human
enteropeptidase (and trypsinogen) may be different [1]. La^{2+}
(10^{-5} M) blocks activation of bovine trypsinogen by human
enteropeptidase, possibly by binding to the -(Asp)$_4$- sequence.

References
[1] Brodrick et al. J. Biol. Chem. 253: 2737-2742, 1978.

Bibliography

1924

Waldschmidt-Leitz, E. Uber Enterokinase und die tryptische
    Wirkung der Pankreasdrüse. Hoppe-Seyler's Z. Physiol. Chem.
    132: 181-237.

1939

Kunitz, M. Formation of trypsin from crystalline trypsinogen by
    means of enterokinase. J. Gen. Physiol. 22: 429-446.
Kunitz, M. Purification and concentration of enterokinase. J.
    Gen. Physiol. 22: 447-450.

1954

Yamashina, I. Studies on enterokinase. I. The purification and
    general properties. Ark. Kemi 7: 539-543.
Yamashina, I. Studies on enterokinase. II. The further
    purification. Ark. Kemi 9: 225-229.

1955

Laskowski, M. Trypsinogen and trypsin. Methods Enzymol. 2: 26-
    36.

1969

Geratz, J. D. Inhibitory effect of aromatic diamidines on trypsin
    and enterokinase. Experientia 25: 1254-1255.
Hadorn, B., Tarlow, M. J., Lloyd, J. K. & Wolff, O. H. Intestinal
    enterokinase deficiency. Lancet 1: 812-813.

1970

Nordström, C. & Dahlqvist, A. The cellular localization of
    enterokinase. Biochim. Biophys. Acta 198: 621-622.
Tarlow, M. J., Hadorn, B., Arthurton, M. W. & Lloyd, J. K.
    Intestinal enterokinase deficiency. A newly-recognized disorder
    of protein digestion. Arch. Dis. Child. 45: 651-655.

1971

Maroux, S., Baratti, J. & Desneulle, P. Purification and
    specificity of porcine enterokinase. J. Biol. Chem. 246: 5031-
    5039.
Nordström, C. Enzymic release of enteropeptidase from isolated
    rat duodenal brush borders. Biochim. Biophys. Acta 268: 711-
    718.

1973

Baratti, J., Maroux, S. & Louvard, D. Effect of ionic strength
    and calcium ions on the activation of trypsinogen by
    enterokinase. Biochim. Biophys. Acta 321: 632-638.
Baratti, J., Maroux, S., Louvard, D. & Desnuelle, P. On porcine
    enterokinase. Further purification and some molecular properties.
    Biochim. Biophys. Acta 315: 147-161.
Louvard, D., Maroux, S., Baratti, J. & Desnuelle, P. On the

distribution of enterokinase in porcine intestine and on its subcellular localization. Biochim. Biophys. Acta 309: 127-137.

1974

Barns, R. J. & Elmslie, R. G. The active site of porcine enteropeptidase. Selective inactivation of the peptidase activity. Biochim. Biophys. Acta 350: 495-498.

Rinderknecht, H., Engeling, E. R., Bunnell, M. J. & Geokas, M. C. A sensitive assay for human enterokinase and some properties of the enzyme. Clin. Chim. Acta 54: 145-160.

Schmitz, J., Preiser, H., Maestracci, D., Crane, R. K., Troesch, V. & Hadorn, B. Subcellular localization of enterokinase in human small intestine. Biochim. Biophys. Acta 343: 435-439.

1975

Grant, D. A. W. & Hermon-Taylor, J. Purification of porcine enterokinase by affinity chromatography. Biochem. J. 147: 363-366.

Preiser, H., Schmitz, J., Maestracci, D. & Crane, R. K. Modification of an assay for trypsin and its application for the estimation of enteropeptidase. Clin. Chim. Acta 59: 169-175.

1976

Baratti, J. & Maroux, S. On the catalytic and binding sites of porcine enteropeptidase. Biochim. Biophys. Acta 452: 488-496.

Barns, R. J. & Elmslie, R. G. Preparation of porcine enteropeptidase free of intestinal aminopeptidase activity with Con A-Sepharose. Biochim. Biophys. Acta 445: 815-818.

Grant, D. A. W. & Hermon-Taylor, J. The purification of human enterokinase by affinity chromatography and immunoadsorption. Some observations on its molecular characteristics and comparisons with the pig enzyme. Biochem. J. 155: 243-254.

Hesford, F., Hadorn, B., Blaser, K. & Schneider, C. H. A new substrate for enteropeptidase. FEBS Lett. 71: 279-282.

Rinderknecht, H. & Friedman, R. M. The effect of lanthanide ions on enteropeptidase-catalyzed activation of trypsinogen. Biochim. Biophys. Acta 452: 497-502.

1977

Anderson, L. E., Walsh, K. A. & Neurath, H. Bovine enterokinase. Purification, specificity, and some molecular properties. Biochemistry 16: 3354-3360.

Antonowicz, I. & Lebenthal, E. Developmental pattern of small intestinal enterokinase and disaccharidase activities in the human fetus. Gastroenterology 72: 1299-1303.

Barns, R. J. & Elmslie, R. G. Identification of a mucosal form of enteropeptidase in Triton X-100 extracts of porcine duodenal mucosa. Biochim. Biophys. Acta 480: 450-460.

Hermon-Taylor, J., Perrin, J., Grant, D. A. W., Appleyard, A., Bubel, M. & Magee, A. I. Immunofluorescent localisation of enterokinase in human small intestine. Gut 18: 259-265.

Lebenthal, E. & Morrissey, G. W.   Subcellular localization of
enterokinase (enteropeptidase EC 3.4.21.9) in rat small
intestine. Biochim. Biophys. Acta 497: 558-566.
Lobley, R. W., Franks, R. & Holmes, R.   Subcellular localization
of enterokinase in human duodenal mucosa.   Clin. Sci. Mol. Med.
53: 551-562.
Magee, A. I., Grant, D. A. W. & Hermon-Taylor, J.   The apparent
molecular weights of human intestinal aminopeptidase,
enterokinase and maltase in native duodenal fluid. Biochem. J.
165: 583-585.

                              1978

Brodrick, J. W., Largman, C., Hsiang, M. W., Johnson, J. H. &
Geokas, M. C.   Structural basis for the specific activation of
human cationic trypsinogen by enteropeptidase. J. Biol. Chem.
253: 2737-2742.
Grant, D. A. W., Magee, A. I. & Hermon-Taylor, J.   Optimisation of
conditions for the affinity chromatography of human enterokinase
on immobilized p-aminobenzamidine. Eur. J. Biochem. 88: 183-
189.
Rinderknecht, H., Nagaraja, M. R. & Adham, N. F.   Enteropeptidase
levels in duodenal juice of normal subjects and patients with
gastrointestinal disease. Am. J. Digest. Dis. 23: 327-331.

                              1979

Grant, D. A. W. & Hermon-Taylor, J.   Hydrolysis of artificial
substrates by enterokinase and trypsin and the development of a
sensitive specific assay for enterokinase in serum. Biochim.
Biophys. Acta 567: 207-215.
Liepnieks, J. J. & Light, A.   The preparation and properties of
bovine enterokinase. J. Biol. Chem. 254: 1677-1683.

Entry 1.02

TRYPSIN

Summary

EC Number: 3.4.21.4

Earlier names: The precursor zymogen is trypsinogen.

Distribution: Trypsinogen is synthesized in the β-cells of the pancreas of mammals (and many other animals), and activated in the intestine.

Source: Pancreas of man and other animals, with or without activation of the zymogen.

Action: Very selective for the cleavage of -Arg-X- or -Lys-X- bonds in polypeptides, and esters and amides of the basic amino acids in synthetic substrates.

Requirements: pH optimum about 8.0. Ca^{2+} (20mM) stabilizes against autolytic inactivation.

Substrate, usual: Tos-Arg-OMe, Bz-Arg-OEt, Bz-DL-Arg-NPhNO$_2$.

Substrate, special: Z-Gly-Gly-Arg-NNap, Z-Phe-Val-Arg-NPhNO$_2$, Z-Val-Gly-Arg-NPhNO$_2$ (Chromozym TRY) or Z-Arg-NMec for maximal sensitivity. Gbz-OPhNO$_2$ or Gbz-OMec for active site titration.

Inhibitors: Dip-F, Pms-F, Tos-Lys-CH$_2$Cl, 4-aminobenzamidine (K_i 8.2×10^{-6} M [1]), Gbz-OPhNO$_2$, leupeptin. Soybean trypsin inhibitor, aprotinin, ovomucoid, α_1-proteinase inhibitor and many others.

Molecular properties: Mol.wt. 23,500. A280,1% 15.4. Trypsinogen is activated to β-trypsin by cleavage of the -Lys$_6$-Ile$_7$- bond, by enteropeptidase or trypsin; α-trypsin is formed by an additional autolytic cleavage, and ψ-trypsin by a second. The major form is a very basic protein in most species, e.g. pI 10.8 (lower for the zymogens); an anionic form also occurs which has an identical sequence at the N-terminus, but many substitutions in the rest of the molecule.

References
[1] Mares-Guia & Shaw J. Biol. Chem. 240: 1579-1585, 1965.

Bibliography

1939

Hofmann, K. & Bergmann, M.   The specificity of trypsin. II.   J. Biol. Chem. 130: 81–86.

Kunitz, M.   Formation of trypsin from crystalline trypsinogen by means of enterokinase. J. Gen. Physiol. 22: 429–446.

1946

McDonald, M. R. & Kunitz, M.   An improved method for the crystallization of trypsin.   J. Gen. Physiol. 29: 155–156.

1951

Kokas, E., Foldes, I. & Banga, I.   The elastase and trypsin contents of the pancreatic secretions in dogs.   Acta Physiol. Acad. Sci. Hung. 2: 333–341.

1955

Laskowski, M.   Trypsinogen and trypsin.   Methods Enzymol. 2: 26–36.

Schwert, G. W. & Takenaka, Y.   A spectrophotometic determination of trypsin and chymotrypsin.   Biochim. Biophys. Acta 16: 570–575.

1959

Hummel, B. C. W.   A modified spectrophotometric determination of chymotrypsin, trypsin, and thrombin.   Can. J. Biochem. 37: 1393–1399.

Riedel, A., Wüensch, E. & Hartl, A.   Nα–Benzoyl-arginin-β–naphthylamid als Substrat zur quantitaven kolorimetrischen Bestimmung von Trypsin. Z. Physiol. Chem. 316: 61–70.

Ronwin, E.   The active center of thrombin and trypsin.   Biochim. Biophys. Acta 33: 326–332.

1960

Hartley, B. S.   Proteolytic enzymes.   Annu. Rev. Biochem. 29: 45–72.

Liener, I. E.   Chromatographic studies on trypsin, trypsinogen and the activation process.   Arch. Biochem. Biophys. 88: 216–221.

Ronwin, E.   The TAC method of tryptic enzyme activity determination.   Can. J. Biochem. Physiol. 38: 57–67.

1961

Blackwood, C. & Mandl, I.   An improved test for the quantitative determination of trypsin, trypsin-like enzymes and enzyme inhibitors.   Anal. Biochem. 2: 370–379.

Erlanger, B. F., Kokowsky, N. & Cohen, W.   The preparation and properties of two new chromogenic substrates of trypsin.   Arch. Biochem. Biophys. 95: 271–278.

1962

Merigan, T. C., Dreyer, W. J. & Berger, A.   A technique for the specific cleavage of arginyl bonds by trypsin. Biochim. Biophys. Acta 62: 122-131.

1963

Charles, M., Rovery, M., Guidoni, A. & Desnuelle, P. Sur le trypsinogène et la trypsine de porc. Biochim. Biophys. Acta 69: 115-129.

Fahrney, D. E. & Gold, A. M. Sulfonyl fluorides as inhibitors of esterases. J. Am. Chem. Soc. 85: 997-1000.

Inagami, T. & Murachi, T. Induced activation of the catalytic site of trypsin. J. Biol. Chem. 238: PC1905-PC1907.

Roth, M. Dosage fluorimetrique de la trypsine. Clin. Chim. Acta 8: 574-578.

1964

Glenner, G. G., Hopsu, V. K. & Cohn, L. A. New substrates of trypsin and some trypsin-like enzymes. J. Histochem. Cytochem. 12: 545-551.

Inagami, T. The mechanism of the specificity of trypsin catalysis. I. Inhibition by alkyl ammonium ions. J. Biol. Chem. 239: 787-791.

Nachlas, M. M., Plapinger, R. E. & Seligman, A. M. Role of some structural features of substrates on trypsin activity. Arch. Biochem. Biophys. 108: 266-274.

1965

Mares-Guia, M. & Shaw, E. Studies on the active center of trypsin. The binding of amidines and guanidines as models of the substrate side chain. J. Biol. Chem. 240: 1579-1585.

Plapinger, R. E., Nachlas, M. M., Seligman, M. L. & Seligman, A. M. Synthesis of chromogenic arginine derivatives as substrates for trypsin. J. Org. Chem. 30: 1781-1785.

Shaw, E., Mares-Guia, M. & Cohen, W. Evidence for an active-center histidine in trypsin through use of a specific reagent 1-chloro-3-tosylamido-7-amino-2-heptanone, the chloromethyl ketone derived from Nα-tosyl-L-lysine. Biochemistry 4: 2219-2224.

Stewart, J. A. & Dobson, J. E. Trypsin-catalyzed hydrolysis of N-benzoyl-L-arginine ethyl ester at low pH. Biochemistry 4: 1086-1091.

Travis, J. & Liener, I. E. The crystallization and partial characterization of porcine trypsin. J. Biol. Chem. 240: 1962-1966.

1966

Bender, M. L., Begue-Canton, M. L., Blakely, R. L., Brubacher, L. J., Feder, J., Gunter, C. R., Kezdy, F. J., Killheffer, J. V., Marshall, T. H., Miller, C. G., Roeske, R. W. & Stoops, J. K. The determination of the concentration of hydrolytic enzyme solutions: α-chymotrypsin, trypsin, papain, elastase, subtilisin, and acetylcholinesterase. J. Am. Chem. Soc. 88: 5890-5913.

Bricteux-Gregoire, S., Schyns, R. & Florkin, M. Structure des peptides libérés au cours de l'activation du trypsinogène de mouton. Biochim. Biophys. Acta 127: 277-279.

Ganrot, P. O. On the heterogeneity and purification of commercial trypsin preparations. Acta Chem. Scand. B 20: 175-182.

Johannin, S. & Yon, J. Some evidence for the existence of two non-covalent complexes before the acylation step during the tryptic hydrolysis of benzoyl-arginine para-nitroanilide. Biochem. Biophys. Res. Commun. 25: 320-325.

1967

Chase, T., Jr. & Shaw, E. p-Nitrophenyl-p'-guanidinobenzoate HCl: a new active site titrant for trypsin. Biochem. Biophys. Res. Commun. 29: 508-514.

Elmore, D. T. & Smyth, J. J. α-N-Methyl-α-N-toluene-p-sulphonyl-L-lysine β-naphthyl ester: a new reagent for determining the absolute molarity of solutions of trypsin. Biochem. J. 103: 36P only.

Elmore, D. T. & Smyth, J. J. p-Nitrophenyl N²-acetyl-N¹-benzylcarbazate ⁻(NPABC): a new reagent for determining the absolute molarity of solutions of trypsin and chymotrypsin. Biochem. J. 103: 37P-38P.

Gorecki, M. & Shalitin, Y. Non cationic substrates of trypsin. Biochem. Biophys. Res. Commun. 29: 189-193.

Miller, J. M., Jr., Opher, A. W. & Miller, J. M. The isoenzymes of trypsin. Johns Hopkins Med. J. 121: 328-332.

Scrimager, S. T. & Hofmann, T. The involvement of the aminoterminal amino acid in the activity of pancreatic proteases. II. The effect of nitrous acid on trypsin. J. Biol. Chem. 242: 2528-2533.

1968

East, E. J. & Trowbridge, C. G. Binding of benzamidine and protons to trypsin as measured by difference spectra. Arch. Biochem. Biophys. 125: 334-343.

Inagami, T. & York, S. S. The effect of alkylguanidines and alkylamines on trypsin catalysis. Biochemistry 7: 4045-4052.

Kramer, S. P., Plapinger, R. E., Bharadwaj, P. D., Patt, H. H. & Seligman, A. M. Assay of trypsin activity of human serum with a chromogenic substrate, CBZ-diglycyl-L-arginyl-2-naphthylamide hydrochloride (GGANA). J. Surg. Res. 8: 253.

Markwardt, F., Landmann, H. & Walsmann, P. Comparative studies on the inhibition of trypsin, plasmin and thrombin by derivatives of benzylamine and benzamidine. Eur. J. Biochem. 6: 502-506.

Schroeder, D. D. & Shaw, E. Chromatography of trypsin and its derivatives. Characterisation of a new active form of bovine trypsin. J. Biol. Chem. 243: 2943-2949.

Tanizawa, K., Ishii, S. & Kanaoka, Y. Trypsin substrates derived from benzamidine: application to active site titration and isolation of acyl-enzyme intermediate. Biochem. Biophys. Res. Commun. 32: 893-897.

Travis, J. Studies on the active site of sheep trypsin. Biochem.

Biophys. Res. Commun. 30: 730-734.

1969

Abita, J. P., Delaage, M., Lazdunski, M. & Savrada, J. The mechanism of activation of trypsinogen: the role of the four N-terminal aspartyl residues. Eur. J. Biochem. 8: 314-324.

Arnon, R. & Neurath, H. An immunological approach to the study of evolution of trypsins. Proc. Natl. Acad. Sci. U.S.A. 64: 1323-1330.

Chase, T. & Shaw, E. Comparison of the esterase activities of trypsin, plasmin, and thrombin on guanidinobenzoate esters. Titration of the enzymes. Biochemistry 8: 2212-2224.

Feeney, R. E., Means, G. E. & Bigler, J. C. Inhibition of human trypsin, plasmin, and thrombin by naturally occurring inhibitors of proteolytic enzymes. J. Biol. Chem. 244: 1957-1960.

Geratz, J. D. Inhibitory effect of aromatic diamidines on trypsin and enterokinase. Experientia 25: 1254-1255.

Smith, R. L. & Shaw, E. Pseudotrypsin. A modified bovine trypsin produced by limited autodigestion. J. Biol. Chem. 244: 4704-4712.

Travis, J. & Roberts, R. C. Human trypsin: isolation and physical-chemical characterization. Biochemistry 8: 2884-2889.

1970

Feinstein, G. Purification of trypsin by affinity chromatography on ovomucoid-Sepharose resin. FEBS Lett. 7: 353-355.

Hartley, B. S. Homologies in serine proteinases. Philos. Trans. Roy. Soc. London Ser. B 257: 77-87.

Jacquot-Armand, Y. & Hill, M. Activité trypsique a l'egard de certains substrats chromogenes. FEBS Lett. 11: 249-253.

Rinderknecht, H., Silverman, P., Geokas, M. C. & Haverback, B. J. Determination of trypsin and chymotrypsin with Remazolbrilliant Blue-hide. Clin. Chim. Acta 28: 239-246.

Shaw, E. & Glover, G. Further observations on substrate-derived chloromethyl ketones that inactivate trypsin. Arch. Biochem. Biophys. 139: 298-305.

Walsh, K. A. Trypsinogens and trypsins of various species. Methods Enzymol. 19: 41-63.

1971

Bricteux-Grégoire, S., Schyns, R. & Florkin, M. Phylogeny of activation peptides of trypsinogen. In: Biochemical Evolution and the Origin of Life (Schoffeniels, E. ed.), pp. 130-149, North-Holland Publishing Co., Amsterdam.

Eyl, A. W. & Inagami, T. Identification of essential carboxyl groups in the specific binding site of bovine trypsin by chemical modification. J. Biol. Chem. 246: 738-746.

Jameson, G. W. & Elmore, D. T. Separation of bovine α- and β-trypsin by affinity chromatograpy. Biochem. J. 124: 66P only.

Keil, B. Trypsin. In: The Enzymes (Boyer, P. D.), 3rd edn., vol. 3, pp. 249-275.

Muramatu, M. & Fujii, S.   Inhibitory effects of ω-amino acid esters on trypsin, plasmin, plasma kallikrein and thrombin. Biochim. Biophys. Acta 242: 203–208.

Ohlsson, K.   Isolation and immunochemical studies of dog trypsin. Biochim. Biophys. Acta 251: 450–455.

Robinson, N. C., Tye, R. W., Neurath, H. & Walsh, K. A.   Isolation of trypsins by affinity chromatography. Biochemistry 10: 2743–2747.

Schroeder, D. D. & Shaw, E.   Active-site-directed phenacyl halides inhibitory to trypsin.   Arch. Biochem. Biophys. 142: 340–350.

Voytek, P. & Gjessing, E. C.   Studies of an anionic trypsinogen and its active enzyme from porcine pancreas.   J. Biol. Chem. 246: 508–516.

1972

Ako, H., Foster, R. J. & Ryan, C. A.   The preparation of anhydro-trypsin and its reactivity with naturally occurring proteinase inhibitors.   Biochem. Biophys. Res. Commun. 47: 1402–1407.

Feinstein, G. & Gertler, A.   The effect of limited proteolysis of chicken ovoinhibitor by bovine chymotrypsin on the inhibitory activities against trypsin, chymotrypsin and elastase.   Eur. J. Biochem. 31: 25–31.

Maki, M. & Sasaki, K.   A new broad spectrum inhibitor of esteroproteases, 4-(2-carboxyethyl)phenyl-trans-4-aminomethyl cyclohexane carboxylate hydrochloride (DV 1006).   Tohoku J. Exp. Med. 106: 233–248.

Morgan, P. H., Robinson, N. C., Walsh, K. A. & Neurath, H. Inactivation of bovine trypsinogen and chymotrypsinogen by diisopropylphosphorofluoridate.   Proc. Natl. Acad. Sci. U.S.A. 69: 3312–3316.

Muramatu, M. & Fujii, S.   Inhibitory effects of ω-guanidino acid esters on trypsin, plasmin, plasma kallikrein and thrombin. Biochim. Biophys. Acta 268: 221–224.

Nakata, H. & Ishii, S.   Substrate activation of trypsin and acetyl-trypsin caused by α-N-benzoyl-L-arginine p-nitroanilide. J. Biochem. 72: 281–290.

Purdie, J. E., Demayo, R. E., Seely, J. H. & Benoiton, N. L.   The trypsin-catalysed hydrolysis of D-lysine and D-arginine ethyl esters.   Biochim. Biophys. Acta 268: 523–526.

Shaw, E.   Site-specific reagents for chymotrypsin, trypsin, and other serine proteases.   Methods Enzymol. 25: 655–671.

Svendsen, L., Blomback, B., Blomback, M. & Olsson, P.   Synthetic chromogenic substrates for determination of trypsin, thrombin and thrombin-like enzymes.   Thromb. Res. 1: 267–278.

Weinstein, M. J. & Doolittle, R. F.   Differential specificities of thrombin, plasmin and trypsin with regard to synthetic and natural substrates and inhibitors.   Biochim. Biophys. Acta 258: 577–590.

1973

Baratti, J., Maroux, S. & Louvard, D.   Effect of ionic strength

and calcium ions on the activation of trypsinogen by
enterokinase. Biochim. Biophys. Acta 321: 632–638.

Beardslee, R. A. & Zahnley, J. C. A simple preparation of
β–trypsin based on a calorimetric study of the thermal
stabilities of α– and β–trypsin. Arch. Biochem. Biohys. 158:
806–811.

Beaven, G. H. & Gratzer, W. B. Modification of the enzymic
activity of trypsin by intramolecular cross-links. Int. J. Pep.
Prot. Res. 5: 215–218.

Gabel, D. & Kasche, V. Autolysis of β–trypsin: influence of
calcium ions and heat. Acta Chem. Scand. B 27: 1971–1981.

Hermodson, M. A., Ericcson, L. H., Neurath, H. & Walsh, K. A.
Determination of the amino acid sequence of porcine trypsin by
sequenator analysis. Biochemistry 12: 3146–3153.

Hixson, H. F. & Nishikawa, A. H. Affinity chromatography:
purification of bovine trypsin and thrombin. Arch. Biochem.
Biophys. 154: 501–509.

Jameson, G. W., Roberts, D. V., Adams, R. W., Kyle, W. S. A. &
Elmore, D. T. Determination of the operational molarity of
solutions of bovine α–chymotrypsin, trypsin, thrombin, and
factor Xa by spectrofluorimetric titration. Biochem. J. 131:
107–117.

Kasai, K. & Ishii, S. Unimportance of histidine and serine
residues of trypsin in the substrate binding function proved by
affinity chromatography. J. Biochem. 74: 631–633.

Kisiel, W. & Hanahan, D. J. The action of factor Xa, thrombin and
trypsin on human factor II. Biochim. Biophys. Acta 329: 221–
232.

Mallory, P. & Travis, J. Human pancreatic enzymes.
Characterization of anionic human trypsin. Biochemistry 12:
2847–2851.

Ohlsson, K. & Tegner, H. Anionic and cationic dog trypsin,
isolation and partial characterization. Biochim. Biophys. Acta
317: 328–337.

Rinderknecht, H. & Geokas, M. C. Anionic and cationic trypsinogens
(trypsins) in mammalian pancreas. Enzyme 14: 116–130.

Rühlmann, A., Kukla, D., Schwager, P., Bartels, K. & Huber, R.
Structure of the complex formed by bovine trypsin and bovine
pancreatic trypsin inhibitor. J. Mol. Biol. 77: 417–436.

1974

Chambers, J. L., Christoph, G. G., Kreiger, M., Kay, L. & Stroud,
R. M. Silver ion inhibition of serine proteases:
crystallographic study of silver trypsin. Biochem. Biophys.
Res. Commun. 59: 70–74.

Feinstein, G., Hofstein, R., Koifmann, J. & Sokolovsky, M. Human
pancreatic proteolytic enzymes and protein inhibitors. Isolation
and molecular properties. Eur. J. Biochem. 43: 569–581.

Kobayashi, R. & Ishi, S. The trypsin-catalyzed hydrolysis of some
L–α–amino–lacking substrates. J. Biochem. 75: 825–835.

Lievaart, P. A. & Stevenson, K. J. The isolation of trypsin and elastase from moose pancreas (Alces alces) by affinity chromatography on lima-bean protease inhibitor-Sepharose resin. Can. J. Biochem. 52: 637-641.

Louvard, M. N. & Puigserver, A.   On bovine and porcine anionic trypsinogens. Biochim. Biophys. Acta 371: 177-185.

Moroi, M. & Yamasaki, M. Mechanism of interaction of bovine trypsin with human α_1-antitrypsin. Biochim. Biophys. Acta 359: 130-141.

Stroud, R. M., Kay, L. M. & Dickerson, R. E.   The structure of bovine trypsin: electron density maps of the inhibited enzyme at 5Å and 2.7 Å resolution. J. Mol. Biol. 83: 185-208.

Sweet, R. M., Wright, H. T., Janin, J., Chothia, C. H. & Blow, D. M. Crystal structure of the complex of porcine trypsin with soybean trypsin inhibitor (Kunitz) at 2.6 Å resolution. Biochemistry 13: 4212-4228.

1975

Bode, W. & Schwager, P.   The single calcium-binding site of crystalline bovine β-trypsin. FEBS Lett. 56: 139-143.

Cohen, A. B.   The interaction of α-1-antitrypsin with chymotrypsin, trypsin and elastase. Biochim. Biophys. Acta 391: 193-200.

Figarella, C., Negri, G. A. & Guy, O. The two human trypsinogens. Inhibition spectra of the two human trypsins derived from their purified zymogens. Eur. J. Biochem. 53: 457-463.

Geratz, J. D., Cheng, M. C.-F. & Tidwell, R. R. New aromatic diamidines with central α-oxyalkane or α, -dioxyalkane chains. Structure-activity relationships for the inhibition of trypsin, pancreatic kallikrein, and thrombin and for the inhibition of the overall coagulation process. J. Med. Chem. 18: 477-481.

Loeffler, L. J., Mar, E.-C., Geratz, J. D. & Fox, L. B. Synthesis of isosteres of p-amidinophenylpyruvic acid. Inhibitors of trypsin, thrombin and pancreatic kallikrein. J. Med. Chem. 18: 287-292.

Mallory, P. A. & Travis, J.   Inhibition spectra of the human pancreatic endopeptidases. Am. J. Clin. Nutr. 28: 823-830.

Preiser, H., Schmitz, J., Maestracci, D. & Crane, R. K. Modification of an assay for trypsin and its application for the estimation of enteropeptidase. Clin. Chim. Acta 59: 169-175.

Stevenson, K. J. & Voordouw, J. K. Characterization of trypsin and elastase from the moose (Alces alces). 1. Amino acid composition and specificity towards polypeptides. Biochim. Biophys. Acta 386: 324-331.

Stroud, R. M., Krieger, M., Koeppe, R. E., II., Kossiakoff, A. A. & Chambers, J. L. Structure-function relationships in the serine proteases. In: Proteases and Biological Control (Reich, E., Rifkin, D. B. & Shaw, E. eds), pp. 13-32, Cold Spring Harbor Laboratory, New York.

1976

Aubry, M. & Bieth, J.   A kinetic study of the human and bovine trypsins and chymotrypsins by the inter-α-inhibiter from human plasma. Biochim. Biophys. Acta 438: 221-230.

Geratz, J. D., Cheng, M. C.-F. & Tidwell, R. R. Novel bis(benzamidine) compounds with an aromatic central link. Inhibitors of thrombin, pancreatic kallikrein, trypsin and complement. J. Med. Chem. 19: 634-639.

Johnson, D. A. Purification and properties of rabbit trypsin. Biochim. Biophys. Acta 452: 482-487.

Lobo, A. P., Wos, J. D., Yu, S. M. & Lawson, W. B. Active site studies of human thrombin and bovine trypsin: peptide substrates. Arch. Biochem. Biophys. 177: 235-244.

Miller, E. J., Finch, J. E., Jr., Chung, E., Butler, W. T. & Robertson, P. B. Specific cleavage of the native type III collagen molecule with trypsin. Similarity of the cleavage products to collagenase-produced fragments and primary structure at the cleavage site. Arch. Biochem. Biophys. 173: 631-637.

Nakaya, K., Ushiwata, A. & Nakamura, Y. Interaction between proteins and detergents which contain a hydrocarbon chain longer than 16 carbon atoms. III. Competitive inhibition of trypsin by cetyldimethylbenzylammonium chloride. Biochim. Biophys. Acta 439: 116-124.

Temler, R. S. & Felber, J.-P. Radioimmunoassay of human plasma trypsin. Biochim. Biophys. Acta 445: 720-728.

Tidwell, R. R., Fox, L. L. & Geratz, J. D. Aromatic tris-amidines. A new class of highly active inhibitors of trypsin-like proteases. Biochim. Biophys. Acta 445: 729-738.

1977

Edy, J. & Collen, D. The interaction in human plasma of antiplasmin, the fast-reacting plasmin inhibitor, with plasmin, thrombin, trypsin and chymotrypsin. Biochim. Biophys. Acta 484: 423-432.

Schechter, Y., Burstein, Y. & Gertler, A. The effect of oxidation of methionine residues in chicken ovoinhibitor on its inhibitory activities against trypsin, chymotrypsin and elastase. Biochemistry 16: 992-997.

Tamura, Y., Hirado, M., Okamura, K., Minato, Y. & Fujii, S. Synthetic inhibitors of trypsin, plasmin, thrombin, C_1r and C_1 esterase. Biochim. Biophys. Acta 484: 417-422.

Walsmann, P. Inhibition of serine proteinases by benzamidine derivatives. Acta Biol. Med. Ger. 36: 1931-1937.

Yokosawa, H. & Ishii, S. Anhydrotrypsin: new features in ligand interactions revealed by affinity chromatography and thionine replacement. J. Biochem. 81: 647-656.

Zimmerman, M., Ashe, B., Yurewicz, E. C. & Patel, G. Sensitive assays for trypsin, elastase, and chymotrypsin using new fluorogenic substrates. Anal. Biochem. 78: 47-51.

1978

Bajusz, S., Barabás, E., Tolnay, P., Széll, E. & Bagdy, D. Inhibition of thrombin and trypsin by tripeptide aldehydes. Int. J. Pept. Protein Res. 12: 217-221.

Brodrick, J. W., Largman, C., Hsiang, M. W., Johnson, J. H. & Geokas, M. C. Structural basis for the specific activation of human cationic trypsinogen by enteropeptidase. J. Biol. Chem. 253: 2737-2742.

Brodrick, J. W., Largman, C., Johnson, J. H. & Geokas, M. C. Human cationic trypsinogen. Purification, characterization, and characteristics of autoactivation. J. Biol. Chem. 253: 2732-2736.

DeTraglia, M. C., Brand, J. S. & Tometsko, A. M. Characterization of azidobenzamidines as photoaffinity labels for trypsin. J. Biol. Chem. 253: 1846-1852.

Figarella, C., Sampaio, M. U. & Greene, L. J. Comparison of the kininogenase activity of human pancreatic trypsins and porcine kallikrein on Met-Lys-bradykinin and human plasma kininogen. Hoppe-Seyler's Z. Physiol. Chem. 359: 1225-1228.

Geratz, J. D. & Tidwell, R. R. Current concepts on action of synthetic thrombin inhibitors. Haemostasis 7: 170-176.

Guy, O., Lombardo, D., Bartelt, D. C., Amic, J. & Figarella, C. Two human trypsinogens. Purification, molecular properties, and N-terminal sequences. Biochemistry 17: 1669-1675.

Huber, R. & Bode, W. Structural basis of the activation and action of trypsin. Acc. Chem. Res. 11: 114-122.

Kakinuma, A., Sugino, H., Moriya, N. & Isono, M. Plasminostreptin, a protein proteinase inhibitor produced by Streptomyces antifibrinolyticus. I. Isolation and characterization. J. Biol. Chem. 253: 1529-1537.

Klausner, Y. S., Rigbi, M., Ticho, T., De Jong, P. J., Neginsky, E. J. & Rindt, Y. The interaction of α-N-(p-toluenesulphonyl)-p-guanidino-L-phenylalanine methyl ester with thrombin and trypsin. Biochem. J. 169: 157-167.

Largman, C., Brodrick, J. W., Geokas, M. C. & Johnson, J. H. Demonstration of human pancreatic anionic trypsinogen in normal serum by radioimmunoassay. Biochem. Biophys. Acta. 543: 450-454.

Lively, M. O. & Powers, J. C. Specificity and reactivity of human granulocyte elastase and cathepsin G, porcine pancreatic elastase, bovine chymotrypsin and trypsin toward inhibition with sulfonyl fluorides. Biochim. Biophys. Acta 525: 171-179.

Petkov, D. D. Detection of a tetrahedral intermediate in the trypsin-catalysed hydrolysis of specific ring-activated anilides. Biochim. Biophys. Acta 523: 538-541.

Vajda, T. & Szabó, T. Effect of methylamine on trypsin catalysis. Eur. J. Biochem. 85: 121-124.

1979

Colomb, E. & Figarella, C. Comparative studies on the mechanism of activation of the two human trypsinogens. Biochim. Biophys. Acta 571: 343-351.

Eddeland, A. Partition of trypsin and Kazal inhibitor in reaction mixtures with human serum. Hoppe-Seyler's Z. Physiol. Chem. 360: 145-149.

Geokas, M. C., Largman, C., Brodrick, J. W. & Johnson, J. H.
Determination of human pancreatic cationic trypsinogen in serum
by radioimmunoassay. Am. J. Physiol. 236: E77–E83.

Grant, D. A. W. & Hermon-Taylor, J. Hydrolysis of artificial
substrates by enterokinase and trypsin and the development of a
sensitive specific assay for enterokinase in serum. Biochim.
Biophys. Acta 567: 207–215.

Green, G. D. J. & Shaw, E. Thiobenzyl benzyloxycarbonyl-L-
lysinate, substrate for a sensitive colorimetric assay for
trypsin-like enzymes. Anal. Biochem. 93: 223–226.

Kuramochi, H., Nakata, H. & Ishii, S. Mechanism of association of
a specific aldehyde inhibitor, leupeptin, with bovine trypsin.
J. Biochem. 86: 1403–1410.

Somorin, O., Tokura, S., Nishi, N. & Noguchi, J. The action of
trypsin on synthetic chromogenic arginine substrates. J.
Biochem. 85: 157–162.

Wyrick, S., Kim, Y.-J., Ishaq, K. & Carter, C.-B. Synthesis of
active site-directed organometallic irreversible protease
inhibitors. Biochim. Biophys. Acta 568: 11–18.

Entry 1.03

CHYMOTRYPSIN

Summary

EC Number: 3.4.21.1

Earlier names: Includes chymotrypsin A and chymotrypsin B. The
chymotrypsinogens are the precursor zymogens.

Distribution: Chymotrypsinogens: zymogen granules of pancreatic
β-cells, probably in all mammals. Active enzymes: intestine.

Source: Autolytically activated pancreas. Most work has been done
with bovine or porcine material.

Action: Endopeptidase with specificity favouring bonds in which
the carbonyl group is that of a hydrophobic residue. Low mol.wt.
substrates also are hydrolyzed.

Requirements: pH 7.5.

Substrate, usual: Bz-Tyr-OEt, Z-Tyr-OPhNO$_2$.

Substrate, special: Z-Gly-Gly-Phe-2-NNap, Suc-Ala-Ala-Pro-Phe-
NPhNO$_2$ [1].

Inhibitors: Dip-F, Pms-F, Tos-Phe-CH$_2$Cl, Ac-Leu-Leu-Phe-H,
chymostatin, hydrocinnamic acid. α_1-chymotrypsin inhibitor, α_1-
proteinase inhibitor, hen ovoinhibitor, soybean trypsin
inhibitor.

Molecular properties: Full sequences and X-ray crystallographic
structures have been determined for chymotrypsins. Tryptic
activation of precursors (chymotrypsinogens A and B in pigs and
cattle) generates active enzymes of mol.wt. about 26,000
containing two polypeptide chains, further heterogeneity
depending upon the bonds cleaved in activation. Forms α, β, γ
and are distinguished by crystal morphology. Chymotrypsin B is
homologous with chymotrypsin A, but has 20% of amino acid
substitutions, with no detected alteration in activity. There
are two or three chromatographically distinct forms of human
chymotrypsin. Chymotrypsin has a pI of about 8.6, and is
generally stable down to pH 3.

Comment: Chymotrypsin-like enzymes, not the same as chymotrypsin,
but sometimes called "chymotrypsins", occur in non-pancreatic
cells including neutrophil granulocytes (cathepsin G, #1.36) and
mast cells (chymases I, #1.42, and II, #1.43).

References
[1] Del Mar et al. Anal. Biochem. 99: 316-320, 1979.

Bibliography

1938

Kunitz, M. Formation of new crystalline enzymes from chymotrypsin.
J. Gen. Physiol. 22: 207-237.

1949

Jensen, E. F., Fellows Nutting, M.-D. & Balls, A. K. Mode of
inhibition of chymotrypsin by diisopropylfluorophosphate. 1.
Introduction of phosphorus. J. Biol. Chem. 179: 201-204.

1952

Balls, A. K. & Jansen, E. F. Stoichiometric inhibition of
chymotrypsin. Adv. Enzymol. 13: 321-343.

1954

Hartley, B. S. & Kilby, B. A. The reaction of p-nitrophenyl
esters with chymotrypsin. Biochem. J. 56: 288-297.
Ravin, H. A., Bernstein, P. & Seligman, A. M. A colorimetric
micromethod for the estimation of chymotrypsin activity. J.
Biol. Chem. 208: 1-15.
Zerner, B. & Bender, M. L. Acyl-enzyme intermediates in the
α-chymotrypsin-catalyzed hydrolysis of "specific" substrates.
The relative rates of hydrolysis of ethyl, methyl and
p-nitrophenyl esters of N-acetyl-L-tryptophan. J. Am. Chem.
Soc. 85: 356-358.

1955

Laskowski, M. Chymotrypsinogens and chymotrypsins. Methods
Enzymol. 2: 8-26.
Schwert, G. W. & Takenaka, Y. A spectrophotometric determination
of trypsin and chymotrypsin. Biochim. Biophys. Acta 16: 570-
575.

1956

Gutfreund, H. & Sturtevant, J. M. The mechanism of chymotrypsin-
catalyzed reactions. Proc. Natl. Acad. Sci. U.S.A. 42: 719-728.

1959

Hummel, B. C. W. A modified spectrophotometric determination of
chymotrypsin, trypsin, and thrombin. Can. J. Biochem. 37: 1393-
1399.
Martin, C. J., Golubow, J. & Axelrod, A. E. A rapid and sensitive
spectrophotometric method for the assay of chymotrypsin. J.
Biol. Chem. 234: 294-298.

1960

Desnuelle, P. Chymotrypsin. In: The Enzymes (Boyer, P. D.,

Lardy, H. & Myrback, K. eds), 2nd edn., vol. 4, pp. 93-118,
Academic Press, New York.

Hartley, B. S. Proteolytic enzymes. Annu. Rev. Biochem. 29: 45-
72.

1961

Schonbaum, G. R., Zerner, B. & Bender, M. L. The
spectrophotometric determination of the operational normality of
an α-chymotrypsin solution. J. Biol. Chem. 236: 2930-2935.

1963

Bundy, H. F. Chymotrypsin-catalyzed hydrolysis of N-acetyl- and
N-benzoyl-L-tyrosine p-nitroanilides. Arch. Biochem. Biophys.
102: 416-422.

Epand, R. M. & Wilson, I. B. Evidence for the formation of
hippuryl chymotrypsin during the hydrolysis of hippuric acid
esters. J. Biol. Chem. 238: 1718-1723.

Fahrney, D. E. & Gold, A. M. Sulfonyl fluorides as inhibitors of
esterases. J. Amer. Chem. Soc. 85: 997-1000.

Schoellmann, G. & Shaw, E. Direct evidence for the presence of
histidine in the active center of chymotrypsin. Biochemistry 2:
252-255.

1964

Bielski, B. H. J. & Freed, S. Fluorometric assay of
α-chymotrypsin. Anal. Biochem. 7: 192-198.

Erlanger, B. F. & Edel, F. The utilization of a specific
chromogenic inactivator in an "all or none" assay for
chymotrypsin. Biochemistry 3: 346-349.

Erlanger, B. F., Cooper, A. G. & Bendich, A. J. On the
heterogeneity of three-times crystallised α-chymotrypsin.
Biochemistry 3: 1880-1883.

Hartley, B. S. Amino-acid sequence of bovine chymotrypsinogen A.
Nature 201: 1284-1287.

Inagami, T. & Sturtevant, J. M. The α-chymotrypsin-catalyzed
hydrolysis and hydroxylaminolysis of acetyltyrosine
p-nitroanilide. Biochem. Biophys. Res. Commun. 14: 69-74.

Lagunoff, D. & Benditt, E. P. Benzosalicylanilide ester
substrates of proteolytic enzymes. Kinetic and histochemical
studies. I. Chymotrypsin. Biochemistry 3: 1427-1431.

1965

Barrett, J. T. & Thompson, L. D. Immunochemial studies with
chymotrypsinogen A. Immunology 8: 136-143.

Blackwood, C. E., Erlarger, B. F. & Mandl, I. A new test for the
quantitative determination of chymotrypsin, chymotrypsin-like
enzymes and their inhibitors. Anal. Biochem. 12: 128-136.

Hartley, B. S., Brown, J. R., Kauffman, D. L. & Smillie, L. B.
Evolutionary similarities between pancreatic proteolytic
enzymes. Nature 207: 1157-1159.

Inagami, T., York, S. S. & Patchornik, A. An electrophilic

mechanism in the chymotrypsin catalyzed hydrolysis of anilide
substrates. J. Am. Chem. Soc. 87: 126-127.

1966

Bender, M. L., Begue-Canton, M. L., Blakely, R. L., Brubacher, L.
J., Feder, J., Gunter, C. R., Kezdy, F. J., Killheffer, J. V.,
Marshall, T. H., Miller, C. G., Roeske, R. W. & Stoops, J. K.
The determination of the concentration of hydrolytic enzyme
solutions: α-chymotrypsin, trypsin, papain, elastase, subtilisin,
and acetylcholinesterase. J. Am. Chem. Soc. 88: 5890-5913.

Bender, M. L., Gibian, M. J. & Whelan, D. J. The alkaline pH
dependence of chymotrypsin reactions: postulation of a pH-
dependent intramolecular competitive inhibition. Proc. Natl.
Acad. Sci. U.S.A. 56: 833-839.

Faller, L. & Sturtevant, J. M. The kinetics of the
α-chymotrypsin-catalyzed hydrolysis of p-nitrophenyl acetate in
organic solvent-water mixtures. J. Biol. Chem. 241: 4825-4834.

Hartley, B. S. & Kauffman, D. L. Corrections to the amino acid
sequence of bovine chymotrypsinogen A. Biochem. J. 101: 229-
231.

Johnson, C. H. & Knowles, J. R. The binding of inhibitors to
α-chymotrypsin. Biochem. J. 101: 56-62.

Matsuoka, Y. & Koide, A. On fin whale chymotrypsins. Arch.
Biochem. Biophys. 114: 422-424.

Meloun, B., Kluh, I., Kostka, V., Morávek, L., Prūsik, Z.,
Vaněcek, J., Keil, B. & Šorm, F. Covalent structure of bovine
chymotrypsinogen A. Biochim. Biophys. Acta 130: 543-546.

Oppenheimer, L. H., Labouesse, B. & Hess, G. P. Implication of an
ionizing group in the control of conformation and activity of
chymotrypsin. J. Biol. Chem. 241: 2720-2730.

Smillie, L. B., Enenkel, A. G. & Kay, C. M. Physicochemical
properties and amino acid composition of chymotrypsinogen B. J.
Biol. Chem. 241: 2097-2102.

1967

Bender, M. L., Kezdy, F. J. & Wedler, F. C. α-Chymotrypsin:
enzyme concentration and kinetics. J. Chem. Ed. 44: 84-88.

Brandt, K. G., Himoe, A. & Hess, G. P. Investigations of the
chymotrypsin-catalyzed hydrolysis of specific substrates. II.
Determination of individual rate constants and enzyme-substrate
binding constants for specific amide and ester substrates. J.
Biol. Chem. 242: 3973-3982.

Himoe, A., Parks, P. C. & Hess, G. P. Investigation of the
chymotrypsin catalyzed hydrolysis of specific substrates. I. The
pH dependency of the catalytic hydrolysis of
N-acetyltryptophanamide by three forms of the enzyme at alkaline
pH. J. Biol. Chem. 242: 919-929.

Mathews, B. M., Sigler, P. B., Henderson, R. & Blow, D. M. Three
dimensional structure of tosyl-α-chymotrypsin. Nature 214: 652-
656.

1968

Baker, B. R. & Hurlbut, J. A. Irreversible enzyme inhibitors.
CXIII. Proteolytic enzymes. III. Active site directed
irreversible inhibitors of α-chymotrypsin derived from
phenoxyacetamides with an N-fluorosulfonylphenyl substituent.
J. Med. Chem. 11: 233-241.

Hofstee, B. H. J. Crystallization and activation of chymotrypsin
"X". Arch. Biochem. Biophys. 125: 1031-1033.

Sigler, P. W., Blow, D. M., Matthews, B. W. & Henderson, R.
Structure of crystalline α-chymotrypsin. II. A preliminary
report including a hypothesis for the activation mechanism. J.
Mol. Biol. 35: 143-164.

Smillie, L. B., Furka, A., Nagabhushan, N., Stevenson, K. J. &
Parkes, C. O. Structure of chymotrypsinogen B compared with
chymotrypsinogen A and trypsinogen. Nature 218: 343-346.

Winzor, D. J., Loke, J. P. & Nichol, L. W. Consideration of
kinetic data on α-chymotrypsin in terms of oligomer activities.
Biochem. J. 108: 159-160.

1969

Blow, D. M., Birktoft, J. J. & Hartley, B. S. Role of a buried
acid group in the mechanism of action of chymotrypsin. Nature
221: 337-340.

Figarella, C., Clemente, F. & Guy, O. On zymogens of human
pancreatic juice. FEBS Lett. 3: 351-353.

Koide, A., Kataoka, T. & Matsuoka, Y. Purification and properties
of ovine cationic chymotrypsin. J. Biochem. 65: 475-477.

Sigman, D. S., Torchia, D. A. & Blout, E. R. Phenacylbromides as
chromophoric reagents for α-chymotrypsin. Biochemistry 8: 4560-
4566.

Steitz, T. A., Henderson, R. & Blow, D. M. Structure of
crystalline α-chymotrypsin. III. Crystallographic studies of
substrates and inhibitors bound to the active site of
α-chymotrypsin. J. Mol. Biol. 46: 337-348.

1970

Birktoft, J. J., Blow, D. M., Henderson, R. & Steitz, T. A. The
structure of α-chymotrypsin. Philos. Trans. Roy. Soc. London
Ser. B 257: 67-76.

Gundlach, H. G. Conformational changes in chymotrypsin detected
with antibodies. I. Antibody response to diisopropylphosphoryl-
chymotrypsin. Hoppe-Seyler's Z. Physiol. Chem. 351: 690-695.

Gundlach, H. G. Conformational changes in chymotrypsin detected
with antibodies. II. Antibody interactions with chemically
modified and denatured chymotrypsins. Hoppe-Seyler's Z.
Physiol. Chem. 351: 696-700.

Gundlach, H. G. Conformational changes in chymotrypsin detected
with antibodies. III. On the antigenic determinant groups of
chymotrypsins. Hoppe-Seyler's Z. Physiol. Chem. 351: 701-704.

Hartley, B. S. Homologies in serine proteinases. Philos. Trans.
Roy. Soc. London Ser. B 257: 77-87.

Hess, G. P., McConn, J., Ku, E. & McConkey, G. Studies of the activity of chymotrypsin. Philos. Trans. R. Soc. London Ser. B 257: 89-104.

Kakade, M. L., Swenson, D. H. & Liener, I. E. Note on the determination of chymotrypsin and chymotrypsin inhibitor activity using casein. Anal. Biochem. 33: 255-258.

Melville, J. C. & Ryan, C. A. Chymotrypsin inhibitor I from potatoes; a multisite inhibitor composed of subunits. Arch. Biochem. Biophys. 138: 700-702.

Rinderknecht, H., Silverman, P., Geokas, M. C. & Haverback, B. J. Determination of trypsin and chymotrypsin with Remazolbrilliant Blue-hide. Clin. Chim. Acta 28: 239-246.

Umezawa, H., Aoyagi, T., Morishima, H., Kunimoto, S., Matsuzaki, M., Hamada, M. & Takeuchi, T. Chymostatin, a new chymotrypsin inhibitor produced by actinomycetes. J. Antibiot. (Tokyo) 23: 425-427.

Wilcox, P. E. Chymotrypsinogens – chymotrypsins. Methods Enzymol. 19: 64-108.

1971

Blow, D. M. The structure of chymotrypsin. In: The Enzymes (Boyer, P. D. ed.), 3rd edn., vol.3, pp. 185-212, Academic Press, New York.

Coan, M. H., Roberts, R. C. & Travis, J. Human pancreatic enzymes. Isolation and properties of a major form of chymotrypsin. Biochemistry 10: 2711-2717.

Haas, E., Elkana, Y. & Kulka, R. G. A sensitive fluorometric assay for α-chymotrypsin. Anal. Biochem. 40: 218-226.

Henderson, R. Catalytic activity of α-chymotrypsin in which histidine-57 has been methylated. Biochem. J. 124: 13-18.

Hess, G. P. Chymotrypsin – chemical properties and catalysis. In: The Enzymes (Boyer, P. D. ed.), 3rd edn., vol.3, pp. 213-248, Academic Press, New York.

Koehler, H. & Lienhard, G. E. 2-Phenylethaneboronic acid, a possible transition state analog for chymotrypsin. Biochemistry 10: 2477-2483.

Kraut, J. Chymotrypsinogen: X-ray structure. In: The Enzymes (Boyer, P. D. ed.), 3rd edn., vol.3, pp. 165-183, Academic Press, New York.

Philipp, M. & Bender, M. L. Inhibition of serine proteases by arylboronic acids. Proc. Natl. Acad. Sci. U.S.A. 68: 478-480.

Segal, D. M., Powers, J. C., Cohen, G. H., Davies, D. R. & Wilcox, P. E. Substrate binding site in chymotrypsin A_γ. A crystallographic study using peptide chloromethylketones as site-specific inhibitors. Biochemistry 10: 3728-3738.

Shaw, E. & Ruscica, J. The reactivity of His-57 in chymotrypsin to alkylation. Arch. Biochem. Biophys. 145: 484-489.

Wünsch, E., Högel-Betz, A. & Jaeger, E. Ein neues Substrat zur quantitativen Bestimmung von Chymotrypsin A. Hoppe-Seyler's Z. Physiol. Chem. 353: 1553-1559.

1972

Ako, H., Ryan, C. A. & Foster, R. J. The purification by affinity chromatography of a protease inhibitor binding species of anhydrochymotrypsin. Biochem. Biophys. Res. Commun. 46: 1639–1645.

Blair, T. T., Marini, M. A. & Martin, C. J. A reexamination of the reaction of TPCK with α-chymotrypsin. FEBS Lett. 20: 41–43.

Coan, M. H. & Travis, J. Human pancreatic enzymes: properties of two minor forms of chymotrypsin. Biochim. Biophys. Acta 268: 207–211.

Feinstein, G. & Gertler, A. The effect of limited proteolysis of chicken ovoinhibitor by bovine chymotrypsin on the inhibitory activities against trypsin, chymotrypsin and elastase. Eur. J. Biochem. 31: 25–31.

Ito, A., Tokawa, K. & Shimizu, B. Peptide aldehydes inhibiting chymotrypsin. Biochem. Biophys. Res. Commun. 49: 343–349.

Melville, J. C. & Ryan, C. A. Chymotrypsin inhibitor I from potatoes. Large scale preparation and characterization of its subunit components. J. Biol. Chem. 247: 3445–3453.

Morgan, P. H., Robinson, N. C., Walsh, K. A. & Neurath, H. Inactivation of bovine trypsinogen and chymotrypsinogen by diisopropylphosphorofluoridate. Proc. Natl. Acad. Sci. U.S.A. 69: 3312–3316.

Pattabiraman, T. N. & Lawson, W. B. Comparative studies of the specificities of α-chymotrypsin and subtilisin BPN'. Studies with flexible substrates. Biochem. J. 126: 645–657.

Pattabiraman, T. N. & Lawson, W. B. Stereochemistry of the active site of α-chymotrypsin. The effect of some tricyclic bromomethyl ketones on α-chymotrypsin. Biochim. Biophys. Acta 258: 548–553.

Segal, D. M. A kinetic investigation of the crystallographically deduced binding subsites of bovine chymotrypsin A_γ. Biochemistry 11: 349–356.

Shaw, E. Site-specific reagents for chymotrypsin, trypsin, and other serine proteases. Methods Enzymol. 25: 655–671.

Yoshida, N. & Ishii, S. Specificity-determining site of α-chymotrypsin. 1. Kinetic investigation of aromatic amino compounds and phenylalanine ester derivatives in α-chymotrypsin-catalysed hydrolysis. J. Biochem. 71: 185–191.

Yoshida, N. & Ishii, S. Specificity-determining site of α-chymotrypsin. II. Inactivation of α-chymotrypsinogen by aromatic diazonium compounds. J. Biochem. 71: 193–199.

1973

Alazard, R., Béchet, J.-J., Dupaix, A. & Yon, J. Inactivation of α-chymotrypsin by a bifunctional reagent, 2-bromomethyl-3,1-benzoxazin-4-one. Biochim. Biophys. Acta 309: 379–396.

Avery, M. J. & Hopkins, T. R. Is γ-chymotrypsin an autolytic product in the activation scheme of chymotrypsinogen A? Biochim. Biophys. Acta 310: 142–147.

Baumann, W. K., Bizzozero, S. A. & Dutler, H. Kinetic

investigation of the α-chymotrypsin-catalyzed hydrolysis of peptide substrates. Eur. J. Biochem. 39: 381-391.

Brown, W. E. & Wold, E. Alkyl isocyanates as active site specific reagents for serine proteases. Identification of the active site serine as the site of reaction. Biochemistry 12: 835-841.

Brown, W. E. & Wold, E. Alkyl isocyanates as active site specific reagents for serine proteases. Reaction properties. Biochemistry 12: 828-834.

Fastrez, J. & Fersht, A. R. Mechanism of chymotrypsin. Structure, reactivity, and non-productive binding relationships. Biochemistry 12: 1067-1074.

Geratz, J. D. & Fox, L. L. B. On the irreversible inhibition of α-chymotrypsin, trypsin, pancreatic kallikrein, thrombin and elastase by N-(β-pyridylmethyl)-3,4-dichlorophenoxyacetamide-p-fluorosulfonylacetanilide bromide. Biochim. Biophys. Acta 321: 296-302.

Grey, S. M., Nurse, G. R. & Visser, L. Implications of the chemical modification of chymotrypsin. Biochem. Biophys. Res. Commun. 52: 687-695.

Imondi, A. R., Stradley, R. P., Butler, E. R. & Wolgemuth, R. L. A method for the assay of chymotrypsin in crude biological materials. Anal. Biochem. 54: 199-204.

Jameson, G. W., Roberts, D. V., Adams, R. W., Kyle, W. S. A. & Elmore, D. T. Determination of the operational molarity of solutions of bovine α-chymotrypsin, trypsin, thrombin, and factor Xa by spectrofluorimetric titration. Biochem. J. 131: 107-117.

Kang, S.-H., & Fuchs, M. S. An improvement in the Hummel chymotrypsin assay. Anal. Biochem. 54: 262-265.

Kurachi, K., Powers, J. C. & Wilcox, P. E. Kinetics of the reaction of chymotrypsin Aα with peptide chloromethyl ketones in relation to its subsite specificity. Biochemistry 12: 771-777.

Tatsuta, K., Mikami, N., Fujimoto, K. & Umezawa, S. The structure of chymostatin, a chymotrypsin inhibitor. J. Antibiot. 26: 625-646.

Yoshida, N., Yamasaki, M. & Ishii, S. Specificity-determining site of α-chymotrypsin. III. Inactivation of α-chymotrypsin by p-diazo-N-acetyl-L-phenylalanine methyl ester. J. Biochem. 73: 1175-1183.

1974

Feinstein, G., Hofstein, R., Koifmann, J. & Sokolovsky, M. Human pancreatic proteolytic enzymes and protein inhibitors. Isolation and molecular properties. Eur. J. Biochem. 43: 569-581.

Gertler, A., Walsh, K. A. & Neurath, H. Catalysis by chymotrypsinogen. Demonstration of an acyl-zymogen intermediate. Biochemistry 13: 1302-1310.

Jones, J. B., Sneddon, D. W. & Lewis, A. J. Inhibition of α-chymotrypsin by hydroxymethyl analogues of D- and L-N-acetylphenylalanine and N-acetyltryptophan of potential affinity labeling value. Biochim. Biophys. Acta 341: 284-290.

Richardson, M. & Cossins, L.   Chymotryptic inhibitor I from potatoes: the amino acid sequences of subunits B, C and D.   FEBS Lett. 45: 11-13.

Tomlinson, G., Shaw, M. C. & Viswanatha, T.   Chymotrypsins. Methods Enzymol. 34: 415-420.

1975

Breaux, E. J. & Bender, M. L.   The binding of specific and non-specific aldehyde substrate analogs to α-chymotrypsin.   FEBS Lett. 56: 81-84.

Caro, A. de., Figarella, C. & Guy, O.   The two human chymotrypsinogens. Purification and characterization.   Biochim. Biophys. Acta 379: 431-443.

Cohen, A. B.   The interaction of α-1-antitrypsin with chymotrypsin, trypsin and elastase.   Biochim. Biophys. Acta 391: 193-200.

Dahlquist, U. & Wåhlby, S.   Formation of a covalent intermediate between α-chymotrypsin and the B-chain of insulin during enzyme-catalyzed hydrolysis.   Biochim. Biophys. Acta 391: 410-414.

De Caro, A., Figarella, C. & Guy, O.   The two human chymotrypsinogens. Purification and characterization.   Biochim. Biophys. Acta 379: 431-443.

Farmer, D. A. & Hageman, J. H.   Use of N-benzoyl-L-tyrosine thiobenzyl ester as a proteinase substrate. Hydrolysis by alpha-chymotrypsin and subtilisin BPN'.   J. Biol. Chem. 250: 7366-7371.

Hageman, J. H., Lara, L. & Farmer, D. A.   Hydrolysis of N-benzoyl-L-tyrosine thiobenzyl ester (BTTBE) by α-chymotrypsin and other serine proteinases: implication of a second hydrophobic site in α-chymotrypsin.   Fed. Proc. Fed. Am. Soc. Exp. Biol. 34: 534 only.

Hamilton, S. E., Dudman, N. P. B., de Jersey, J., Stoops, J. K. & Zerner, B.   Organophosphate inhibitors: the reactions of bis(p-nitrophenyl)methyl phosphate with liver carboxylesterases and α-chymotrypsin.   Biochim. Biophys. Acta 377: 282-296.

Mallory, P. A. & Travis, J.   Inhibition spectra of the human pancreatic endopeptidases.   Am. J. Clin. Nutr. 28: 823-830.

Powers, J. C. & Carroll, D. L.   Reaction of acyl carbazates with proteolytic enzymes.   Biochem. Biophys. Res. Commun. 67: 639-644.

Rao, K. N.   Substrate solubilization for the Hummel α-chymotrypsin assay.   Anal. Biochem. 65: 548-551.

Rinderknecht, H. & Fleming, R. M.   A new highly sensitive and specific assay for chymotrypsin.   Clin. Chim. Acta 59: 139-146.

Stroud, R. M., Krieger, M., Koeppe, R. E., II., Kossiakoff, A. A. & Chambers, J. L.   Structure-function relationships in the serine proteases.   In: Proteases and Biological Control (Reich, E., Rifkin, D. B. & Shaw, E. eds), pp. 13-32, Cold Spring Harbor Laboratory, New York.

1976

Aubry, M. & Bieth, J.   A kinetic study of the human and bovine trypsins and chymotrypsins by the inter-α-inhibiter from human plasma. Biochim. Biophys. Acta 438: 221-230.

Bauer, C. A., Thompson, R. C. & Blout, E. R. The active centers of Streptomyces griseus protease 3, α-chymotrypsin and elastase. Enzyme substrate interactions close to the scissile bond. Biochemistry 15: 1296-1299.

Bauer, C. A., Thompson, R. C. & Blout, E. R. The active centers of Streptomyces griseus protease 3, α-chymotrypsin and elastase. Enzyme substrate interactions remote from the scissile bond. Biochemistry 15: 1291-1295.

Erlanger, B. F., Wasserman, N. H., Cooper, A. G. & Monk, R. J. Allosteric activation of the hydrolysis of specific substrates by chymotrypsin. Eur. J. Bioch. 61: 287-295.

Powers, J. C., Tuhy, P. M. & Witter, F. Modification of chymotrypsin with the specific cyanate reagent N-acetyl-p-cyanato-L-phenylalanine ethyl ester. Biochim. Biophys. Acta 445: 426-436.

Smith, C. M. & Hanahan, D. J. The activation of factor V by factor Xa or α-chymotrypsin and comparison with thrombin and RVV-V action. Biochemistry 15: 1830-1838.

Warren, J. R. & Gordon, J. A. The nature of alkylurea and urea denaturation of α-chymotrypsinogen. Biochim. Biophys. Acta 420: 397-405.

Zimmerman, M., Yurewicz, E. & Patel, G. A new fluorogenic substrate for chymotrypsin. Anal. Biochem. 70: 258-262.

1977

Birnbaum, E. R., Abbott, F., Gomez, J. E. & Darnall, D. W. The calcium ion binding site in bovine chymotrypsinogen A. Arch. Biochem. Biophys. 179: 469-476.

Edy, J. & Collen, D. The interaction in human plasma of antiplasmin, the fast-reacting plasmin inhibitor, with plasmin, thrombin, trypsin and chymotrypsin. Biochim. Biophys. Acta 484: 423-432.

Fastrez, J. & Houyet, N. Mechanism of chymotrypsin: acid/base catalysis and transition-state solvation by the active site. Eur. J. Biochem. 81: 515-522.

Powers, J. C. & Gupton, B. F. Reaction of serine proteases with aza-amino acid and aza-peptide derivatives. Methods Enzymol. 46: 208-216.

Schechter, Y., Burstein, Y. & Gertler, A. The effect of oxidation of methionine residues in chicken ovoinhibitor on its inhibitory activities against trypsin, chymotrypsin and elastase. Biochemistry 16: 992-997.

Zimmerman, M., Ashe, B., Yurewicz, E. C. & Patel, G. Sensitive assays for trypsin, elastase, and chymotrypsin using new fluorogenic substrates. Anal. Biochem. 78: 47-51.

1978

Brock, J. H. & Esparza, I. A simple colorimetric assay of enhanced sensitivity for chymotrypsin. Clin. Chim. Acta 85: 99-

100.

Klausner, Y. S., Rigbi, M., Ticho, T., De Jong, P. J., Neginsky, E. J. & Rindt, Y. The interaction of α-N-(p-toluenesulphonyl)-p-guanidino-L-phenylalanine methyl ester with thrombin and trypsin. Biochem. J. 169: 157-167.

Lively, M. O. & Powers, J. C. Specificity and reactivity of human granulocyte elastase and cathepsin G, porcine pancreatic elastase, bovine chymotrypsin and trypsin toward inhibition with sulfonyl fluorides. Biochim. Biophys. Acta 525: 171-179.

Moffitt, M. J. & Means, G. E. Subsite interactions in chymotrypsin as reflected by acyl enzyme stability. Biochem. Biophys. Res. Commun. 83: 1415-1421.

Oka, T. & Morihara, K. Peptide bond synthesis catalyzed by α-chymotrypsin. J. Biochem. 84: 1277-1283.

Petkov, D., Christova, E. & Stoineva, I. Catalysis and leaving group binding in anilide hydrolysis by chymotrypsin. Biochim. Biophys. Acta 527: 131-141.

Stoesz, J. D. & Lumry, R. W. Refolding transition of α-chymotrypsin: pH and salt dependence. Biochemistry 17: 3693-3699.

Travis, J., Bowen, J. & Baugh, R. Human α-1-antichymotrypsin: interaction with chymotrypsin-like proteinases. Biochemistry 17: 5651-5656.

1979

Bauer, R.S. & Berliner, L. J. Spin label investigations of α-chymotrypsin active site structure in single crystals. J. Mol. Biol. 128: 1-19.

Blumberg, S. & Katchalski-Katzir, E. Inhibition of α-chymotrypsin by D-tryptophan amide covalently bound to macromolecular carriers: specific, steric and electrostatic effects. Biochemistry 18: 2126-2133.

Chen, R., Gorenstein, D. G., Kennedy, W. P., Lowe, G., Nurse, D. & Schultz, R. M. Evidence for hemiacetal formation between N-acyl-L-phenylalaninals and α-chymotrypsin by cross-saturation nuclear magnetic resonance spectroscopy. Biochemistry 18: 921-926.

Del Mar, E. G., Largman, C., Brodrick, J. W. & Geokas, M. C. A sensitive new substrate for chymotrypsin. Anal. Biochem. 99: 316-320.

Dunn, B. M. & Gilbert, W. A. Quantitative affinity chromatography of α-chymotrypsin. Arch. Biochem. Biophys. 198: 533-540.

Geokas, M. C., Largman, C., Brodrick, J. W., Johnson, J. H. & Fassett, M. Immunoreactive forms of human pancreatic chymotrypsin in normal plasma. J. Biol. Chem. 254: 2775-2781.

Kennedy, W. P. & Schultz, R. M. Mechanism of association of a specific aldelyde "transition state analogue" to the active site of α-chymotrypsin. Biochemistry 18: 349-356.

Obara, M., Karasaki, Y. & Ohno, M. α-Chymotrypsin-catalyzed hydrolysis of peptide substrates. Effect on the reactivity of the secondary interaction due to the peptide moiety C-terminal to the cleaved bond. J. Biochem. 86: 461–468.

Tellam, R., de Jersey, J. & Winzor, D. J. Evaluation of
  equilibrium constants for the binding of N–acetyl–L–tryptophan
  to monomeric and dimeric forms of α–chymotrypsin. Biochemistry
  18: 5316–5321.

Thompson, R. C. & Bauer, C.-A. Reaction of peptide aldehydes with
  serine proteases. Implications for the entropy changes
  associated with enzymic catalysis. Biochemistry 18: 1552–1558.

Watanabe, H., Green, G. D. J. & Shaw, E. A comparison of the
  behavior of chymotrypsin and cathepsin B towards peptidyl
  diazomethyl ketones. Biochem. Biophys. Res. Commun. 89: 1354–
  1360.

Wyrick, S., Kim, Y.-J., Ishaq, K. & Carter, C.-B. Synthesis of
  active site–directed organometallic irreversible protease
  inhibitors. Biochim. Biophys. Acta 568: 11–18.

Entry 1.04

CHYMOTRYPSIN C

Summary

EC Number: 3.4.21.2

Earlier names: Precursor is known from porcine and bovine
pancreas. Active enzyme from intestine.

Distribution: Autolytically activated pancreas.

Source: An endopeptidase preferentially cleaving Leu-, Tyr-, Phe-,
Met-, Trp-, Gln- or Asn-X bonds. Broader specificity than
chymotrypsins A and B, resembling cathepsin G (#1.36). Also
cleaves synthetic substrates based on hydrophobic amino acids.

Action: pH optimum 7.8.

Requirements: Bz-Leu-OEt.

Substrate, usual: Z-Leu-2-ONap (chromogenic, with a diazonium
salt).

Substrate, special: As for chymotrypsin, but Tos-Leu-CH_2Cl more
effective than Tos-Phe-CH_2Cl.

Inhibitors: Formed from porcine chymotrypsinogen C and from bovine
subunit II of procarboxypeptidase A. Mol.wt. of active enzyme
(porcine) about 23,800. pI considerably below that of
chymotrypsin and pancreatic serine proteinases generally.

Molecular properties: Probably a single polypeptide chain, unlike
chymotrypsin A.

Bibliography

1965

Folk, J. E. & Cole, P. W. Chymotrypsin C. II. Enzymatic
specificity toward several polypeptides. J. Biol. Chem. 240:
193-197.

Folk, J. E. & Schirmer, E. W. Chymotrypsin C. 1. Isolation of the
zymogen and the active enzyme: preliminary structure and
specificity studies. J. Biol. Chem. 240: 181-192.

1966

McConnell, B. & Gjessing, E. C. Isolation and crystallization of
an esteroproteolytic zymogen from porcine pancreas. J. Biol.

Chem. 241: 573-579.

1967

Tobita, T. & Folk, J. E. Chymotrypsin C. III. Sequence of amino acids around an essential histidine residue. Biochim. Biophys. Acta 147: 15-25.

1969

Peanasky, R. J., Gratecos, D., Baratti, J. & Rovery, M. Mode of activation and N-terminal sequence of subunit II in bovine procarboxypeptidase A and of porcine chymotrypsinogen C. Biochim. Biophys. Acta 181: 82-92.

1970

Baumstark, J. S. Isolation and partial characterization of an elastase-associated acidic endopeptidase and its interaction with elastase at low ionic strength. Biochim. Biophys. Acta 220: 534-551.

Folk, J. E. Chymotrypsin C (porcine pancreas). Methods Enzymol. 19: 109-112.

Neurath, H., Bradshaw, R. A., Pétra, P. H. & Walsh, K. A. Bovine carboxypeptidase A - activation, chemical structure and molecular heterogeneity. Philos. Trans. R. Soc. London Ser. B 257: 159-176.

Wilcox, P. E. Chymotrypsinogens - chymotrypsins. Methods Enzymol. 19: 64-108.

1975

Puigserver, A. & Desnuelle, P. Dissociation of bovine 6S procarboxypeptidase A by reversible condensation with 2,3-dimethyl maleic anhydride: application to the partial characterization of subunit III. Proc. Natl. Acad. Sci. U.S.A. 72: 2442-2445.

1976

Thomson, A. & Denniss, I. S. The identity of the elastase-associated acidic endopeptidase and chymotrypsin C from porcine pancreas. Biochim. Biophys. Acta 429: 581-590.

Entry 1.05

PANCREATIC ELASTASE

Summary

EC Number: 3.4.21.11 (shared by all mammalian elastases).

Earlier names: Pancreatopeptidase E, pancreatic elastase I, elastase 1.

Distribution: Best known from pig, but occurs in other mammals. Precursor, proelastase, in zymogen granules of pancreatic β-cells. Active enzyme is generated in the intestine by the action of trypsin.

Source: Pancreas of pig and other species, with or without autolysis to activate the zymogen.

Action: Solubilizes elastin and hydrolyzes bonds in polypeptides and synthetic ester and amide substrates. Preference for the residue contributing the carboxyl group may be Ala, Leu, Gly, Val, Ile. Catalytic efficiency is greatly enhanced by increase of peptide chain length to 4 residues, but site S_3 cannot accommodate Pro.

Requirements: pH optimum about 8.5. Low ionic strength, for elastolysis.

Substrate, usual: Boc-Ala-OPhNO$_2$, Ac-(Ala)$_3$-OPhNO$_2$, Z-Ala-2-ONap. Elastin.

Substrate, special: MeOSuc-Ala-Ala-Pro-Val-NMec and -SBzl, for maximal sensitivity [1]. Ac-Ala-Ala-Aala-NPhNO$_2$ for active site titration.

Inhibitors: Dip-F, Ac-Ala-Ala-Pro-Ala-CH$_2$Cl, Ac-Pro-Ala-Pro-Ala-H, elastatinal. CF$_3$CO-Ala-Ala-Ala, Cu^{2+} (10^{-5} M) (not Zn^{2+}, 10^{-3} M). Turkey ovomucoid. (Not oleic acid, soybean trypsin inhibitor or aprotinin). α_1-Proteinase inhibitor, α_2-macroglobulin (not other plasma inhibitors).

Molecular properties: A single polypeptide chain without carbohydrate. Mol.wt. 25,900, pI 9.5 (or higher), A280,1% 20.2 (pH 5). Stable in the range pH 4-12.

Comment: It seems the human pancreas produces much less elastinolytic activity than that of the pig, and that the human elastase resembles pig pancreatic elastase II (#1.07) rather than pig elastase I.

References
[1] Castillo et al. Anal. Biochem. 99: 53-64, 1979.

Bibliography

1949

Balo, J. & Banga, I. Elastase and elastase inhibitor. Nature
164: 491 only.

1951

Kokas, E., Foldes, I. & Banga, I. The elastase and trypsin
contents of the pancreatic secretions in dogs. Acta Physiol.
Acad. Sci. Hung. 2: 333-341.

1952

Banga, I. Isolation and crystallization of elastase from the
pancreas of cattle. Acta Physiol. Acad. Sci. Hung. 3: 317-324.

1953

Balo, J. & Banga, I. Changes in elastase content of the human
pancreas in relation to atherosclerosis. Acta Physiol. Acad.
Sci. Hung. 4: 187-193.

1955

Grant, N. H. & Robbins, K. C. Occurrence and activation of an
elastase precursor in pancreas. Proc. Soc. Exp. Biol. Med. 90:
264-265.
Partridge, S. M. & Davis, H. F. Soluble proteins derived from
partial hydrolysis of elastin. Biochem. J. 61: 21-33.
Sachar, L. A., Winter, K. K., Sicher, N., & Frankel, S.
Photometric method for estimation of elastase activity. Proc.
Soc. Exp. Biol. Med, 90: 323-326.

1956

Lewis, U. J., Williams, D. E. & Brink, N. G. Pancreatic elastase:
purification, properties and function. J. Biol. Chem. 222: 705-
720.

1957

Bagdy, D. & Banga, I. Extraction and purification of elastase
from dried pancreas. Acta Physiol. Acad. Sci. Hung. 11: 371-
376.
Hall, D. A. The complex nature of the enzyme elastase. Arch.
Biochem. Biophys. 67: 366-377.

1959

Banga, I., Balő, J. & Horváth, M. Nephelometric determination
elastase activity and method for elastoproteolytic measurements.
Biochem. J. 71: 544-551.
Hall, D. A. & Czerkawsky, J. W. The purification of the
proteolytic component of elastase. Biochem. J. 73: 356-361.

Hartley, B. S., Naughton, M. A. & Sanger, F. The amino acid sequence around the reactive serine of elastase. Biochim. Biophys. Acta 34: 243–244.

McIvor, B. & Moon, H. D. Antigenic studies on elastase. J. Immunol. 82: 328–331.

Tolnay, P. & Bagdy, D. On the elastase-inhibiting action of sera. Biochim. Biophys. Acta 31: 566–568.

1960

Moon, H. D. & McIvor, B. C. Elastase in the exocrine pancreas: localization with fluorescent antibody. J. Immunol. 85: 78–80.

1961

Hall, D. A. & Czerkawski, J. W. The reaction between elastase and elastic tissue. V. Groupings essential for elastolysis. Biochem. J. 80: 128–136.

Hall, D. A. & Czerkawski, J. W. The reaction between elastase and elastic tissue. VI. The mechanism of elastolysis. Biochem. J. 80: 134–136.

Lamy, F., Craig, C. P. & Tauber, S. Studies on elastase and elastin. I. Assay and properties of elastase. J. Biol. Chem. 236: 86–91.

Naughton, M. A. & Sanger, F. Purification and specificity of pancreatic elastase. Biochem. J. 78: 156–163.

1962

Bagdy, D., Falk, M. & Tolnay, P. Inhibition of elastase by trypsin inhibitors. Acta Physiol. Acad. Sci. Hung. 21: 123–126.

Walford, R. L. & Kickhöfen, B. Selective inhibition of elastolytic and proteolytic properties of elastase. Arch. Biochem. Biophys. 98: 191–196.

1963

Banga, I. Determination of elastase and elastase inhibitor by means of orcein-elastin. Acta Physiol. Acad. Sci. Hung. 24: 1–9.

Baumstark, J. S., Bardawil, W. A., Sbarra, A. J. & Hayes, N. Purification of pancreatopeptidase E by batch seperation on DEAE cellulose. Biochim. Biophys. Acta 77: 676–679.

Lamy, F. & Tauber, S. Studies on elastase and elastin. II. Partial purification and properties of activation of proelastase. J. Biol. Chem. 238: 939–945.

1966

Bender, M. L., Begue-Canton, M. L., Blakely, R. L., Brubacher, L. j., Feder, J., Gunter, C. R., Kezdy, F. J., Killheffer, J. V., Marshall, T. H., Miller, C. G., Roeske, R. W. & Stoops, J. K. The determination of the concentration of hydrolytic enzyme solutions: α-chymotrypsin, trypsin, papain, elastase, subtilisin, and acetylcholinesterase. J. Am. Chem. Soc. 88: 5890–5913.

Hall, D. A. The identification and estimation of elastase in serum and plasma. Biochem. J. 101: 29–36.

1967

Banga, I. & Ardelt, W. Studies on the elastolytic activity of serum. Biochim. Biophys. Acta 146: 284–286.

Baumstark, J. S. Studies on the elastase–serum protein interaction. 1. Molecular identity of the inhibitors in human serum and direct demonstration of inhibitor-elastase complexes by zone and immunoelectrophoresis. Arch. Biochem. Biophys. 118: 619–630.

Brown, J. R., Kauffman, D. L. & Hartley, B. S. The primary structure of porcine pancreatic elastase. The N-terminus and disulphide bridges. Biochem J. 103: 497–507.

Geokas, M., Wilding, P., Rinderknecht, H. & Haverback, B. J. Determination of elastase in mammalian plasma and serum. J. Clin. Invest. 46: 1059–1064.

Gertler, A. & Hofmann, T. The involvement of the aminoterminal amino acid in the activity of pancreatic proteases. I. The effects of nitrous acid on elastase. J. Biol. Chem. 242: 2522–2527.

1968

Bender, M. L. & Marshall, T. H. The elastase catalysed hydrolysis of p-nitrophenyl acetate. J. Am. Chem. Soc. 90: 201–208.

Rinderknecht, H., Geokas, M. C., Silverman, P., Lillard, Y. & Haverback, B. J. New methods for the determination of elastase. Clin. Chim. Acta 19: 327–339.

Shotton, D. M., Hartley, B. S., Cameman, N., Hofmann, T. & Rao, L. Crystalline porcine pancreatic elastase. J. Mol. Biol. 32: 155–156.

Wasi, S. & Hofmann, T. A pH-dependent conformational change in porcine elastase. Biochem. J. 106: 926–930.

1969

Geneste, P. & Bender, M. L. Elastolytic activity of elastase. Proc. Natl. Acad. Sci. U.S.A. 64: 683–685.

Kaplan, H. & Dugas, H. Evidence that the activity of elastase is not dependent on the ionization of its N-terminal amino group. Biochem. Biophys. Res. Commun. 34: 681–685.

Marshall, T., Whitaker, J. & Bender, M. L. Porcine elastase. II. Properties of the tyrosinate-splitting enzymes and the specificity of elastase. Biochemistry 8: 4671–4677.

Robert, B. & Robert, L. Determination of elastolytic activity with ^{125}I and ^{131}I labelled elastin. Eur. J. Biochem. 11: 62–67.

Sampath Narayanan, A. & Anwar, R. A. The specificity of purified porcine pancreatic elastase. Biochem. J. 144: 11–17.

Visser, L. & Blout, E. R. Elastase substrates and inhibitors. Fed. Proc. Fed. Am. Soc. Exp. Biol. 28: 407 only.

1970

Ardelt, W., Ksienzy, S. & Nidzwiecka, N. Spectrophotometric method for the determination of pancreatopeptidase E activity.

Anal. Biochem. 34: 180–187.

Atlas, D., Levit, S., Schechter, I. & Berger, A. On the active site of elastase. Partial mapping by means of peptide substrates. FEBS Lett. 11: 281–283.

Baumstark, J. S. Isolation and partial characterization of an elastase-associated acidic endopeptidase and its interaction with elastase at low ionic strength. Biochim. Biophys. Acta 220: 534–551.

Baumstark, J. S. Studies on the elastase-serum protein interaction. II. On the digestion of human α_2-macroglobulin, an elastase inhibitor, by elastase. Biochim. Biophys. Acta 207: 318–330.

Bieth, J., Pichoir, M. & Metais, P. The influence of α_2-macroglobulin on the elastolytic and esterolytic activity of elastase. FEBS Lett. 8: 319–321.

Geokas, M. C., Silverman, P. & Rinderknecht, H. An elastase inhibitor from canine mandibulary gland. Experientia 25: 942–943.

Gertler, A. & Birk, Y. Isolation and characterization of porcine proelastase. Eur. J. Biochem. 12: 170–176.

Gertler, A. & Hofman, T. Acetyl-L-alanyl-L-alanyl-L-alanine methyl ester: a new highly specific elastase substrate. Can. J. Biochem. 48: 384–386.

Hartley, B. S. Homologies in serine proteinases. Philos. Trans. Roy. Soc. London Ser. B 257: 77–87.

Kaplan, H., Symonds, V. B., Dugas, H. & Whitaker, D. R. A comparison of properties of the α-lytic protease of Sorangium sp. and porcine elastase. Can. J. Biochem. 48: 649–658.

Quinn, R. S. & Blout, E. R. Spectrofluorometric assay for elastolytic enzymes. Biochem. Biophys. Res. Commun. 40: 328–333.

Shotton, D. M. Elastase. Methods Enzymol. 19: 113–140.

Shotton, D. M. & Hartley, B. S. Amino acid sequence of porcine pancreatic elastase and its homologies with other serine proteinases. Nature 225: 802–806.

Shotton, D. M. & Watson, H. C. The three-dimensional structure of crystalline porcine pancreatic elastase. Philos. Trans. R. Soc. London Ser. B 257: 111–118.

Shotton, D. M. & Watson, H. C. Three dimensional structure of tosyl-elastase. Nature 225: 811–817.

Thompson, R. C. & Blout, E. R. Evidence for an extended active center in elastase. Proc. Natl. Acad. Sci. U.S.A. 67: 1734–1740.

Watson, H. C., Shotton, D. M., Cox, J. C. & Muirhead, H. Three dimensional Fourier synthesis of tosyl-elastase at 3.5Å resolution. Nature 225: 806–811.

1971

Gertler, A. Selective, reversible loss of elastolytic activity of elastase and sublilisin resulting from electrostatic changes due to maleylation. Eur. J. Biochem. 23: 36–40.

Gertler, A. The non-specific electrostatic nature of the adsorption of elastase and other basic proteins on elastin. Eur. J. Biochem. 20: 541-546.

Gertler, A. & Feinstein, G. Inhibition of porcine elastase by turkey ovomucoid and chicken ovoinhibitor. Eur. J. Biochem. 20: 547-552.

Hartley, B. S. & Shotton, D. M. Pancreatic elastase. In: The Enzymes (Boyer, P. D. ed.), 3rd edn., vol. 3, pp. 323-373, Academic Press, New York.

Kaplan, H., Stevenson, K. J. & Hartley, B. S. Competitive labelling: a method for determining the reactivity of individual groups in proteins. The amino groups of porcine elastase. Biochem. J. 124: 289-299.

Katayama, K. & Fujita, T. Preparation of ^{131}I-labelled elastase. J. Labelled Compounds 7: 87-89.

Keller, S. & Mandl, I. Solubilized elastin as a substrate for elastase and elastase inhibitor determinations. Biochem. Med. 5: 342-347.

Rinderknecht, H., Silverman, P. & Geokas, M. C. Investigations on the standardization, assay and action of elastase. Enzymologia 40: 345-359.

Senior, R. M., Huebner, P. F. & Pierce, J. A. Measurement of elastase activity by elastin agar and its use in the detection of antitrypsin deficiency. J. Lab. Clin. Med. 77: 510-516.

Visser, L. & Blout, E. R. Elastase. II. Optical properties and the effects of sodium dodecyl sulfate. Biochemistry 10: 743-752.

Visser, L., Sigman, D. S. & Blout, E. R. Elastase. I. A new inhibitor, 1-bromo-4-(2,4-dinitrophenyl)butan-2-one. Biochemistry 10: 735-742.

1972

Atlas, D. & Berger, A. On the specificity of elastase. Hydrolysis of peptide p-nitrobenzyl esters. Biochemistry 11: 4719-4723.

Feinstein, G. & Gertler, A. The effect of limited proteolysis of chicken ovoinhibitor by bovine chymotrypsin on the inhibitory activities against trypsin, chymotrypsin and elastase. Eur. J. Biochem. 31: 25-31.

Gorbunoff, M. J. & Timashef, S. N. The role of tyrosines in elastase. Arch. Biochem. Biophys. 152: 413-422.

Kagan, H. M., Crombie, G., Jordan, R. & Franzblau, C. Proteolysis of elastin-ligand complexes. Stimulation of elastase digestion of insoluble elastin by sodium dodecyl sulfate. Biochemistry 11: 3412-3418.

Katayama, K. & Fujita, T. Studies on biotransformation of elastase. I. Transport of ^{131}I-labelled elastase across rat intestine in vitro. Biochim. Biophys. Acta 288: 172-180.

Katayama, K. & Fujita, T. Studies on biotransformation of elastase. II. Intestinal absorption of ^{131}I-labelled elastase in vivo. Biochim. Biophys. Acta 288: 181-189.

Powers, J. C. & Tuhy, P. M. Active site specific inhibitors of elastase. J. Am. Chem. Soc. 94: 6544–6545.

Trowbridge, J. O. & Moon, H. D. Purification of human elastase. Proc. Soc. Exp. Biol. Med. 141: 928–931.

Visser, L. & Blout, E. R. The use of p-nitrophenyl N-tert-butyloxycarbonyl-L-alaninate as substrate for elastase. Biochim. Biophys. Acta 268: 257–260.

1973

Atlas, D. & Berger, A. Size and sterospecificity of the active site of porcine elastase. Biochemistry 12: 2573–2577.

Baumstark, J. S. Studies on the elastase–serum protein interaction. III. The elastase inhibitors of swine serum with emphasis on the elastase–α_2-macroglobulin interaction. Biochim. Biophys. Acta 309: 181–195.

Bieth, J. & Meyer, J. F. A colorimetric method for measuring the esterolytic activity of elastase. Anal. Biochem. 51: 121–126.

Bieth, J. & Wermuth, C. G. The action of elastase on p-nitroanilide substrates. Biochem. Biophys. Res. Commun. 53: 383–390.

Brown, W. E. & Wold, E. Alkyl isocyanates as active site specific reagents for serine proteases. Identification of the active site serine as the site of reaction. Biochemistry 12: 835–841.

Brown, W. E. & Wold, E. Alkyl isocyanates as active site specific reagents for serine proteases. Reaction properties. Biochemistry 12: 828–834.

Dzialoszynski, L. & Hofmann, T. Competitive inhibition of pancreatic elastase. Biochim. Biophys. Acta 302: 406–410.

Feinstein, G., Kupfer, A. & Sokolovsky, M. N-Acetyl-(L-Ala)$_3$-p-nitroanilide as a new chromogenic substrate for elastase. Biochem. Biophys. Res. Commun. 50: 1020–1027.

Jori, G., Galiazzo, G. & Buso, O. Photosensitized oxidation of elastase. Selective modification of the tryptophyl residues. Arch. Biochem. Biophys. 158: 116–125.

Powers, J. C. & Tuhy, P. M. Active-site specific inhibitors of elastase. Biochemistry 12: 4767–4774.

Sawyer, L., Shotton, D. M. and Watson, H. C. Atomic co-ordinates for tosyl-elastase. Biochem. Biophys. Res. Commun. 53: 944–952.

Shaw, M. C. & Whitaker, D. R. A comparison of inhibition constants for the inhibition of α-lytic protease and porcine elastase by acetylated alanine peptides. Can. J. Biochem. 51: 112–114.

Shotton, D. M. & Hartley, B. S. Evidence for the amino acid sequence of porcine pancreatic elastase. Biochem. J. 131: 643–647.

Takahashi, S., Seifter, S. & Yang, F.-C. A new radioactive assay for enzymes with elastolytic activity using reduced tritiated elastin. The effect of sodium dodecyl sulphate on elastolysis. Biochim. Biophys. Acta 327: 138–145.

Thompson, R. C. Use of peptide aldehydes to generate transition-state analogs of elastase. Biochemistry 12: 47–51.

Thompson, R. C. & Blout, E. R. Dependence of the kinetic

parameters for elastase-catalysed amide hydrolysis on the length of peptide substrates. Biochemistry 12: 57-65.

Thompson, R. C. & Blout, E. R.   Elastase-catalysed amide hydrolysis of tri- and tetrapeptide amides. Biochemistry 12: 66-71.

Thompson, R. C. & Blout, E. R.   Peptide chloromethyl ketones as irreversible inhibitors of elastase. Biochemistry 12: 44-47.

Thompson, R. C. & Blout, E. R.   Restrictions on the binding of proline-containing peptides to elastase. Biochemistry 12: 51-57.

Thomson, A.   Denniss, I. S.   The reaction of active-site inhibitors with elastase using a new assay substrate. Eur. J. Biochem. 3b: 1-5.

Umezawa, H., Aoyagi, T., Okura, A., Morishima, H., Takeuchi, T. & Okami, Y.   Elastatinal, a new elastase inhibitor produced by actinomycetes. J. Antibiot. (Tokyo) 26: 787-789.

1974

Atlas, D.   The active site of pancreatic porcine elastase: specificity, size and sterospecificity. Israel J. Chem. 12: 455-469.

Bieth, J. & Frechin, J.-C.   Elastase inhibitors as impurities in commercial preparations of Kunitz soybean trypsin inhibitor. Biochim. Biophys. Acta 364: 97-102.

Bieth, J., Spiess, B. & Wermuth, C. G.   The synthesis and analytical use of a highly sensitive and convenient substrate of elastase. Biochem. Med. 11: 350-357.

Carballo, J., Kasahara, K., Appert, H. E. & Howard, J. M.   Radioimmunoassay of plasma elastase. Proc. Soc. Exp. Biol. Med. 146: 997-1002.

Feinstein, G., Hofstein, R., Koifmann, J. & Sokolovsky, M.   Human pancreatic proteolytic enzymes and protein inhibitors. Isolation and molecular properties. Eur. J. Biochem. 43: 569-581.

Jordan, R. E., Hewitt, N., Lewis, W., Kagan, H. & Franzblau, C.   Regulation of elastase-catalyzed hydrolysis of insoluble elastin by synthetic and naturally occurring hydrophobic ligands. Biochemistry 13: 3497-3503.

Karibian, D., Jones, C., Gertler, A., Dorrington, K. J. & Hofmann, T.   On the reaction of acetic and maleic anhydride with elastase. Evidence for a role of the NH_2-terminal valine. Biochemistry 13: 2891-2897.

Kasafírek, E., Frič, P. & Mališ, F.   The significance of the N-acyl residue of L-alanyl-L-alanyl-L-alanine p-nitroanilide for cleavage by pancreatic elastase. FEBS Lett. 40: 353-356.

Katayama, K. & Fujita, T.   Studies on biotransformation of elastase. III. Effects of elastase-binding proteins in serum on the disappearance of ^{131}I-labelled elastase from blood. Biochim. Biophys. Acta 336: 165-177.

Katayama, K. & Fujita, T.   Studies on biotransformation of elastase. IV. Tissue distribution of ^{131}I-labelled elastase and intracellular distribution in liver after intravenous administration in rats. Biochim. Biophys. Acta **336**: 178-190.

Katayama, K. & Fujita, T.   Studies on biotransformation of elastase. V. Degradation of injected ^{131}I-labelled elastase by subcellular particles of rat liver. Biochim. Biophys. Acta **336**: 191-200.

Lievaart, P. A. & Stevenson, K. J.   The isolation of trypsin and elastase from moose pancreas (Alces alces) by affinity chromatography on lima-bean protease inhibitor-Sepharose resin. Can. J. Biochem. **52**: 637-641.

Robert, B., Hornebeck, W., & Robert, L.   Cinétique hétérogène de l'interaction élastine-élastase. Biochimie **56**: 239-244.

Thompson, R. C.   Binding of peptides to elastase: implications for the mechanism of substrate hydrolysis. Biochemistry **13**: 5495-5501.

1975

Atlas, D.   The active site of porcine elastase. J. Mol. Biol. **93**: 39-53.

Bielefeld, D. R., Senior, R. M. & Yu, S. Y.   A new method for determination of elastolytic activity using [^{14}C]labelled elastin and its application to leukocyte elastase. Biochem. Biophys. Res. Commun. **67**: 1553-1559.

Cohen, A. B.   The interaction of α-1-antitrypsin with chymotrypsin, trypsin and elastase. Biochim. Biophys. Acta **391**: 193-200.

Gold, R. & Shalitin, Y.   On the specificity of porcine elastase. Biochim. Biophys. Acta **410**: 421-425.

Ledoux, M. & Lamy, F.   Electrophoretic characterization of porcine pancreatic (pro)elastases A and B. Can. J. Biochem. **53**: 421-432.

Mallory, P. A. & Travis, J.   Human pancreatic enzymes: purification and characterisation of a non-elastolytic enzyme, protease E, resembling elastase. Biochemistry **14**: 722-730.

Meyer, J.-F., Beith, J. & Métais, P.   On the inhibition of elastase by serum. Some distinguishing properties of α_1-antitrypsin and α_2-macroglobulin. Clin. Chim. Acta **62**: 43-53.

Okura, A., Morishima, H., Takita, T., Aoyogi, T., Takeuchi, T. & Umezawa, H.   The structure of elastatinal, an elastase inhibitor of microbial origin. J. Antibiot. (Tokyo) **28**: 337-339.

Powers, J. C. & Carroll, D. L.   Reaction of acyl carbazates with proteolytic enzymes. Biochem. Biophys. Res. Commun. **67**: 639-644.

Powers, J. C., Carroll, D. L. & Tuhy, P. M.   Synthetic active site-directed inhibitors of elastolytic proteases. Ann. N. Y. Acad. Sci. **256**: 420-425.

Stevenson, K. J. & Voordouw, J. K.   Characterization of trypsin and elastase from the moose (Alces alces). 1. Amino acid composition and specificity towards polypeptides. Biochim. Biophys. Acta **386**: 324-331.

Stroud, R. M., Krieger, M., Koeppe, R. E., II., Kossiakoff, A. A. & Chambers, J. L.   Structure-function relationships in the serine proteases. In: Proteases and Biological Control (Reich,

E., Rifkin, D. B. & Shaw, E. eds), Cold Spring Harbor Laboratory,
New York, pp. 13-32.

1976

Ardelt, W., Koj, A., Chudzik, J. & Dubin, A.   Inactivation of some
pancreatic and leucocyte elastases by peptide chloromethyl
ketones and alkyl isocyanates. FEBS Lett. 67: 156-160.

Bauer, C. A., Thompson, R. C. & Blout, E. R.   The active centers
of Streptomyces griseus protease 3, α-chymotrypsin and elastase.
Enzyme substrate interactions close to the scissile bond.
Biochemistry 15: 1296-1299.

Bauer, C. A., Thompson, R. C. & Blout, E. R.   The active centers
of Streptomyces griseus protease 3, α-chymotrypsin and elastase.
Enzyme substrate interactions remote from the scissile bond.
Biochemistry 15: 1291-1295.

Bieth, J., Pichoir, M. & Metais, P.   Pseudo-zero order
solubilization of Remazol-Brilliant-Blue elastin by elastase.
Anal. Biochem. 70: 430-433.

Breaux, E. J. & Bender. M. L.   Direct spectrophotometric
observation of an acyl-enzyme intermediate in elastase
catalysis. Biochem. Biophys. Res. Commun. 70: 235-240.

Dimicoli, J.-L., Bieth, J. & Lhoste, J. M.   Trifluoroacetylated
peptides as substrates and inhibitors of elastase: a nuclear
magnetic resonance study. Biochemistry 15: 2230-2236.

El-Ridi, S. S. & Hall, D. A.   Isolation of elastase from human
serum. Biochem. Soc. Trans. 4: 336-337.

Kasafírek, E., Frič, P., Slabý, J. & Mališ, F.   p-Nitroanilides of
3-carboxypropionyl-peptides. Their cleavage by elastase,
trypsin, and chymotrypsin. Eur. J. Biochem. 69: 1-13.

Largman, C., Brodrick, J. W. & Geokas, M. C.   Purification and
characterization of two human pancreatic elastases. Biochemistry
15: 2491-2500.

Ohlsson, K. & Olsson, A.S.   Purification and partial
characterization of human pancreatic elastase. Hoppe-Seyler's
Z. Physiol. Chem. 357: 1153-1161.

1977

Aubry, M. & Bieth, J.   Kinetics of the inactivation of human and
bovine trypsins and chymotrypsins by α_1-proteinase inhibitor and
of their reactivation by α_2-macroglobulin. Clin. Chim. Acta 78:
371-380.

Dimicoli, J. L. & Bieth, J.   Location of the calcium ion binding
site in porcine pancreatic elastase using a lanthanide ion
probe. Biochemistry 16: 5532-5537.

Dorn, C. P., Zimmerman, M., Yang, S. S., Yurewicz, E. C., Ashe, B.
M., Frankshun, R. & Jones, H.   Proteinase inhibitors. 1.
Inhibitors of elastase. J. Med. Chem. 20: 1464-1468.

Geokas, M. C., Brodrick, J. W., Johnson, J. H. & Largman, C.
Pancreatic elastase in human serum. Determination by
radioimmunoassay. J. Biol. Chem. 252: 61-67.

Gnosspelius, G.   Assay of elastase activity using trypsin as amplifying agent. Anal. Biochem. 81: 315-319.

Lestienne, P., Dimicoli, J.-L. & Bieth, J. Trifluoroacetyl tripeptides as potent inhibitors of human leukocyte elastase. J. Biol. Chem. 252: 5931-5933.

Powers, J. C. & Gupton, B. F. Reaction of serine proteases with aza-amino acid and aza-peptide derivatives. Methods Enzymol. 46: 208-216.

Powers, J. C., Gupton, B. F., Harley, A. D., Nishino, N. & Whitley, R. J. Specificity of porcine pancreatic elastase, human leucocyte elastase and cathepsin G. Inhibition with peptide chloromethyl ketones. Biochim. Biophys. Acta 485: 156-166.

Rifkin, D. B. & Crowe, R. M. A sensitive assay for elastase employing radioactive elastin coupled to Sepharose. Anal. Biochem. 79: 268-275.

Schechter, Y., Burstein, Y. & Gertler, A. The effect of oxidation of methionine residues in chicken ovoinhibitor on its inhibitory activities against trypsin, chymotrypsin and elastase. Biochemistry 16: 992-997.

Stone, P. J., Crombie, G. & Franzblau, C. The use of tritiated elastin for the determination of subnanogram amounts of elastase. Anal. Biochem. 80: 572-577.

Twumasi, D. Y., Liener, I. E., Galdstone, M. & Levytska, V. Activation of human leukocyte elastase by human αμ-macroglobulin. Nature 267: 61-63.

Zimmerman, M., Ashe, B., Yurewicz, E. C. & Patel, G. Sensitive assays for trypsin, elastase, and chymotrypsin using new fluorogenic substrates. Anal. Biochem. 78: 47-51.

1978

Bieth, J. Elastases: structure, function and pathological role. Front. Matrix Biol. 6: 1-82.

Christner, P., Weinbaum, G., Sloan, B. & Rosenbloom, J. Degradation of tropoelastin by proteases. Anal. Biochem. 88: 682-688.

Katagiri, K., Takeuchi, T., Taniguchi, K. & Sasaki, M. One step purification procedure of elastase from pancreatic powder by affinity chromatography. Anal. Biochem. 86: 159-165.

Lively, M. O. & Powers, J. C. Specificity and reactivity of human granulocyte elastase and cathepsin G, porcine pancreatic elastase, bovine chymotrypsin and trypsin toward inhibition with sulfonyl fluorides. Biochim. Biophys. Acta 525: 171-179.

Winninger, C., Lestienne, P., Dimicoli, J.-L. & Bieth, J. NMR and enzymatic investigation of the interaction between elastase and sodium trifluoroacetate. Biochim. Biophys. Acta 526: 227-234.

1979

Castillo, M. J., Nakajima, K., Zimmerman, M. & Powers, J. C. Sensitive substrates for human leukocyte and porcine pancreatic elastase: a study of the merits of various chromophoric and fluorogenic leaving groups in assays for serine proteases.

Anal. Biochem. **99**: 53-64.

Satoh, S., Kurecki, T., Kress, L. F. & Laskowski, M., Jr. The dual nature of the reaction between porcine elastase and human α_1-proteinase inhibitor. Biochem. Biophys. Res. Commun. **86**: 130-137.

Thompson, R. C. & Bauer, C.-A. Reaction of peptide aldehydes with serine proteases. Implications for the entropy changes associated with enzymic catalysis. Biochemistry **18**: 1552-1558.

Entry 1.06

PROTEASE E

Summary

EC Number: (3.4.21.-)

Earlier names: -

Distribution: Human pancreas.

Source: Human pancreas.

Action: Hydrolyzes Boc-Ala-OPhNO$_2$, Ac-(Ala)$_3$-OMe and Suc-(Ala)$_3$-NPhNO$_2$ (substrates of elastase); also casein and hemoglobin, but has little or no action on elastin.

Requirements: Neutral pH.

Substrate, usual: Boc-Ala-OPhNO$_2$.

Inhibitors: Dip-F, Ac-Ala-Ala-Pro-Ala-CH$_2$Cl. α_1-Proteinase inhibitor (but poor inhibition by several other natural inhibitors).

Molecular properties: Mol.wt. 31,000; anionic (unlike most true elastases).

Comment: It is suggested that the low pI of this enzyme prevents the electrostatic interaction with elastin that contributes to elastolysis by elastase I (#1.05). There is a weak elastolytic activity in autolyzed human pancreas, however (see #1.07).

Bibliography

1975

Mallory, P. A. & Travis, J. Inhibition spectra of the human pancreatic endopeptidases. Am. J. Clin. Nutr. 28: 823-830.

Mallory, P. A. & Travis, J. Human pancreatic enzymes: purification and characterization of a nonelastolytic enzyme, protease E, resembling elastase. Biochemistry 14: 722-730.

1976

Largman, C., Brodrick, J. W. & Geokas, M. C. Purification and characterization of two human pancreatic elastases. Biochemistry 15: 2491-2500.

1978

Kobayashi, R., Kobayashi, Y. & Hirs, C. H. W. Identification of a
   binary complex of procarboxypeptidase A and a precursor of
   protease E in porcine pancreatic secretion. J. Biol. Chem. 253:
   5526–5530.

Entry 1.07

PANCREATIC ELASTASE II

Summary

EC Number: (3.4.21.-)

Earlier names: Pancreatic elastase 2.

Distribution: Porcine and human pancreas (as proelastase II).

Source: Pig pancreas.

Action: Digests elastin by cleavage of Leu-, Phe- and Tyr- bonds. Also active against synthetic substrates of chymotrypsin, but not of elastase I (#1.05).

Requirements: Alkaline pH.

Substrate, usual: Elastin. Ac-Tyr-OEt.

Inhibitors: Tos-Phe-CH$_2$Cl. (Not turkey ovomucoid, unlike elastase I.)

Molecular properties: Mol.wt. 26,500; A280,1% 27.4; pI 10.7. N-terminal isoleucine.

Comment: More like chymotrypsin than elastase in specificity, apart from solubilization of elastin. The human pancreatic elastase is an enzyme with weak elastinolytic activity resembling elastase II rather than elastase I [1].

References
[1] Largman et al. Biochemistry 15: 2491-2500, 1976.

Bibliography

1957

Grant, N. H. & Robbins, K. C. Studies on porcine elastase and proelastase. Arch. Biochem. Biophys. 66: 396-403.

1965

Avrameas, S. & Uriel, J. Systematic fractionation of swine pancreatic hydrolases. II. Fractionation of enzymes insoluble in ammonium sulfate solution at 0.40 saturation. Biochemistry 4: 1750-1757.

1969

Uram, M. & Lamy, F. Purification of two proeleastase from porcine pancreas. Biochim. Biophys. Acta 194: 102–111.

1972

Clemente, F., de Caro, A. & Figarella, C. Composition du suc pancréatique humain. Étude immunoenzymologique. Eur. J. Biochem. 31: 186–193.

1974

Ardelt, W. Partial purification and properties of porcine pancreatic elastase II. Biochim. Biophys. Acta 341: 318–326.

1975

Ardelt, W. Physical parameters and chemical composition of porcine pancreatic elastase II. Biochim. Biophys. Acta 393: 267–273.

Ledoux, M. & Lamy, F. Electrophoretic characterization of porcine pancreatic (pro)elastasès A and B. Can. J. Biochem. 53: 421–432.

1976

Ardelt, W., Koj, A., Chudzik, J. & Dubin, A. Inactivation of some pancreatic and leucocyte elastases by peptide chloromethyl ketones and alkyl isocyanates. FEBS Lett. 67: 156–160.

Largman, C., Brodrick, J. W. & Geokas, M. C. Purification and characterization of two human pancreatic elastases. Biochemistry 15: 2491–2500.

1977

Gertler, A., Weiss, Y. & Burstein, Y. Purification and characterization of porcine elastase II and investigation of its elastolytic specificity. Biochemistry 16: 2709–2716.

1979

Del Mar, E. G., Brodrick, J. W., Geokas, M. C. & Largmann, C. Effect of oxidation of methionine in a peptide substrate for human elastases: a model for inactivation of α_1-protease inhibitor. Biochem. Biophys. Res. Commun. 88: 346–350.

Largman, C., Brodrick, J. W., Geokas, M. C., Sischo, W. M. & Johnson, J. H. Formation of a stable complex between human proelastase 2 and human α_1-protease inhibitor. J. Biol. Chem. 254: 8516–8523.

1980

Del Mar, E. G., Largman, C., Brodrick, J. W., Fassett, M. & Geokas, M. C. Substrate specificity of human pancreatic elastase 2. Biochemistry 19: 468–472.

Entry 1.08

DOG PANCREATIC COLLAGENASE

Summary

EC Number: (3.4.21.-)

Earlier names: -

Distribution: Dog pancreatic secretion. The limits of species and tissues distribution are not defined.

Source: Dog pancreatic secretion stimulated by pancreozymin.

Action: Degrades soluble native collagen to small peptides. Also active with insoluble collagen, but not casein.

Requirements: Neutral pH.

Substrate, usual: Collagen.

Inhibitors: Partially inhibited by 10mM Dip-F or cysteine.

Molecular properties: -

Comment: "Serine" collagenases have been reported previously from crab hepatopancreas [1] and lower organisms [2].

References
[1] Eisen & Jeffrey Biochim. Biophys. Acta 191: 517-526, 1969.
[2] Keil In: Enzyme Regulation and Mechanism of Action (Mildner & Ries, eds), pp. 351-362, Pergamon, Oxford, 1980.

Bibliography

1974

Takahashi, S. & Seifter, S. An enzyme with collagenolytic activity from dog pancreatic juice. Israel J. Chem. 12: 557-571.

Entry 1.09

COAGULATION FACTOR XII$_a$

Summary

EC Number: (3.4.21.-)

Earlier names: Hageman factor (activated), contact factor. May
  well be identical with plasma factor PF/dil (#1.19).

Distribution: Plasma, as the precursor.

Source: Human and bovine plasma (about 30 µg/ml in human plasma).

Action: Activates prekallikrein, coagulation factor XI, factor VII
  and plasma plasminogen proactivator (often cleaving -Arg-Ile-
  bonds). Shows restricted trypsin-like esterase activity.

Requirements: Neutral pH.

Substrate, usual: Prekallikrein, activation being followed by
  increase of activity against Bz-Arg-OEt. Bz-Arg-OMe, Tos-Arg-OMe,
  Ac-Gly-Lys-OMe.

Substrate, special: Bz-Phe-Val-Arg-NPhNO$_2$.

Inhibitors: Dip-F, Pms-F, benzamidine. Lima bean trypsin inhibitor
  (not soybean trypsin inhibitor or ovomucoid). C$\overline{\text{I}}$ inhibitor,
  antithrombin-heparin.

Molecular properties: Bovine factor XII is a single chain
  glycoprotein, mol.wt. 74,000, A280,1% 14.2, with sequences
  resembling inhibitor and active site at N- and C-termini,
  respectively.
    Factor XII$_a$ is the product of cleavage of an -Arg-Val- bond
  within a disulfide loop, so that the two chains (mol.wt. 46,000
  and 28,000) are disulfide-linked. The light chain contains the
  catalytic residues. Human factor XII, mol.wt. 76,000, is
  activatable by plasmin, kallikrein or factor XII$_a$ to form
  fragments of 52,000, 40,000 and 28,000 mol.wt., the last again
  containing the catalytic residues. Normally, the fragments
  separate without reduction, but there may be a form in which the
  light chain is disulfide-bonded, as in the bovine enzyme.

Comment: Factor XII adheres readily to glass, and tends to be
  activated spontaneously (perhaps without proteolytic cleavage)
  in contact with negatively charged and other surfaces (e.g.
  kaolin, barium carbonate, charcoal); the process is stimulated
  by high molecular weight kininogen, and inhibited by
  hexadimethrine bromide (Polybrene). Traces of ellagic acid (10^{-8}

M) and bacterial endotoxin also activate. Physiological
activation may follow adsorption to collagen and vascular
basement membrane. Catalytic activation by kallikrein probably
completes a positive feed-back loop.

Bibliography

1960

Becker, E. L. Inactivation of Hageman factor by
diisopropylfluorophosphate. J. Lab. Clin. Med. 56: 136-138.

1961

Ratnoff, O. D., Davie, E. W. & Mallett, D. L. Studies on the
action of Hageman factor: evidence that activated Hageman factor
in turn activates plasma thromboplastin antecedent. J. Clin.
Invest. 40: 803-819.

Webster, M. E. & Ratnoff, O. D. Role of Hageman factor in the
activation of vasodilator activity in human plasma. Nature 192:
180-181.

1964

Ratnoff, O. D. & Crum, J. D. Activation of Hageman factor by
solutions of ellagic acid. J. Lab. Clin. Med. 63: 359-377.

Schoenmakers, J., Matze, R., Haanen, C. & Zilliken, F. Proteolytic
activity of purified bovine Hageman factor. Biochim. Biophys.
Acta 93: 433-436.

1965

Donaldson, V. H. & Ratnoff, O. D. Hageman factor: alterations in
physical properties during activation. Science 150: 754-756.

Schoenmakers, J. G. G., Matze, R., Haanen, C. & Zilliken, F.
Hageman factor, a novel sialoglycoprotein with esterase
activity. Biochim. Biophys. Acta 101: 166-176.

1968

Esnouf, M. P. & Macfarlane, R. G. Enzymology and the blood
clotting mechanism. Advan. Enzymol. 30: 255-315.

Nagasawa, S., Takahashi, H., Koida, M. & Suzuki, T. Partial
purification of bovine plasma kallikreinogen, its activation by
the Hageman factor. Biochem. Biophys. Res. Commun. 32: 644-649.

1970

Forbes, C. D., Pensky, J. & Ratnoff, O. D. Inhibition of
activated Hageman factor and activated plasma thromboplastin
antecedent by purified serum C$\bar{1}$ inactivator. J. Lab. Clin.
Med. 76: 809-815.

Kaplan, A. P. & Austen, K. F. A pre-albumin activator of
prekallikrein. J. Immunol. 105: 802-811.

1971

Cochrane, C. G. & Wuepper, K. D. The first component of the

kinin-forming system in human and rabbit plasma. Its relationship to clotting factor XII (Hageman factor). J. Exp. Med. 134: 986-1004.

Kaplan, A. P. & Austen, K. F. A prealbumin activator of prekallikrein. II. Derivation of activators of prekallikrein from active Hageman factor by digestion with plasmin. J. Exp. Med. 133: 696-712.

Soltay, M. J., Movat, H. Z. & Ozge-Anwar, A. H. The kinin system of human plasma. V. The probable derivation of prekallikrein activator from activated Hageman factor (XIIa). Proc. Soc. Exp. Biol. Med. 138: 952-958.

1972

Kaplan, A. P. & Austen, K. F. The fibrinolytic pathway of human plasma. Isolation and characterization of the plasminogen proactivator. J. Exp. Med. 136: 1378-1393.

Komiya, M., Nagasawa, S. & Suzuki, T. Bovine prekallikrein activator with functional activity as Hageman factor. J. Biochem. 72: 1205-1218.

Takahashi, N., Nagasawa, S. & Suzuki, T. Conversion of bovine prekallikrein to kallikrein. Evidence of limited proteolysis of prekallikrein by bovine Hageman factor (factor XII). FEBS Lett. 24: 98-100.

1973

Bagdasarian, A., Lahiri, B. & Colman. R. W. Origin of the high molecular weight activator of prekallikrein. J. Biol. Chem. 248: 7742-7747.

Cochrane, C. G., Revak, S. D. & Wuepper, K. D. Activation of Hageman factor in solid and fluid phases. A critical role of kallikrein. J. Exp. Med. 138: 1564-1583.

Laake, K. & Venneröd, A. M. Determination of factor XII in human plasma with arginine proesterase (prekallikrein). I. Preparation and properties of the substrate. Thromb. Res. 2: 393-407.

Laake, K. & Venneröd, A. M. Determination of factor XII in human plasma with arginine proesterase (prekallikrein). II. Studies on the method. Thromb. Res. 2: 409-421.

Stormorken, H., Gjönnaess, H. & Laake, K. Interrelations between the clotting and kinin systems. Haemostasis 2: 245-252.

1974

Heck, L. W. & Kaplan, A. P. Substrates of Hageman factor. 1. Isolation and characterization of human factor XI (PTA) and inhibition of the activated enzyme by α_1-antitrypsin. J. Exp. Med. 140: 1615-1630.

Johnston, A. R., Cochrane, C. G. & Revak, S. D. The relationship between PF/dil and activated human Hageman factor. J. Immunol. 113: 103-109.

Laake, K. & Venneröd, A. M. Factor XII induced fibrinolysis: studies on the separation of prekallikrein, plasminogen proactivator, and factor XI in human plasma. Thromb. Res. 4:

285-302.
Laake, K., Vennceröd, A. M., Haugen, G. & Gjönnaess, H. Cold-
promoted activation of factor VII in human plasma: studies on
the associated acyl-arginine esterase activity. Thromb. Res. 4:
769-785.
McMillin, C. R., Saito, H., Ratnoff, O. D. & Walton, A. G. The
secondary structure of human Hageman factor (factor XII) and its
alteration by activating agents. J. Clin. Invest. 54: 1312-
1322.
Morrison, D. C. & Cochrane, C. G. Direct evidence for Hageman
factor (Factor XII) activation by bacterial lipopolysaccharides
(endotoxins). J. Exp. Med. 140: 797-811.
Movat, H. Z. & Özge-Anwar, A. H. The contact phase of blood
coagulation: clotting factor XI and XII, their isolation and
interaction. J. Lab. Clin. Med. 84: 861-878.
Revak, S. D., Cochrane, C. G., Johnston, A. R. & Hugli, T. E.
Structural changes accompanying enzymatic activation of human
Hageman factor. J. Clin. Invest. 54: 619-627.
Saito, H., Ratnoff, O. D. & Donaldson, V. H. Defective activation
of clotting, fibrinolytic, and permeability-enhancing systems in
human Fletcher trait plasma. Circ. Res. 34: 641-651.
Vennceröd, A. M. & Laake, K. Isolation and characterization of a
prealbumin activator of prekallikrein from acetone-activated
human plasma. Thromb. Res. 4: 103-118.

                                    1975

Davie, E. W. & Fujikawa, K. Basic mechanisms in blood coagulation.
Annu. Rev. Biochem. 44: 799-829.
Davie, E. W., Fujikawa, K., Legaz, M. E. & Kato, H. Role of
proteases in blood coagulation. In: Proteases in Biological
Control (Reich, E., Rifkin, D. B. & Shaw, E. eds), pp. 65-77,
Cold Spring Harbor Laboratory, New York.
Oh-ishi, S. & Webster, M. E. Vascular permeability factors
(PF/Nat and PF/Dil) - their relationship to Hageman factor and
the kallikrein-kinin system. Biochem. Pharmacol. 24: 591-598.
Saito, H. & Ratnoff, O. D. Alteration of factor VII activity by
activated Fletcher factor (a plasma kallikrein): a potential
link between the intrinsic and extrinsic blood-clotting systems.
J. Lab. Clin. Med. 85: 405-415.
Saito, H., Ratnoff, O. D., Waldman, R. & Abraham, J. P.
Fitzgerald trait. Deficiency of a hitherto unrecognized agent,
Fitzgerald factor, participating in surface-mediated reactions
of clotting, fibrinolysis, generation of kinins, and the
property of diluted plasma enhancing vascular permeability
(PF/Dil). J. Clin. Invest. 55: 1082-1089.
Ulevitch, R. J., Cochrane, C. G., Revak, S. D., Morrison, D. C. &
Johnston, A. R. The structural and enzymatic properties of the
components of the Hageman factor-activated pathways. In:
Proteases and Biological Control (Reich, E., Rifkin, D. B. &
Shaw, E. eds), pp.85-93, Cold Spring Harbor Laboratory, New
York.

Wuepper, K. D., Miller, D. R. & Lacombe, M. J. Flaujeac trait. Deficiency of human plasma kininogen. J. Clin. Invest. 56: 1663-1672.

1976

Chan, J. Y. C. & Movat, H. Z. Purification of factor XII (Hageman factor) from human plasma. Thromb. Res. 8: 337-349.

Chan, J. Y. C., Habal, F. M., Burrows, C. E. & Movat, H. Z. Interaction between factor XII (Hageman factor), high molecular weight kininogen and prekallikrein. Thromb. Res. 9: 423-433.

Cochrane, C. G., Revak, S. D., Ulevitch, R., Johnston, A. & Morrison, D. Hageman factor: characterization and mechanism of activation. In: Chemistry and Biology of the Kallikrein-Kinin System in Health and Disease, DHEW Publication No. (NIH) 76-791 (Pisano, J. J. & Austen, K. F. eds), pp. 17-35, U.S. Government Printing Office, Washington.

Griffin, J. H. & Cochrane, C. G. Human factor XII (Hageman factor). Methods Enzymol. 45: 56-65.

Griffin, J. H. & Cochrane, C. G. Mechanisms for the involvement of high molecular weight kininogen in surface-dependent reactions of Hageman factor. Proc. Natl. Acad. Sci. U.S.A. 73: 2554-2558.

Kaplan, A. P., Meier, H. L., Yecies, L. D. & Heck, L. W. Hageman factor and its substrates: the role of factor XI, prekallikrein, and plasminogen proactivator in coagulation, fibrinolysis, and kinin generation. In: Chemistry and Biology of the Kallikrein-Kinin System in Health and Disease, DHEW Publication No. (NIH) 76-791 (Pisano, J. J. & Austen, K. F. eds), pp. 237-254, U.S. Government Printing Office, Washington.

Revak, S. D. & Cochrane, C. G. Hageman factor: its structure and modes of activation. Thromb. Heamostas. 35: 570-575.

Revak, S. D. & Cochrane, C. G. The relationship of structure and function in human Hageman factor. The association of enzymatic and binding activities with separate regions of the molecule. J. Clin. Invest. 57: 852-860.

Spragg, J. Immunological assays for components of the human plasma kinin-forming system. In: Chemistry and Biology of the Kallikrein-Kinin System in Health and Disease, DHEW Publication No. (NIH) 76-791 (Pisano, J. J. & Austen, K. F. eds), pp. 195-204, U.S. Government Printing Office, Washington.

Stead, N., Kaplan, A. P. & Rosenberg, R. D. Inhibition of activated factor XII by antithrombin-heparin cofactor. J. Biol. Chem. 251: 6481-6488.

Suzuki, T. & Kato, H. Protein components of the bovine kallikrein-kinin system. In: Chemistry and Biology of the Kallikrein-Kinin System in Health and Disease, DHEW Publication No. (NIH) 76-791 (Pisano, J. J. & Austen, K. F. eds), pp. 107-119, U.S. Government Printing Office, Washington.

Webster, M. E. & Oh-ishi, S. Activation of Hageman factor (factor XII): requirement for activators other than prekallikrein. In: Chemistry and Biology of the Kallikrein-Kinin System in Health

and Disease, DHEW Publication No. (NIH) 76-791 (Pisano, J. J. &
Austen, K. F. eds), pp. 55-60, U.S. Government Printing Office,
Washington.

1977

Chan, J. Y. C., Burrowes, C. E. & Movat, H. Z. Activation of
    factor XII (Hageman factor): enhancing effect of a potentiator.
    Thromb. Res. 10: 309-313.
Chan, J. Y. C., Burrowes, C. E., Habal, F. M. & Movat, H. Z. The
    inhibition of activated factor XII (Hageman factor) by
    antithrombin III; the effect of other plasma proteinase
    inhibitors. Biochem. Biophys. Res. Commun. 74: 150-158.
Fujikawa, K., Kurachi, K. & Davie, E. W. Characterization of
    bovine factor XII_a (activated Hageman factor). Biochemistry 16:
    4182-4188.
Fujikawa, K., Walsh, K. A. & Davie, E. W. Isolation and
    characterization of bovine factor XII (Hageman factor).
    Biochemistry 16: 2270-2278.
Griffin, J. H. Molecular mechanism of surface-dependent
    activation of Hageman factor (HF) (coagulation factor XII).
    Fed. Proc. Fed. Am. Soc. Exp. Biol. 36: 329 only.
Kisiel, W., Fujikawa, K. & Davie, E. W. Activation of bovine
    factor VII (proconvertin) by factor XIIa (activated Hageman
    factor). Biochemistry 16: 4189-4194.
Kurachi, K. & Davie, E. W. Activation of human factor XI (plasma
    thromboplastin antecedent) by factor XII_a (activated Hageman
    factor). Biochemistry 16: 5831-5839.
Mandle, R., Jr. & Kaplan, A. P. Hageman factor substrates. Human
    plasma prekallikrein: mechanism of activation by Hageman factor
    and participation in Hageman factor-dependent fibrinolysis. J.
    Biol. Chem. 252: 6097-6104.
Meier, H. L., Pierce, J. V., Colman, R. W. & Kaplan, A. P.
    Activation and function of human Hageman factor. The role of
    high molecular weight kininogen and prekallikrein. J. Clin.
    Invest. 60: 18-31.
Radcliffe, R., Bagdasarian, A., Colman, R. & Nemerson, Y.
    Activation of bovine factor VII by Hageman factor fragments.
    Blood 50: 611-617.
Revak, S. D., Cochrane, C. G. & Griffin, J. H. The binding and
    cleavage characteristics of human Hageman factor during contact
    activation. J. Clin. Invest. 59: 1167-1175.
Schiffman, S., Pecci, R. & Lee, P. Contact activation of factor
    Xl: evidence that the primary role of contact activation
    cofactor (CAC) is to facilitate the activation of factor Xll.
    Thromb. Res. 10: 319-323.

1978

Chan, J. Y. C., Burrowes, C. E. & Movat, H. Z. Surface activation
    of factor XII (Hageman factor) - critical role of high molecular
    weight kininogen and another potentiator. Agents Actions 8: 66-
    72.

Claeys, H. & Collen, D. Purification and characterization of bovine coagulation factor XII (Hageman factor). Eur. J. Biochem. 87: 69–74.

Goldsmith, G. H., Jr., Saito, H. & Ratnoff, O. D. Activation of plasminogen by Hageman factor (Factor-XII) and Hageman factor fragments. J. Clin. Invest. 62: 54–60.

Griffin, J.H. Role of surface in surface-dependent activation of Hageman Factor (blood coagulation Factor XII). Proc. Natl. Acad. Sci. U.S.A. 75: 1998–2002.

Revak, S. D., Cochrane, C. G., Bouma, B. N. & Griffin, J. H. Surface and fluid phase activities of two forms of activated Hageman factor produced during contact activation of plasma. J. Exp. Med. 147: 719–729.

Seligsohn, U., Østerud, B., Griffin, J. H. & Rapaport, S. I. Evidence for the participation of both activated factor XII and activated factor IX in cold-promoted activation of factor VII. Thromb. Res. 13: 1049–1056.

Yecies, L., Meier, H., Mandle, R., Jr. & Kaplan, A. P. The Hageman factor-dependent fibrinolytic pathway. In: Progress in Chemical Fibrinolysis and Thrombolysis (Davidson, J. F., Rowan, R. M., Samama, M. M. & Desnoyers, P. C. eds), vol. 3, pp. 121–139, Raven Press, New York.

1979

Beretta, G. & Griffin, J. H. DFP studies of the mechanism of surface dependent reactions of Hageman factor (HF, factor XII). Fed. Proc. Fed. Am. Soc. Exp. Biol. 38: 811 only.

Colman, R. W., Shames, P. V., Liu, C. Y., Scott, C. F. & Kirby, E. P. Von Willebrand's factor (VWF) potentiates prekallikrein activation by Hageman factor fragments. Fed. Proc. Fed. Am. Soc. Exp. Biol. 38: 1272 only.

Davie, E. W., Fujikawa, K., Kurachi, K. & Kisiel, W. The role of serine proteases in the blood coagulation cascade. Adv. Enzymol. 48: 277–318.

Ratnoff, O. D. & Saito, H. Amidolytic properties of single-chain activated Hageman factor. Proc. Natl. Acad. Sci. U.S.A. 76: 1461–1463.

Ratnoff, O. D. & Saito, H. Interactions among Hageman factor, plasma prekallikrein, high molecular weight kininogen, and plasma thromboplastin antecedent. Proc. Natl. Acad. Sci. U.S.A. 76: 958–961.

Saito, H., Goldsmith, G. H., Moroi, M. & Aoki, N. Inhibitory spectrum of alpha-2-plasmin inhibitor. Proc. Natl. Acad. Sci. U.S.A. 76: 2013–2017.

Vinazzer, H. Assay of total factor XII and of activated factor XII in plasma with a chromogenic substrate. Thromb. Res. 14: 155–166.

Entry 1.10

PLASMA KALLIKREIN

Summary

EC Number: 3.4.21.8 (tissue kallikrein inappropriately bears the same number).

Earlier names: Kininogenin, kininogenase. The precursor has been called kallikreinogen, Fletcher factor and prekallikrein, but the best name may be prokallikrein.

Distribution: Plasma of man and other mammals (as prekallikrein).

Source: Bovine, rabbit or human plasma (prekallikrein)(10 mg/100 ml in man).

Action: Liberates bradykinin from plasma high mol.wt. kininogen by cleavage of -Lys-Arg- and -Arg-Ser- bonds. Hydrolyzes polyarginine, salmine and arginine-rich histones, but not polylysine or casein. Activates coagulation factor XII, factor VII and plasminogen.

Requirements: pH 7-9.

Substrate, usual: Tos-Arg-OMe, Bz-Arg-OEt, Z-Lys-OPhNO$_2$.

Substrate, special: Bz-Pro-Phe-Arg-NPhNO$_2$ (Chromozym PK), D-Pro-Phe-Arg-NPhNO$_2$ (S-2302), Z-Phe-Arg-NMec. Gbz-OPhNO$_2$ for active site titration.

Inhibitors: Dip-F, Pro-Phe-Arg-CH$_2$Cl, Boc-Lys-CH$_2$Cl (but not Tos-Lys-CH$_2$Cl), benzamidine, Gbz-OPhNO$_2$, Cu^{2+} (10^{-6}M). Aprotinin, soybean (not lima bean) trypsin inhibitor. $\overline{C1}$ inhibitor, α_2-macroglobulin [1] (cf. tissue kallikreins), antithrombin III-heparin [2] (not α_1-proteinase inhibitor).

Molecular properties: Prekallikrein, a single chain glycoprotein of apparent mol.wt. 115,000, is converted to kallikrein, mol.wt. 90,000, by factor XII$_a$ or XII$_f$. Kallikrein has light and heavy chains (mol.wt. 36,000 and 52,000), and the light chain contains the catalytic Ser. The heavy chain is autolytically degraded to smaller fragments. Kallikrein has pI 8.3-9.1.

Comment: Plasma kallikrein has a much higher mol.wt. than the tissue kallikreins. Prekallikrein exists in the plasma as a non-covalent complex with high mol.wt. kininogen, apparent mol.wt. 285,000 [3]. After activation, part of the kallikrein is bound in a complex (800,000 mol.wt.) with α_2-macroglobulin.

References
[1] Harpel J. Exp. Med. 132: 329-352, 1970.
[2] Venneröd et al. Thrombos. Res. 9: 457-466, 1976.
[3] Mandle et al. Proc. Natl. Acad. Sci. U.S.A. 73: 4179-4183, 1976.

Bibliography

1962

Landerman, N. S., Webster, M. E., Becker, E. L. & Ratcliffe, H. D. Hereditary angioneurotic edema. II. Deficiency of inhibitor for serum globulin permeability factor and/or plasma kallikrein. J. Allergy 33: 330-341.

Siegelman, A. M., Carlson, A. S. & Robertson, T. Investigation of serum trypsin and related substances. 1. The quantitative demonstration of trypsinlike activity in human blood serum by a micromethod. Arch. Biochem. Biophys. 97: 159-163.

1964

Kagen, L. J. Some biochemical and physical properties of the human permeability globulins. Br. J. Exp. Pathol. 45: 604-611.

1968

Nagasawa, S., Takahashi, H., Koida, M. & Suzuki, T. Partial purification of bovine plasma kallikreinogen, its activation by the Hageman factor. Biochem. Biophys. Res. Commun. 32: 644-649.

Paskhina, T. S., Zukova, V. P., Egorova, T. P., Narticova, V. F., Karpova, A. V. & Guseva, M. V. [Reaction of N-α-tosyl-L-lysyl-chloromethane with kallikrein from blood plasma of rabbit and with kallikreins from urine and saliva of man.] Biokhimia 33: 745-752.

1969

Colman, R. W., Mason, J. W. & Sherry, S. The kallikreinogen-kallikrein enzyme system of human plasma: assay of components and observations in disease states. Ann. Intern. Med. 71: 763-773.

Colman, R. W., Mattler, L. & Sherry, S. Studies on the prekallikrein (kallikreinogen)-kallikrein enzyme system of human plasma. 1. Isolation and purification of plasma kallikreins. J. Clin. Invest. 48: 11-22.

Colman, R. W., Mattler, L. & Sherry, S. Studies on the prekallikrein (kallikreinogen)-kallikrein enzyme system of human plasma. 2. Evidence relating the kaolin-activated arginine esterase to plasma kallikrein. J. Clin. Invest. 48: 23-32.

Ratnoff, O. D., Pensky, J., Ogston, D. & Naff, G. B. The inhibition of plasmin, plasma kallikrein, plasma permeability factor, and the C'1r subcomponent of the first component of complement by serum C'1 esterase inhibitor. J. Exp. Med. 129: 315-331.

1970

Gigli, I., Mason, J. W., Colman, R. W. & Austen, K. F. Interaction
of plasma kallikrein with the C$\overline{1}$ inhibitor. J. Immunol. 104:
574-581.

Harpel, P. C. Human plasma α_2-macroglobulin. An inhibitor of
plasma kallikrein. J. Exp. Med. 132: 329-352.

Kaplan, A. P. & Austen, K. F. A pre-albumin activator of
prekallikrein. J. Immunol. 105: 802-811.

McConnell, D. J. & Mason, B. The isolation of human plasma
prekallikrein. Br. J. Pharmacol. 38: 490-502.

Wuepper, K. D., Tucker, E. S., III & Cochrane, C. G. Plasma kinin
system: proenzyme components. J. Immunol. 105: 1307-1311.

1971

Harpel, P. C. Separation of plasma thromboplastin antecedent from
kallikrein by the plasma α_2-macroglobulin, kallikrein inhibitor.
J. Clin. Invest. 50: 2084-2090.

Kaplan, A. P. & Austen, K. F. A prealbumin activator of
prekallikrein. II. Derivation of activators of prekallikrein
from active Hageman factor by digestion with plasmin. J. Exp.
Med. 133: 696-712.

Muramatu, M. & Fujii, S. Inhibitory effects of ω-amino acid
esters on trypsin, plasmin, plasma kallikrein and thrombin.
Biochim. Biophys. Acta 242: 203-208.

Soltay, M. J., Movat, H. Z. & Özge-Anwar, A. H. The kinin system
of human plasma. V. The probable derivation of prekallikrein
activator from activated Hageman factor (XIIa). Proc. Soc. Exp.
Biol. Med. 138: 952-958.

1972

Fritz, H. & Förg-Brey, B. Zur Isolierung von Organ- und
Harnkallikreinen durch Affinitätschromatography. Spezifische
Bindung an wasserunlösliche Inhiborderivate und Dissoziation der
Komplex mit kompetitiven Hemmstoffen (Benzamidin). Hoppe-
Seyler's Z. Physiol. Chem. 353: 901-905.

Fritz, H., Wunderer, G. & Dittmann, B. Zur Isolierung von
Schweine- und Human-Serumkallikrein durch
Affinitätschromatographie. Spezifische Bindung an
wasserunlösliche Kunitz-Sojabohnen-Inhibitor-Cellulosen und
Dissoziation mit kompetitiven Hemmstoffen (Benzamidin). Hoppe-
Seyler's Z. Physiol. Chem. 353: 893-900.

Fritz, H., Wunderer, G., Kummer, K., Heimberger, N. & Werle, E.
α_1-Antitrypsin und C$\overline{1}$-Inactivator: Progressiv-Inhibitoren für
Serumkallikreine von Mensch und Schwein. Hoppe-Seyler's Z.
Physiol. Chem. 353: 906-910.

Fujimoto, Y., Moriya, H., Yamaguchi, K. & Moriwaki, C. Detection
of aginine esterase of various kallikrein preparations on
gellified electrophoretic media. J. Biochem. 71: 751-754.

Harpel, P. C. Studies on the interaction between collagen and a
plasma kallikrein-like activity. J. Clin. Invest. 51: 1813-
1822.

Komiya, M., Nagasawa, S. & Suzuki, T. Bovine prekallikrein activator with functional activity as Hageman factor. J. Biochem. 72: 1205-1218.

Muramatu, M. & Fujii, S. Inhibitory effects of ω-guanidino acid esters on trypsin, plasmin, plasma kallikrein and thrombin. Biochim. Biophys. Acta 268: 221-224.

Takahashi, H., Nagasawa, S. & Suzuki, T. Studies on prekallikrein of bovine plasma. I. Purification and properties. J. Biochem. 71: 471-483.

Takahashi, H., Nagasawa, S. & Suzuki, T. Conversion of bovine prekallikrein to kallikrein. Evidence of limited proteolysis of prekallikrein by bovine Hageman factor (factor XII). FEBS Lett. 24: 98-100.

Wuepper, K. D. & Cochrane, C. G. Effect of plasma kallikrein on coagulation in vitro. Proc. Soc. Exp. Biol. Med. 141: 271-276.

Wuepper, K. D. & Cochrane, C. G. Plasma prekallikrein: isolation, characterization and mechanism of activation. J. Exp. Med. 135: 1-20.

1973

Bagdasarian, A., Lahiri, B. & Colman. R. W. Origin of the high molecular weight activator of prekallikrein. J. Biol. Chem. 248: 7742-7747.

Bagdasarian, A., Talamo, R. C. & Colman, R. W. Isolation of high molecular weight activators of human plasma kallikrein. J. Biol. Chem. 248: 3456-3463.

Cochrane, C. G., Revak, S. D. & Wuepper, K. D. Activation of Hageman factor in solid and fluid phases. A critical role of kallikrein. J. Exp. Med. 138: 1564-1583.

Haberland, G. L. & Rohen, J. W. (eds) Kininogenases. Kallikrein: 1st Symposium on Physiological Properties and Pharmacological Rational, Schattauer Verlag, Stuttgart.

Laake, K. & Venneröd, A. M. Determination of factor XII in human plasma with arginine proesterase (prekallikrein). I. Preparation and properties of the substrate. Thromb. Res. 2: 393-407.

Laake, K. & Venneröd, A. M. Determination of factor XII in human plasma with arginine proesterase (prekallikrein). II. Studies on the method. Thromb. Res. 2: 409-421.

Stormorken, H., Gjönnaess, H. & Laake, K. Interrelations between the clotting and kinin systems. Haemostasis 2: 245-252.

Wuepper, K. D. Prekallikrein deficiency in man. J. Exp. Med. 138: 1345-1355.

1974

Colman, R. W. Formation of human plasma kinin. New England J. Med. 291: 509-515.

Gallin, J. I. & Kaplan, A. P. Mononuclear cell chemotactic activity of kallikrein and plasminogen activator and its inhibition by CĪ inhibitor and α_2-macroglobulin. J. Immunol. 113: 1928-1934.

Laake, K. & Vennerod, A. M. Factor XII induced fibrinolysis:

studies on the separation of prekallikrein, plasminogen proactivator, and factor XI in human plasma. Thromb. Res. 4: 285-302.

Laake, K., Venneröd, A. M., Haugen, G. & Gjönnaess, H. Cold-promoted activation of factor VII in human plasma: studies on the associated acyl-arginine esterase activity. Thromb. Res. 4: 769-785.

Lahiri, B., Rosenberg, R., Talamo, R. C., Mitchell, B., Bagdasarian, A. & Colman, R. W. Antithrombin III: an inhibitor of human plasma kallikrein. Fed. Proc. Fed. Am. Soc. Exp. Biol. 33: 642 only.

McConnell, D. J. & Loeb, J. N. Kallikrein inhibitory capacity of α_2-macroglobulin subunits. Proc. Soc. Exp. Biol. Med. 147: 891-896.

Moriwaki, C., Hojima, Y. & Moriya, H. A proposal – use of combined assays of kallikrein measurement. Chem. Pharm. Bull. 22: 975-983.

Saito, H. & Ratnoff, O. D. Inhibition of normal clotting and Fletcher factor activity by rabbit anti-kallikrein antiserum. Nature 248: 597-598.

Saito, H., Ratnoff, O. D. & Donaldson, V. H. Defective activation of clotting, fibrinolytic, and permeability-enhancing systems in human Fletcher trait plasma. Circ. Res. 34: 641-651.

Sampaio, C., Wong, S.-C. & Shaw, E. Human plasma kallikrein. Purification and preliminary characterization. Arch. Biochem. Biophys. 165: 133-139.

Sumi, H. & Fujii, S. Purification of a new plasma inhibitor of kallikrein. J. Biochem. 75: 541-551.

Venneröd, A. M. & Laake, K. Isolation and characterization of a prealbumin activator of prekallikrein from acetone-activated human plasma. Thromb. Res. 4: 103-118.

1975

Burrowes, C. E., Habal, F. M. & Movat, H. Z. The inhibition of human plasma kallikrein by antithrombin III. Thromb. Res. 7: 175-183.

Davie, E. W., Fujikawa, K., Legaz, M. E. & Kato, H. Role of proteases in blood coagulation. In: Proteases in Biological Control (Reich, E., Rifkin, D. B. & Shaw, E. eds), pp. 65-77, Cold Spring Harbor Laboratory, New York.

Oh-ishi, S. & Webster, M. E. Vascular permeability factors (PF/Nat and PF/Dil) – their relationship to Hageman factor and the kallikrein-kinin system. Biochem. Pharmacol. 24: 591-598.

Pisano, J. J. Chemistry and biology of the kallikrein-kinin system. In: Proteases and Biological Control (Reich, E., Rifkin, D. B. & Shaw, E. eds), pp. 199-222, Cold Spring Harbor Laboratory, New York.

Venneröd, A. M. & Laake, K. Inhibition of purified kallikrein by antithrombin III and heparin. Thromb. Res. 7: 223-226.

1976

Amundsen, E., Svendsen, L., Venerød, A. M. & Laake, K. Determination of plasma kallikrein with a new chromogenic tripeptide derivative. In: Chemistry and Biology of the Kallikrein-Kinin System in Health and Disease, DHEW Publication No. (NIH) 76-791 (Pisano, J. J. & Austen, K. F. eds), pp. 215-220, U.S. Government Printing Office, Washington.

Chan, J. Y. C., Habal, F. M., Burrows, C. E. & Movat, H. Z. Interaction between factor XII (Hageman factor), high molecular weight kininogen and prekallikrein. Thromb. Res. 9: 423-433.

Colman, R. W. & Bagdasarian, A. Human kallikrein and prekallikrein. Methods Enzymol. 45: 303-322.

Griffin, J. H. & Cochrane, C. G. Mechanisms for the involvement of high molecular weight kininogen in surface-dependent reactions of Hageman factor. Proc. Natl. Acad. Sci. U.S.A. 73: 2554-2558.

Han, Y. N., Kato, S. & Iwanaga, S. Identification of Ser-Leu-Met-Lys-bradykin isolated from chemically modified high-molecular weight bovine kininogen. FEBS Lett. 71: 45-48.

Imanari, T., Kaizu, T., Yoshida, H., Yates, K., Pierce, J. V. & Pisano, J. J. Radiochemical assays for human urinary, salivary, and plasma kallikreins. In: Chemistry and Biology of the Kallikrein-Kinin System in Health and Disease, DHEW Publication No. (NIH) 76-791 (Pisano, J. J. & Austen, K. F. eds), pp. 205-213, U.S. Government Printing Office, Washington.

Kaplan, A. P., Meier, H. L., Yecies, L. D. & Heck, L. W. Hageman factor and its substrates: the role of factor XI, prekallikrein, and plasminogen proactivator in coagulation, fibrinolysis, and kinin generation. In: Chemistry and Biology of the Kallikrein-Kinin System in Health and Disease, DHEW Publication No. (NIH) 76-791 (Pisano, J. J. & Austen, K. F. eds), pp. 237-254, U.S. Government Printing Office, Washington.

Lahiri, B., Bagdasarian, A., Mitchell, B., Talamo, R. C. Colman, R. W. & Rosenberg, R. D. Antithrombin-heparin cofactor: an inhibitor of plasma kallikrein. Arch. Biochem. Biophys. 175: 737-747.

Mandle, R. J., Colman, R. W. & Kaplan, A. P. Identification of prekallikrein and high-molecular-weight kininogen as a complex in human plasma. Proc. Natl. Acad. Sci. U.S.A. 73: 4179-4183.

Matsuda, Y., Moriya, H., Moriwaki, C., Fujimoto, Y. & Matsuda, M. Fluorimetric method for assay of kallikrein-like arginine esterase. J. Biochem. 79: 1197-1200.

Oh-ishi, S. & Webster, M. E. Formation of prekallikrein activator and TAME esterase activity by dilution of human plasma. In: Chemistry and Biology of the Kallikrein-Kinin System in Health and Disease, DHEW Publication No. (NIH) 76-791 (Pisano, J. J. & Austen, K. F. eds), pp. 61-64, U.S. Government Printing Office, Washington.

Spragg, J. Immunological assays for components of the human plasma kinin-forming system. In: Chemistry and Biology of the

Kallikrein-Kinin System in Health and Disease, DHEW Publication No. (NIH) 76-791 (Pisano, J. J. & Austen, K. F. eds), pp. 195-204, U.S. Government Printing Office, Washington.

Suzuki, T. & Kato, H. Protein components of the bovine kallikrein-kinin system. In: Chemistry and Biology of the Kallikrein-Kinin System in Health and Disease, DHEW Publication No. (NIH) 76-791 (Pisano, J. J. & Austen, K. F. eds), pp. 107-119, U.S. Government Printing Office, Washington.

Tamura, Y. & Fujii, S. Purification of human plasma kallikrein and urokinase by affinity chromatography. J. Biochem. 80: 507-511.

Tidwell, R. R., Fox, L. L. & Geratz, J. D. Aromatic Tris-amidines. A new class of highly active inhibitors of trypsin-like proteases. Biochim. Biophys. Acta 445: 729-738.

Vennerőd, A. M., Laake, K., Solberg, A. K. & Strőmland, S. Inactivation and binding of human plasma kallikrein by antithrombin III and heparin. Thromb. Res. 9: 457-466.

Vogt, W. Formation of a kininogenase-α_2-macroglobulin complex during activation of human plasma with acetone. In: Chemistry and Biology of the Kallikrein-Kinin System in Health and Disease, DHEW Publication No. (NIH) 76-791 (Pisano, J. J. & Austen, K. F. eds), p. 75 only, U.S. Government Printing Office, Washington.

Vogt, W. & Dugal, B. Generation of an esterolytic and kinin-forming kallikrein-α_2-macroglobulin complex in human serum by treatment with acetone. Naunyn-Schmiederberg's Arch. Pharmacol. 294: 75-84.

Wuepper, K. D. Plasma prekallikrein: its characterization, mechanism of action, and inherited deficiency in man. In: Chemistry and Biology of the Kallikrein-Kinin System in Health and Disease, DHEW Publication No. (NIH) 76-791 (Pisano, J. J. & Austen, K. F. eds), pp. 37-51, U.S. Government Printing Office, Washington.

1977

Amundsen, E. & Svendsen, L. A new assay for plasma kallikrein activity utilizing a synthetic chromogenic substrate. In: New Methods for the Analysis of Coagulation using Chromogenic Substrates (Witt, I., ed.), pp. 211-219, Walter de Gruyter, Berlin.

Claeson, G., Aurell, L., Karlsson, G. & Friberger, P. Substrate structure and activity relationship. In: New Methods for the Analysis of Coagulation using Chromogenic Substrates (Witt, I., ed.), pp. 37-54, Walter de Gruyter, Berlin.

Halabi, M. & Back, N. Kinetic studies with serine proteases and protease inhibitors utilizing a synthetic nitro-anilide chromogenic substrate. In: Kininogenases. Kallikrein. 4th Symposium on Physiological Properties and Pharmacological Rationale. (Haberland, G. L., Rohen, J. W. & Suzuki, T. eds), pp. 35-46, Schattauer Verlag, Stuttgart.

Mandle, R., Jr. & Kaplan, A. P. Hageman factor substrates. Human

plasma prekallikrein: mechanism of activation by Hageman factor
and participation in Hageman factor-dependent fibrinolysis. J.
Biol. Chem. 252: 6097-6104.

Mattler, L. E. & Bang, N. U. Serine protease specificity for
peptide chromogenic substrates. Thromb. Haemost. 38: 776-792.

Meier, H. L., Pierce, J. V., Colman, R. W. & Kaplan, A. P.
Activation and function of human Hageman factor. The role of
high molecular weight kininogen and prekallikrein. J. Clin.
Invest. 60: 18-31.

Morita, T., Kato, H., Iwanago, S., Takada, K., Kimura, T. &
Sakakibara, S. New fluorogenic substrates for α-thrombin,
factor X, kallikreins and urokinase. J. Biochem. 82: 1495-1498.

Tamura, Y., Hirado, M., Okamura, K., Minato, Y. & Fujii, S.
Synthetic inhibitors of trypsin, plasmin, thrombin, C_1r and C_1
esterase. Biochim. Biophys. Acta 484: 417-422.

1978

Amundsen, E., Gallimore, M. J., Aasen, A. O., Larsbraaten, M. &
Lyngaas, K. Activation of human plasma prekallikrein: influence
of activators, activation time and temperature and inhibitors.
Thromb. Res. 13: 625-636.

Colman, R. W., Edelman, R., Scott, C. F. & Gilman, R. H. Plasma
kallikrein activation and inhibition during typhoid fever, J.
Clin. Invest. 61: 287-296.

Gallimore, M. J., Fareid, E. & Stormorken, H. The purification of
a human plasma kallikrein with weak plasminogen activator
activity. Thromb. Res. 12: 409-420.

Heber, H., Geiger, R. & Heimburger, N. Human plasma kallikrein:
purification, enzyme characterization and quantitative
determination in plasma. Hoppe-Seyler's Z. Physiol. Chem. 359:
659-669.

Heimark, R. L., Fujikawa, K. & Davie, E. W. Isolation and
characterization of bovine plasma prekallikrein. Fed. Proc.
Fed. Am. Soc. Exp. Biol. 37: 1587 only.

Kettner, C. & Shaw, E. Synthesis of peptides of arginine
chloromethyl ketone. Selective inactivation of human plasma
kallikrein. Biochemistry 17: 4778-4784.

Sampaio, C. A. M. & Grisolia, D. Human plasma kallikrein.
Preliminary studies on hydrolysis of proteins and peptides.
Agents Actions 8: 125-131.

Sumi, H., Takasugi, S. & Toki, N. Studies on kallikrein-kinin
system in plasma of patients with acute pancreatitis. Clin.
Chim. Acta 87: 113-118.

Yecies, L., Meier, H., Mandle, R., Jr. & Kaplan, A. P. The
Hageman factor-dependent fibrinolytic pathway. In: Progress in
Chemical Fibrinolysis and Thrombolysis (Davidson, J. F., Rowan,
R. M., Samama, M. M. & Desnoyers, P. C. eds), vol. 3, pp. 121-
139, Raven Press, New York.

1979

Colman, R. W., Shames, P. V., Liu, C. Y., Scott, C. F. & Kirby, E.

P. Von Willebrand's factor (VWF) potentiates prekallikrein activation by Hageman factor fragments. Fed. Proc. Fed. Am. Soc. Exp. Biol. 38: 1272 only.

Davie, E. W., Fujikawa, K., Kurachi, K. & Kisiel, W. The role of serine proteases in the blood coagulation cascade. Adv. Enzymol. 48: 277-318.

Heimark, R. L. & Davie, E. W. Isolation and characterization of bovine plasma prekallikrein (Fletcher factor). Biochemistry 18: 5743-5750.

Oh-ishi, S. & Katori, M. Fluorometric assay for plasma prekallikrein using peptidylmethylcoumarinyl-amide as substrate. Thromb. Res. 14: 551-559.

Ratnoff, O. D. & Saito, H. Interactions among Hageman factor, plasma prekallikrein, high molecular weight kininogen, and plasma thromboplastin antecedent. Proc. Natl. Acad. Sci. U.S.A. 76: 958-961.

Saito, H., Goldsmith, G. H., Moroi, M. & Aoki, N. Inhibitory spectrum of alpha-2-plasmin inhibitor. Proc. Natl. Acad. Sci. U.S.A. 76: 2013-2017.

Scott, C. F., Liu, C. Y. & Colman, R. W. Human plasma kallikrein: a rapid high-yield method of purification. Eur. J. Biochem. 100: 77-83.

Soulier, J.-P. & Gozin, D. Assay of Fletcher factor (plasma prekallikrein) using an artificial clotting reagent and a modified chromogenic assay. Thromb. Haemostas. 42: 538-547.

Yarovaya, G. A., Dotsenko, V. L., Orekhovich, V. N., Kawiak, J. & Kawalec, M. Preparation and some properties of highly purified human serum kallikrein. Clin. Chim. Acta 93: 321-327.

Yokosawa, N., Takahashi, N. & Inagami, T. Isolation of completely inactive plasma prorenin and its activation by kallikreins: a possible new link between renin and kallikrein. Biochim. Biophys. Acta 569: 211-219.

1980

Takahashi, H., Nagasawa, S. & Suzuki, T. Studies on prekallikrein of bovine plasma. II. Activation of prekallikrein with proteinases and properties of kallikrein activated by bovine Hageman factor. J. Biochem. 87: 23-34.

COAGULATION FACTOR XI$_a$

Summary

EC Number: 3.4.21.27

Earlier names: Plasma thromboplastin. Precursor: plasma thromboplastin antecendent (PTA), antihemophilic factor.

Distribution: Precursor zymogen: plasma of man and other mammals.

Source: Bovine plasma.

Action: Activates factor IX by cleavage of −Arg−Ala− and −Arg−Val− bonds. Also acts on factor VII, factor XIII and plasminogen. Cleaves Bz−Arg−OEt, Tos−Arg−OMe and Bz−Phe−Val−Arg−NPhNO$_2$.

Requirements: Neutral pH.

Substrate, usual: (Assayed by clotting-time with deficient plasma.)

Inhibitors: Dip−F, Pms−F, N−acetylimidazole. Antithrombin III, α_1−proteinase inhibitor, C$\bar{\text{I}}$ inhibitor. Aprotinin, soybean trypsin inhibitor.

Molecular properties: The precursor (bovine and human) is a protein of apparent mol.wt. 124,000, comprising two identical subunits (mol.wt. 60,000−80,000, according to method), containing 11% carbohydrate. The protein also contains Gla residues, so biosynthesis is vitamin K−dependent.

In the activation by factor XII$_a$, each subunit is cleaved to fragments of 30,000 and 50,000 approximate mol.wt., without detected liberation of any free peptide. Each of the 30,000 mol.wt. chains contains an active site, and is highly homologous with other serine proteinases.

Comment: It is still not understood how three zymogens (factors XII, XI and prekallikrein) and the non−proteolytic high mol.wt. kininogen interact to generate the active proteinases.

Bibliography

1953

Rosenthal, R. L., Dreskin, O. H. & Rosenthal, N. New hemophilia-like disease caused by deficiency of a third plasma thromboplastin factor. Proc. Soc. Exp. Biol. Med. 82: 171-174.

1961

Ratnoff, O. D., Davie, E. W. & Mallett, D. L. Studies on the action of Hageman factor: evidence that activated Hageman factor in turn activates plasma thromboplastin antecedent. J. Clin. Invest. 40: 803-819.

1962

Ratnoff, O. D. & Davie, E. W. The activation of Christmas Factor (Factor IX) by activated plasma thromboplastin antecedent (activated Factor XII). Biochemistry 1: 677-685.

1964

Kingdon, H. S., Davie, E. W. & Ratnoff, O. D. The reaction between activated plasma thromboplastin antecedent and diisopropylphosphofluoridate. Biochemistry 3: 166-173.

1968

Esnouf, M. P. & Macfarlane, R. G. Enzymology and the blood clotting mechanism. Advan. Enzymol. 30: 255-315.

1969

Kociba, G. J., Ratnoff, O. D., Loeb, W. F., Wall, R. L. & Heider, L. E. Bovine plasma thromboplastin antecedent (Factor XI) deficiency. J. Lab. Clin. Med. 74: 37-41.

1970

Forbes, C. D., Pensky, J. & Ratnoff, O. D. Inhibition of activated Hageman factor and activated plasma thromboplastin antecedent by purified serum CĪ inactivator. J. Lab. Clin. Med. 76: 809-815.

1971

Harpel, P. C. Separation of plasma thromboplastin antecedent from kallikrein by the plasma α_2-macroglobulin, kallikrein inhibitor. J. Clin. Invest. 50: 2084-2090.

1972

Forbes, C. D. & Ratnoff, O. D. Studies on plasma thromboplastin antecedent (factor XI), PTA deficiency and inhibition of PTA by plasma: pharmacologic inhibitors and specific antiserum. J. Lab. Clin. Med. 79: 113-127.

1973

Cochrane, C. G., Revak, S. D. & Wuepper, K. D. Activation of Hageman factor in solid and fluid phases. J. Exp. Med. 138: 1564-1583.
Damus, P. S., Hicks, M. & Rosenberg, R. D. Anticoagulant action of heparin. Nature 246: 355-357.

1974

Heck, L. W. & Kaplan, A. P. Substrates of Hageman factor. 1.

Isolation and characterization of human factor XI (PTA) and inhibition of the activated enzyme by α_1-antitrypsin. J. Exp. Med. 140: 1615-1630.

Laake, K. & Vennerod, A. M. Factor XII induced fibrinolysis: studies on the separation of prekallikrein, plasminogen proactivator, and factor XI in human plasma. Thromb. Res. 4: 285-302.

Lundblad, R. L. & Kingdon, H. S. Biochemistry of the interactions of bovine factors XIa and IX. Thromb. Diath. Haemorrh. Suppl. 57: 315-333.

Movat, H. Z. & Özge-Anwar, A. H. The contact phase of blood coagulation: clotting factor XI and XII, their isolation and interaction. J. Lab. Clin. Med. 84: 861-878.

1975

Connellan, J. M. & Castaldi, P. A. Partial purification of factors XI and XIa: a comparison of their molecular size and structure. Thromb. Res. 7: 717-728.

Davie, E. W. & Fujikawa, K. Basic mechanisms in blood coagulation. Annu. Rev. Biochem. 44: 799-829.

Davie, E. W., Fujikawa, K., Legaz, M. E. & Kato, H. Role of proteases in blood coagulation. In: Proteases in Biological Control (Reich, E., Rifkin, D. B. & Shaw, E. eds), pp. 65-77, Cold Spring Harbor Laboratory, New York.

Schiffman, S. & Lee, P. Partial purification and characterization of contact activation cofactor. J. Clin. Invest. 56: 1082-1092.

1976

Griffin, J. H. & Cochrane, C. G. Mechanisms for the involvement of high molecular weight kininogen in surface-dependent reactions of Hageman factor. Proc. Natl. Acad. Sci. U.S.A. 73: 2554-2558.

Kaplan, A. P., Meier, H. L., Yecies, L. D. & Heck, L. W. Hageman factor and its substrates: the role of factor XI, prekallikrein, and plasminogen proactivator in coagulation, fibrinolysis, and kinin generation. In: Chemistry and Biology of the Kallikrein-Kinin System in Health and Disease, DHEW Publication No. (NIH) 76-791 (Pisano, J. J. & Austen, K. F. eds), pp. 237-254, U.S. Government Printing Office, Washington.

Koide, T., Kato, H. & Davie, E. W. Bovine factor XI (plasma thromboplastin antecedent). Methods Enzymol. 45: 65-73.

1977

Bouma, B. N. & Griffin, J. H. Human blood coagulation Factor XI. Purification, properties, and mechanism of activation by activated Factor XII. J. Biol. Chem. 252: 6432-6437.

Koide, T., Hermodson, M. A. & Davie, E. W. Active site of bovine factor XI (plasma thromboplastin antecedent). Nature 266: 729-730.

Koide, T., Kato, H. & Davie, E. W. Isolation and characterization of bovine factor XI (plasma thromboplastin antecedent).

Biochemistry 16: 2279–2286.

Kurachi, K. & Davie, E. W. Activation of human factor XI (plasma thromboplastin antecedent) by factor XII_a (activated Hageman factor). Biochemistry 16: 5831–5839.

Saitio, H. Purification of high molecular weight kininogen and the role of this agent in blood coagulation. J. Clin. Invest. 60: 584–594.

Schiffman, S., Pecci, R. & Lee, P. Contact activation of factor Xl: evidence that the primary role of contact activation cofactor (CAC) is to facilitate the activation of factor Xll. Thromb. Res. 10: 319–323.

Thompson, R. E., Mandle, R. & Kaplan, A. P. Association of factor Xl and high molecular weight kininogen in human plasma. J. Clin. Invest. 60: 1376–1380.

1979

Lipscomb, M. S. & Walsh, P. N. Human platelets and factor-XI: localization in platelet membranes of factor-XI-like activity and its functional distinction from plasma factor-XI. J. Clin. Invest. 63: 1006–1014.

Mannhalter, C., Slattery, C. & Schiffman, S. Adsorption properties of factor Xl. Fed. Proc. Fed. Am. Soc. Exp. Biol. 38: 810 only.

Ratnoff, O. D. & Saito, H. Interactions among Hageman factor, plasma prekallikrein, high molecular weight kininogen, and plasma thromboplastin antecedent. Proc. Natl. Acad. Sci. U.S.A. 76: 958–961.

Saito, H., Goldsmith, G. H., Moroi, M. & Aoki, N. Inhibitory spectrum of alpha-2-plasmin inhibitor. Proc. Natl. Acad. Sci. U.S.A. 76: 2013–2017.

Tuszynski, G. P., Camp, E., Bevacqua, S. & Walsh, P. N. Factor Xl antigen in human platelets. Fed. Proc. Fed. Am. Soc. Exp. Biol. 38: 1307 only.

Entry 1.12

COAGULATION FACTOR IX_a

Summary

EC Number: 3.4.21.22

Earlier names: Activated Christmas factor, antihemophilic factor
  B, plasma thromboplastin component (PTC).

Distribution: Plasma, as the inactive precursor.

Source: The precursor is isolated from plasma, usually bovine.

Action: Cleaves an -Arg-Ile- bond to activate factor X, as part of
  a complex with phospholipid, Ca^{2+} and factor VIII_a. Shows weak
  esterase activity toward Tos-Arg-OMe and Bz-Arg-OMe.

Requirements: Neutral pH.

Substrate, usual: (Clotting time assays are used).

Inhibitors: Antithrombin III-heparin, hirudin. (Not Dip-F).

Molecular properties: The precursor is a glycoprotein (26%
  carbohydrate), single chain, mol.wt. about 58,000, pI 4.2. It
  contains Gla residues, so biosynthesis is vitamin K-dependent.
    Two-stage activation (cleavage of -Arg-Ala- and -Arg-Val-
  bonds) by thrombin or factor XI_a with Ca^{2+} releases a 9000
  mol.wt. fragment leaving two-chain, disulfide-bonded active
  factor (27,000 plus 17,000 mol.wt.). The heavy chain is
  homologous with the pancreatic serine proteinases, and the light
  chain contains the Gla residues. An enzyme from Russell's viper
  venom cleaves only the -Arg-Val- bond to form an active factor
  with the same mol.wt. as the zymogen.

Comment: Deficiency in Christmas disease (hemophilia B) causes a
  severe bleeding disorder.

Bibliography

1962

Ratnoff, O. D. & Davie, E. W.   The activation of Christmas Factor
  (Factor IX) by activated plasma thromboplastin antecedent
  (activated Factor XII).   Biochemistry 1: 677-685.

1963

Schiffman, S., Rapaport, S. I. & Patch, M. J.   The identification

and synthesis of activated plasma thromboplastin component
(PTC'). Blood 22: 733-749.

1965

Biggs, R., Macfarlane, R. G., Denson, K. W. E. & Ash, B. J.
Thrombin and the interaction of factors VIII and IX. Br. J.
Haematol. 11: 276-295.

1968

Esnouf, M. P. & Macfarlane, R. G. Enzymology and the blood
clotting mechanism. Adv. Enzymol. 30: 255-315.

1972

De Lange, J. A. & Hemker, H. C. Inhibition of blood coagulation
factors by serine esterase inhibitors. FEBS Lett. 24: 265-268.
Reekers, P. & Hemker, H. C. Purification of blood coagulation
factors II, VII, IX and X from bovine citrated plasma.
Haemostasis 1: 2-22.

1973

Chandra, S. & Pechet, L. The separation of clotting factors. II.
Purification of human Factor IX by isoelectric focusing.
Biochim. Biophys. Acta 328: 456-463.
Fujikawa, K., Thompson, A. R., Legaz, M. E., Meyer, R. G. & David,
E. W. Isolation and characterization of bovine factor IX
(Christmas factor). Biochemistry 12: 4938-4945.

1974

Enfield, D. L., Ericsson, L. H., Fujikawa, F., Titani, K., Walsh,
K. A. & Neurath, H. Bovine factor IX (Christmas factor).
Further evidence of homology with factor X (Stuart factor) and
prothrombin. FEBS Lett. 47: 132-135.
Fujikawa, K., Coan, M. H., Enfield, D. L., Titani, K., Ericsson,
L. H. & Davie, E. W. A comparison of bovine prothrombin, factor
IX (Christmas factor), and factor X (Stuart factor). Proc.
Natl. Acad. Sci. U.S.A. 71: 427-430.
Fujikawa, K., Legaz, M. E., Kato, H. & Davie, E. W. The mechanism
of activation of bovine factor IX (Christmas factor) by bovine
factor XIa (activated plasma thromboplastin antecedent).
Biochemistry 13: 4508-4516.
Lundblad, R. L. & Kingdon, H. S. Biochemistry of the interactions
of bovine factors XIa and IX. Thromb. Diath. Haemorrh. Suppl.
57: 315-333.

1975

Davie, E. W. & Fujikawa, K. Basic mechanisms in blood coagulation.
Annu. Rev. Biochem. 44: 799-829.
Davie, E. W., Fujikawa, K., Legaz, M. E. & Kato, H. Role of
proteases in blood coagulation. In: Proteases in Biological
Control (Reich, E., Rifkin, D. B. & Shaw, E. eds), pp. 65-77,
Cold Spring Harbor Laboratory, New York.
Kalousek, F., Konigsberg, W. & Nemerson, Y. Activation of factor

IX by activated factor X: a link between the extrinsic and
intrinsic coagulation systems. FEBS Lett. 50: 382–385.
Rosenberg, J. S., McKenna, P. W. & Rosenberg, R. D. Inhibition of
human factor IX$_a$ by human antithrombin. J. Biol. Chem. 250:
8883–8888.

1976

Bucher, D., Nebelin, E., Thomsen, J. & Stenflo, J. Identification
of γ-carboxyglutamic acid residues in bovine factors IX and X,
and in a new vitamin K-dependent protein. FEBS Lett. 68: 293–
296.
Fryklund, L., Borg, H. & Anderson, L.-O. Amino-terminal sequence
of human factor IX: presence of γ-carboxyl glutamic acid
residues. FEBS Lett. 65: 187–189.
Fujikawa, K. & Davie, E. W. Bovine factor IX (Christmas factor).
Methods Enzymol. 45: 74–83.
Kurachi, K., Fujikawa, K., Schmer, G. & Davie, E. W. Inhibition
of bovine factor IX$_a$ and factor X$_{a\beta}$ by antithrombin III.
Biochemistry 15: 373–377.
Østerud, B., Miller-Andersson, M., Abildgaard, U. & Prydz, H. The
effect of antithrombin III on the activity of the coagulation
factors VII, IX, and X. Thromb. Haemostas. 35: 295–304.
Váradi, K. & Elödi, S. Purification and properties of human
factor IX$_a$. Thromb. Haemostas. 35: 576–585.

1977

Di Scipio, R. G., Hermodson, M. A., Yates, S. G. & Davie, E. W. A
comparison of human prothrombin, factor IX (Christmas factor),
factor X (Stuart factor), and protein S. Biochemistry 16: 698–
706.
Østerud, B. & Rapaport, S. I. Activation of factor IX by the
reaction product of tissue factor and factor VII: additional
pathway for initiating blood coagulation. Proc. Natl. Acad.
Sci. U.S.A. 74: 5260–5264.
Thompson, A. R. Factor IX antigen by radioimmunoassay. Abnormal
factor IX protein in patients on warfarin therapy and with
hemophilia B. J. Clin. Invest. 59: 900–910.
Titani. K., Enfield, D. L., Katayama, K., Ericsson, L. H.,
Fujikawa, K., Walsh, K. A. & Neurath, H. Primary structure of
bovine factor IX. Thromb. Haemostas. 38: 116 only.

1978

Byrne, R. & Castellino, F. J. The influence of metal ions on the
activation of bovine factor IX by the coagulant protein of
Russell's viper venom. Arch. Biochem. Biophys. 190: 687–692.
Di Scipio, R. G., Kurachi, K. & Davie, E. W. Activation of human
factor IX (Christmas factor). J. Clin. Invest. 61: 1528–1538.
Hultin, M. B. & Nemerson, Y. Activation of factor X by factors
IXa and VIII; a specific assay for factor IXa in the presence of
thrombin-activated factor VIII. Blood 52: 928–940.
Lindquist, P. A., Fujikawa, K. & Davie, E. W. Activation of

bovine factor IX (Christmas factor) by factor XI$_a$ (activated plasma thromboplastin antecent) and a protease from Russell's viper venom. J. Biol. Chem. 253: 1902-1909.

Østerud, B., Bouma, B. N. & Griffin, J. H. Human blood coagulation Factor IX. Purification, properties, and mechanism of activation by activated Factor XI. J. Biol. Chem. 253: 5946-5951.

Seligsohn, U., Østerud, B., Griffin, J. H. & Rapaport, S. I. Evidence for the participation of both activated factor XII and activated factor IX in cold-promoted activation of factor VII. Thromb. Res. 13: 1049-1056.

1979

Amphlett, G. W., Byrne, R. & Castellino, F. J. The binding of calcium to the activation products of bovine factor IX. J. Biol. Chem. 254: 6333-6336.

Davie, E. W., Fujikawa, K., Kurachi, K. & Kisiel, W. The role of serine proteases in the blood coagulation cascade. Adv. Enzymol. 48: 277-318.

COAGULATION FACTOR VIII$_a$

Summary

EC Number: (3.4.21.-)

Earlier names: Antihemophilic factor A, platelet cofactor I.

Distribution: Plasma as the precursor, factor VIII.

Source: Plasma.

Action: Participates in activation of factor X together with factor IX$_a$, Ca^{2+} and phospholipid. No proteolytic or esterase activity has been directly demonstrated.

Requirements: Neutral pH. Acts in conjunction with other factors (see above).

Substrate, usual: Factor X, in complex system.

Inhibitors: Dip-F. Antithrombin III.

Molecular properties: Factor VIII occurs in plasma as a very high mol.wt. (up to 2,000,000) complex with the von Willebrand platelet aggregating factor, but the two may be separated in 1M NaCl or 0.25M CaCl$_2$. SDS gel electrophoresis shows components of 93,000, 88,000 and 85,000 mol.wt. [1]. The coagulant activity of factor VIII$_a$ appears following the action of thrombin, factor X$_a$, plasmin or trypsin.

Comment: Factor VIII is absent from the plasma of people with "classic" hemophilia. If the existence of a serine active center in factor VIII$_a$ is confirmed, the supposition that the cleavage of $-Arg_{51}-Ile_{52}-$ in factor X is mediated by factor IX$_a$, while factor VIII$_a$ acts as a regulator, may need re-examination.

References
[1] Vehar & Davie Biochemistry 19: 401-410, 1980.

Bibliography

1965

Biggs, R., Macfarlane, R. G., Denson, K. W. E. & Ash, B. J. Thrombin and the interaction of factors VIII and IX. Br. J. Haematol. 11: 276-295.

Lundblad, R. L. & Davie, E. W. The activation of Stuart factor (factor X) by activated antihemophilic factor (activated factor

VIII). Biochemistry 4: 113-120.

1972

Bouma, B. N., Wiegernick, Y., Sixma, J. J., van Mourik, J. A. & Mochtar, I. A. Immunological characterization of purified anti-haemophilic factor A (factor VIII) which corrects abnormal platelet retention in von Willebrand's disease. Nature New Biol. 236: 104-106.

1974

Brown, J. E., Baugh, R. F., Sargent, R. B. & Hougie, C. Separation of bovine factor VIII-related antigen (platelet aggregating factor) from bovine antihemophilic factor. Proc. Soc. Exp. Biol. Med. 147: 608-611.

1975

Davie, E. W. & Fujikawa, K. Basic mechanisms in blood coagulation. Annu. Rev. Biochem. 44: 799-829.

Legaz, M. E., Weinstein, M. J., Heldebrant, C. M. & Davie, E. W. Isolation, subunit structure, and proteolytic modification of bovine factor VIII. Ann. N.Y. Acad. Sci. 240: 43-61.

McKee, P. A., Andersen, J. C. & Switzer, M. E. Molecular structural studies of human factor VIII. Ann N.Y. Acad. Sci. 240: 8-33.

1976

Legaz, M. E. & Davie, E. W. Bovine factor VIII (antihaemophilic factor). Methods Enzymol. 45: 83-89.

1977

Brown, J. E. & Hougie, C. The effect of DFP on factor VIII preparations. Fed. Proc. Fed. Am. Soc. Exp. Biol. 36: 287 only.

Rick, M. E. & Hoyer, L. W. Thrombin activation of factor VIII: the effect of inhibitors. Br. J. Haematol. 36: 585-597.

Tan, H. K., & Andersen, J. C. Human factor VIII: morphometric analysis of purified material in solution. Science 198: 932-934.

Vehar, G. A. & Davie, E. W. Formation of a serine enzyme in the presence of bovine factor VIII (antihaemophilic factor) and thrombin. Science 197: 374-376.

1978

Hultin, M. B. & Nemerson, Y. Activation of factor X by factors IXa and VIII; a specific assay for factor IXa in the presence of thrombin-activated factor VIII. Blood 52: 928-940.

1979

Beck, E. A., Tranqui-Pouit, L., Chapel, A., Perret, B. A., Furlan, M., Hudry-Clergeon, G. & Suscillon, M. Studies on factor VIII-related protein. I. Ultrastructural and electrophoretic heterogeneity of human factor VIII-related protein. Biochim. Biophys. Acta 578: 155-163.

Davie, E. W., Fujikawa, K., Kurachi, K. & Kisiel, W. The role of serine proteases in the blood coagulation cascade. Adv. Enzymol. 48: 277-318.

Gadarowski, J. J. & Iatridis, P. G. Effects of 2,3-diphosphoglycerate (2,3-DPG) on factor VIII activities. Fed. Proc. Fed. Am. Soc. Exp. Biol. 38: 1409 only.

Johnson, G. S. Ristocetin cofactor activity of purified canine factor VIII. Fed. Proc. Fed. Am. Soc. Exp. Biol. 38: 997 only.

Kao, K. J., Pizzo, S. V. & McKee, P. A. The importance of carbohydrate residues in the binding of factor VIII/von Willebrand factor to human platelets. Fed. Proc. Fed. Am. Soc. Exp. Biol. 38: 810 only.

Thuy, L. P., Baugh, R. F., Brown, J. E. & Hougie, C. Succinylation and dodecyl succinylation of bovine von Willebrand factor/factor VIII. Fed. Proc. Fed. Am. Soc. Exp. Biol. 38: 997 only.

von Kaulla, E. & von Kaulla, K. N. In vitro factor VIII activity of certain synthetic fibrinolysis-inducing compounds. Fed. Proc. Fed. Am. Soc. Exp. Biol. 38: 758 only.

1980

Vehar, G. A. & Davie, E. W. Preparation and properties of bovine factor VIII (antihemophilic factor). Biochemistry 19: 401-410.

Entry 1.14

COAGULATION FACTOR X_a

Summary

EC Number: 3.4.21.6

Earlier names: Precursor: autoprothrombin III, Stuart (-Prower) factor. Enzyme: autoprothrombin C, thrombokinase, prothrombinase.

Distribution: Precursor: plasma of man and other mammals.

Source: Activation of precursor from plasma.

Action: Activates prothrombin (#1.15), factor VII (#1.16), factor V and factor IX (#1.12). Also cleaves Tos-Arg-OMe, Bz-Arg-OMe, etc.

Requirements: Neutral pH.

Substrate, usual: Prothrombin (assaying by correction of coagulation of deficient plasma).

Substrate, special: Bz-Ile-Glu-Gly-Arg-NPhNO$_2$ (S-2222), Boc-Ile-Glu-Gly-Arg-NMec, Boc-Ser-Gly-Arg-NMec.

Inhibitors: Dip-F (5-10 mM for several hours), Pms-F, Gbz-NPhNO$_2$. (Not Tos-Phe-CH$_2$Cl or Tos-Lys-CH$_2$Cl.) Antithrombin III, soybean trypsin inhibitor.

Molecular properties: Bovine factor X is a glycoprotein (10% carbohydrate), mol.wt. 55,000, composed of a light chain (16,500) and a heavy chain (39,300) linked by a disulfide bond. Human factor X has a slightly larger heavy chain, and thus a mol.wt. of 58,900.
  Cleavage of an -Arg-Ile- bond in the heavy chain of factor X produces the active factor $X_{a\alpha}$, and further proteolysis can liberate a 2700 mol.wt. peptide, giving factor $X_{a\beta}$ without loss of activity.

Comment: Factor X is activated in the presence of factor IX$_a$, factor VIII$_a$, Ca^{2+} and phospholipid, in the intrinsic pathway of coagulation, or factor VII$_a$ and tissue factor in the extrinsic route. Activation can also be produced by Russell's viper venom factor, and trypsin. Hereditary deficiency of factor X causes Stuart disease, a bleeding disorder.

Bibliography

1956

Miller, K. D. & van Vunakis, H. The effect of diisopropyl
fluorophosphate on the proteinase and esterase activities of
thrombin and on prothrombin and its activators. J. Biol. Chem.
223: 227-237.

1957

Bachmann, F., Duckert, F., Fluckiger, P. & Hitzig, W. H.
Hämorrhagische Diathesen verusacht durch Mangel des Stuart-
Prower-Faktors. Schweiz Med. Wochenschr. 87: 1221-1223.

1960

Milstone, J. H. TAMe esterase activity of blood thrombokinase
after repeated electrophoretic fractionations. Proc. Soc. Exp.
Biol. Med. 103: 361-363.

1961

Denson, K. W. E. The specific assay of Prower-Stuart factor and
factor VII. Acta Haematol. 25: 105-120.

1962

Cole, E. R., Marciniak, E. & Seegers, W. H. Procedures for the
quantitative determination of autoprothrombin C. Thromb. Diath.
Haemorrh. 8: 434-441.
Esnouf, M. P. & Williams, W. J. The isolation and purification of
a bovine-plasma protein which is a substrate for the coagulant
fraction of Russell's-viper venom. Biochem. J. 84: 62-71.

1963

Seegers, W. H., Cole, E. R., Harmison, C. R. & Marciniak, E.
Purification and some properties of autoprothrombin C. Can. J.
Biochem. Physiol. 41: 1047-1063.

1964

Milstone, J. H. Thrombokinase as prime activator of prothrombin:
historical perspectives and present status. Fed. Proc. Fed. Am.
Soc. Exp. Biol. 23: 742-761.
Papahadjopoulos, D., Yin, E. T. & Hanahan, D. J. Purification and
properties of bovine factor X: molecular changes during
activation. Biochemistry 3: 1931-1939.
Seegers, W. H., Cole, E. R., Harmison, C. R. & Monkhouse, F. C.
Neutralization of autoprothrombin C activity with antithrombin.
Can. J. Biochem. 42: 359-364.

1965

Lundblad, R. L. & Davie, E. W. The activation of Stuart factor
(factor X) by activated antihemophilic factor (activated factor
VIII). Biochemistry 4: 113-120.

1966

Nemerson, Y. The reaction between bovine brain tissue factor and factors VII and X. Biochemistry 5: 601–608.

1967

Seegers, W. H., Marciniak, E., Kipfer, R. K. & Yasunaga, K. Isolation and some properties of prethrombin and autoprothrombin III. Arch. Biochem. Biophys. 121: 372–383.

1968

Aronson, D. L. & Ménaché, D. Action of human thrombokinase on human prothrombin and p-tosyl-L-arginine methyl ester. Biochim. Biophys. Acta 167: 378–387.

Esnouf, M. P. & Macfarlane, R. G. Enzymology and the blood clotting mechanism. Adv. Enzymol. 30: 255–315.

Jackson, C. M. & Hanahan, D. J. Studies on bovine factor X. II. Characterization of purified factor X. Observations on some alterations in zone electrophoretic and chromatographic behaviour occurring during purification. Biochemistry 7: 4506–4517.

Jackson, C. M., Johnson, T. F. & Hanahan, D. J. Studies on bovine factor X. Large-scale purification of the bovine plasma protein possessing factor X activity. Biochemistry 7: 4492–4505.

1969

Aronson, D. L., Mustafa, A. J. & Mushinski, J. F. Purification of human factor X and comparison of peptide maps of human factor X and prothrombin. Biochim. Biophys. Acta 188: 25–30.

Leveson, J. E. & Esnouf, M. P. The inhibition of activated factor X with di:isopropyl fluorophosphate. Br. J. Haematol. 17: 173–178.

1970

Barton, P. G., Yin, E. T. & Wessler, S. Reactions of activated factor X-phosphatide mixtures in vitro and in vivo. J. Lipid Res. 11: 87–95.

Biggs, R., Denson, K. W. E., Akman, N., Borrett, R. & Hadden, M. Antithrombin III, antifactor Xa and heparin. Br. J. Haematol. 19: 283–305.

Gladhaug, Å. & Prydz, H. Purification of the coagulation factors VII and X from human serum. Some properties of factor VII. Biochim. Biophys. Acta 215: 105–111.

1971

Adams, R. W. & Elmore, D. T. The kinetics of hydrolysis of synthetic substrates by bovine factor Xa. Biochem. J. 124: 66P–67P.

Aronson, D. L. & Mustafa, A. J. Two forms of the human prothrombin converting enzyme (active factor X). Proc. Soc. Exp. Biol. Med. 137: 1262–1266.

Dombrose, F. A., Seegers, W. H. & Sedensky, J. A. Antithrombin.

Inhibition of thrombin and antiprothrombin C (F-Xa) as a mutual depletion system. Thromb. Diath. Haemorrh. 26: 103-123.

Yin, E. T., Wessler, S. & Stoll, P. J. Biological properties of the naturally occurring plasma inhibitor to activated factor X. J. Biol. Chem. 246: 3703-3711.

Yin, E. T., Wessler, S. & Stoll, P. J. Identity of plasma-activated factor X inhibitor with antithrombin III and heparin cofactor. J. Biol. Chem. 246: 3712-3719.

Yin, E. T., Wessler, S. & Stoll, P. J. Rabbit plasma inhibitor of the activated species of blood coagulation factor X. J. Biol. Chem. 246: 3694-3702.

1972

Chuang, T. F., Sargeant, R. B. & Hougie, C. The intrinsic activation of factor X in blood coagulation. Biochim. Biophys. Acta 273: 287-291.

De Lange, J. A. & Hemker, H. C. Inhibition of blood coagulation factors by serine esterase inhibitors. FEBS Lett. 24: 265-268.

Fujikawa, K., Legaz, M. E. & Davie, E. W. Bovine factor X_1 (Stuart factor). Mechanism of activation by a protein from Russell's viper venom. Biochemistry 11: 4892-4898.

Fujikawa, K., Legaz, M. E. & Davie, E. W. Bovine factors X_1 and X_2 (Stuart factor). Isolation and characterization. Biochemistry 11: 4882-4891.

Jackson, C. M. Characterization of two glycoprotein variants of bovine factor X and demonstration that the factor X zymogen contains two polypeptide chains. Biochemistry 11: 4873-4881.

Kandall, C., Akinbami, T. K. & Colman, R. W. Determinants of prothrombinase activity and modification of prothrombin conversion by thrombin-treated factor V. Br. J. Haematol. 23: 655-668.

Radcliffe, R. D. & Barton, P. G. The purification and properties of activated factor X. Bovine factor X activated with Russell's viper venom. J. Biol. Chem. 247: 7735-7742.

Reekers, P. & Hemker, H. C. Purification of blood coagulation factors II, VII, IX and X from bovine citrated plasma. Haemostasis 1: 2-22.

Roberts, P. S. & Fleming, P. B. The effects of salts and of ionic strength on the hydrolysis of TAME (p-toluenesulfonyl-L-arginine methyl ester) by thrombin and thrombokinase. Thromb. Diath. Haemorrh. 27: 573-583.

Seegers, W. H., McCoy, L. E., Reuterby, J., Sakuragawa, N., Murano, G. & Agrawal, B. B. L. Further observations on the purification and properties of autoprothrombin III (factor X). Thromb. Res. 1: 209-220.

Titani, K., Hermodson, M. A., Fujikawa, K., Ericsson, L. H., Walsh, K. A., Neurath, H. & Davie, E. W. Bovine factor X_{1a} (activated Stuart factor). Evidence of homology with mammalian serine proteases. Biochemistry 11: 4899-4903.

1973

Bajaj, S. P. & Mann, K. G. Simultaneous purification of bovine prothrombin and factor X. Activation of prothrombin by trypsin-activated factor X. J. Biol. Chem. 248: 7729-7741.

Dombrose, F. A. & Seegers, W. H. Evidence for multiple molecular forms of autoprothrombin C (factor X_a). Thromb. Res. 3: 737-743.

Elmore, D. T. The enzymatic properties of thrombin and factor Xa. Biochem. Soc. Trans. 1: 1191-1194.

Esnouf, M. P., Lloyd, P. H. & Jesty, J. A method for simultaneous isolation of factor X and prothrombin from bovine plasma. Biochem. J. 131: 781-789.

Jameson, G. W., Roberts, D. V., Adams, R. W., Kyle, W. S. A. & Elmore, D. T. Determination of the operational molarity of solutions of bovine α-chymotrypsin, trypsin, thrombin, and factor Xa by spectrofluorimetric titration. Biochem. J. 131: 107-117.

Jesty, J. & Esnouf, M. P. The preparation of activated factor X and its action on prothrombin. Biochem. J. 131: 791-797.

Kisiel, W. & Hanahan, D. J. The action of factor Xa, thrombin and trypsin on human factor II. Biochim. Biophys. Acta 329: 221-232.

Marciniak, E. Factor-Xa inactivation by antithrombin III: evidence for biological stabilization of factor Xa by factor V-phospholipid complex. Br. J. Haematol. 24: 391-400.

Mattock, P. & Esnouf, M. P. A form of bovine factor X with a single polypeptide chain. Nature New Biol. 242: 90-92.

McCoy, L. E., Walz, D. A., Agrawal, B. B. & Seegers, W. H. Isolation of L-chain polypeptide of autoprothrombin III (factor X): homology with prothrombin indicated. Thromb. Res. 2: 293-296.

Radcliffe, R. D. & Barton, P. G. Molecular characteristics of bovine factor X activated in concentrated sodium citrate solution. Arch. Biochem. Biophys. 155: 381-390.

Smith, R. L. Titration of activated bovine factor X. J. Biol. Chem. 248: 2418-2423.

1974

Esmon, C. T. & Jackson, C. M. The conversion of prothrombin to thrombin. III. The factor Xa-catalyzed activation of prothrombin. J. Biol. Chem. 249: 7782-7790.

Esmon, C. T., Owen, W. G. & Jackson, C. M. A plausible mechanism for prothrombin activation by factor X_a, factor V_a, phospholipid and calcium ions. J. Biol. Chem. 249: 8045-8047.

Esmon, C. T., Owen, W. G. & Jackson, C. M. The conversion of prothrombin to thrombin. II. Differentiation between thrombin and factor X_a-catalyzed proteolysis. J. Biol. Chem. 249: 606-611.

Fujikawa, K., Coan, M. H., Enfield, D. L., Titani, K., Ericsson, L. H. & Davie, E. W. A comparison of bovine prothrombin, factor

IX (Christmas factor), and factor X (Stuart factor). Proc. Natl. Acad. Sci. U.S.A. 71: 427-430.

Fujikawa, K., Coan, M. H., Legaz, M. E. & Davie, E. W. The mechanism of activation of bovine factor X (Stuart factor) by intrinsic and extrinsic pathways. Biochemistry 13: 5290-5299.

Jackson, C. M. Biochemistry of bovine factor X. Thromb. Diath. Haemorrh. Suppl. 57: 197-216.

Jesty, J. & Nemerson. Purification of factor VII from bovine plasma. Reaction with tissue factor and activation of factor X. J. Biol. Chem. 249: 509-515.

Jesty, J., Spencer, A. K. & Nemerson, Y. The mechanism of activation of factor X. Kinetic control of alternative pathways leading to the formation of activated factor X. J. Biol. Chem. 249: 5614-5622.

Kisiel, W. & Hanahan, D. J. Proteolysis of human factor II by factor Xa in the presence of hirudin. Biochem. Biophys. Res. Commun. 59: 570-577.

Morita, T., Nishibe, H., Iwanaga, S. & Suzuki, T. Studies on the activation of bovine prothrombin. Isolation and characterization of the fragments released from the prothrombin by activated factor X. J. Biochem. 76: 1031-1048.

1975

Davie, E. W., Fujikawa, K., Legaz, M. E. & Kato, H. Role of proteases in blood coagulation. In: Proteases in Biological Control (Reich, E., Rifkin, D. B. & Shaw, E. eds), pp. 65-77, Cold Spring Harbor Laboratory, New York.

Enfield, D. L., Ericsson, L. H., Walsh, K. A., Neurath, H. & Titani, K. Bovine factor X_1 (Stuart factor). Primary structure of the light chain. Proc. Natl. Acad. Sci. U.S.A. 72: 16-19.

Fujikawa, K., Titani, K. & Davie, E. W. Activation of bovine factor X (Stuart factor): conversion of factor $X_{a\alpha}$ to factor $X_{a\beta}$. Proc. Natl. Acad. Sci. U.S.A. 72: 3359-3363.

Henrikson, R. A. & Jackson, C. M. The chemistry and enzymology of bovine factor X. Seminars Thromb. Haemostas. 1: 284-309.

Kalousek, F., Konigsberg, W. & Nemerson, Y. Activation of factor IX by activated factor X: a link between the extrinsic and intrinsic coagulation systems. FEBS Lett. 50: 382-385.

Radcliffe, R. & Nemerson, Y. Activation and control of factor VII by activated factor X and thrombin. J. Biol. Chem. 250: 388-395.

Titani, K., Fujikawa, K., Enfield, D. L., Ericsson, L. H., Walsh, K. A. & Neurath, H. Bovine factor X_1 (Stuart factor): amino acid sequence of heavy chain. Proc. Natl. Acad. Sci. U.S.A. 72: 3082-3086.

Ulevitch, R. J., Cochrane, C. G., Revak, S. D., Morrison, D. C. & Johnston, A. R. The structural and enzymatic properties of the components of the Hageman factor-activated pathways. In: Proteases and Biological Control (Reich, E., Rifkin, D. B. & Shaw, E. eds), pp. 85-93, Cold Spring Harbor Laboratory, New York.

1976

Bucher, D., Nebelin, E., Thomsen, J. & Stenflo, J. Identification of γ-carboxyglutamic acid residues in bovine factors IX and X, and in a new vitamin K-dependent protein. FEBS Lett. 68: 293-296.

Fujikawa, K. & Davie, E. W. Bovine factor X (Stuart factor). Methods Enzymol. 45: 89-95.

Jesty, J. & Nemerson, Y. The activation of bovine coagulation factor X. Methods Enzymol. 45: 95-107.

Kosow, D. P. Purification and activation of human factor X: cooperative effect of Ca^{2+} on the activation reaction. Thromb. Res. 9: 565-573.

Kurachi, K., Fujikawa, K., Schmer, G. & Davie, E. W. Inhibition of bovine factor IX_a and factor $X_{a\beta}$ by antithrombin III. Biochemistry 15: 373-377.

Østerud, B., Miller-Andersson, M., Abildgaard, U. & Prydz, H. The effect of antithrombin III on the activity of the coagulation factors VII, IX, and X. Thromb. Haemost. 35: 295-304.

Smith, C. M. & Hanahan, D. J. The activation of factor V by factor Xa or α-chymotrypsin and comparison with thrombin and RVV-V action. Biochemistry 15: 1830-1838.

Stürzebecher, J., Markwardt, F. & Walsmann, P. Synthetic inhibitors of serine proteinases. XIV. Inhibition of factor Xa by derivatives of benzamidine. Thromb. Res. 9: 637-646.

Vician, L. & Tishkoff, G. H. Purification of human blood clotting factor X by blue dextran agarose affinity chromatography. Biochim. Biophys. Acta 434: 199-208.

1977

Aurell, L., Friberger, P., Karlsson, G. & Claeson, G. A new sensitive and highly specific chromogenic peptide substrate for factor X_a. Thromb. Res. 11: 595-609.

Claeson, G., Aurell, L., Karlsson, G. & Friberger, P. Substrate structure and activity relationship. In: New Methods for the Analysis of Coagulation using Chromogenic Substrates (Witt, I., ed.), pp. 37-54, Walter de Gruyter, Berlin.

Di Scipio, R. G., Hermodson, M. A. & Davie, E. W. Activation of human factor X (Stuart factor) by a protease from Russell's viper venom. Biochemistry 16: 5253-5260.

Di Scipio, R. G., Hermodson, M. A., Yates, S. G. & Davie, E. W. A comparison of human prothrombin, factor IX (Christmas factor), factor X (Stuart factor), and protein S. Biochemistry 16: 698-706.

Miletich, J. P., Jackson, C. M. & Majerus, P. W. Interaction of coagulation factor Xa with human platelets. Proc. Natl. Acad. Sci. U.S.A. 74: 4033-4036.

Morita, T., Kato, H., Iwanaga, S., Takada, K., Kimura, T. & Sakakibara, S. New fluorogenic substrates for α-thrombin, factor X, kallikreins and urokinase. J. Biochem. 82: 1495-1498.

Silverberg, S. A., Nemerson, Y. & Zur, M. Kinetics of the

activation of bovine coagulation factor X by components of the extrinsic pathway. Kinetic behavior of two-chain factor VII in the presence and absence of tissue factor. J. Biol. Chem. 252: 8481-8488.

Suomela, H., Blombäck, M. & Blombäck, B. The activation of factor X evaluated by using synthetic substrates. Thromb. Res. 1: 267-281.

Vinazzer, H. Assay of Factor X_a with a chromogenic substrate. In: New Methods for the Analysis of Coagulation using Chromogenic Substrates (Witt, I., ed.), pp. 203-209, Walter de Gruyter, Berlin.

Walsmann, P. Inhibition of serine proteinases by benzamidine derivatives. Acta Biol. Med. Ger. 36: 1931-1937.

Wylie, A. R. G., Lonsdale-Eccles, J. D., Blumson, N. L. & Elmore, D. T. The effect of the proteinase from rat peritoneal mast cells on prothrombin and Factor X. Biochem. Soc. Trans. 5: 1449-1452.

Yue, R. H. & Gertler, M. M. Isolation of the high molecular form of bovine factor X and some of its physical properties. Biochim. Biophys. Acta 490: 350-362.

1978

Aurell, L., Simonsson, R., Arielly, S., Karlsson, G., Friberger, P. & Claeson, G. New chromogenic peptide substrates for Factor X_a. Haemostasis 7: 92-94.

Geratz, J. D. & Tidwell, R. R. Current concepts on action of synthetic thrombin inhibitors. Haemostasis 7: 170-176.

Hultin, M. B. & Nemerson, Y. Activation of factor X by factors IXa and VIII; a specific assay for factor IXa in the presence of thrombin-activated factor VIII. Blood 52: 928-940.

Jesty, J. The inhibition of activated bovine coagulation factors X and VII by antithrombin III. Arch. Biochem. Biophys. 185: 165-173.

Kerr, M. A., Grahn, D. T., Walsh, K. A. & Neurath, H. Activation of bovine factor X (Stuart factor) - analogy with pancreatic zymogen-enzyme systems. Biochemistry 17: 2645-2648.

Miletich, J. P., Jackson, C. M. & Majerus, P. W. Properties of the factor X_a binding site on human platelets. J. Biol. Chem. 253: 6908-6916.

Morris, S., Robey, F. A. & Kosow, D. P. Kinetic studies on the activation of human Factor X. The role of metal ions on the reaction catalyzed by the venom coagulant protein of Vipera russelli. J. Biol. Chem. 253: 4604-4608.

Orthner, C. L. & Kosow, D. P. The effect of metal ions on the amidolytic activity of human factor X_a (activated Stuart-Prower factor). Arch. Biochem. Biophys. 185: 400-406.

Walker, F. J. & Esmon, C. T. The molecular mechanisms of heparin action. III. The anticoagulant properties of polyanetholesulfonate. Biochem. Biophys. Res. Commun. 83: 1339-1346.

Wu, V.-Y. & Seegers, W. H. Preservation of factor Xa activity.

Thromb. Res. 13: 701 only.

1979

Davie, E. W., Fujikawa, K., Kurachi, K. & Kisiel, W. The role of serine proteases in the blood coagulation cascade. Adv. Enzymol. 48: 277-318.

Jesty, J. Dissociation of complexes and their derivatives formed during inhibition of bovine thrombin and activated Factor X by antithrombin III. J. Biol. Chem. 254: 1044-1049.

Marciniak, E. Factor Xa adsorption on fibrin. Fed. Proc. Fed. Am. Soc. Exp. Biol. 38: 1272 only.

Marlar, R. A., Kleiss, A. J. & Griffin, J. H. Development of a new assay for the activation of human coagulation factor X. Fed. Proc. Fed. Am. Soc. Exp. Biol. 38: 1272 only.

Morita, T. & Jackson, C. M. Bovine factor X_1 and X_2 activation peptide based chromatographic differences. Fed. Proc. Fed. Am. Soc. Exp. Biol. 38: 811 only.

Robison, D. J., Bing, D. H., Furie, B., Laura, R. & Furie, B. C. Active site-directed labeling of bovine activated factor X. Fed. Proc. Fed. Am. Soc. Exp. Biol. 38: 793 only.

Thomas, D. P., Merton, R. E., Barrowcliffe, T. W., Mulloy, B. & Johnson, E. A. Anti-factor Xa activity of heparan sulphate. Thromb. Res. 14: 501-506.

Walker, F. J. & Esmon, C. T. Factors that influence the rate of antithrombin III inhibition of thrombin and factor Xa. Fed. Proc. Fed. Am. Soc. Exp. Biol. 38: 809 only.

Walker, F. J. & Esmon, C. T. Interactions between heparin and factor Xa. Inhibition of prothrombin activation. Biochim. Biophys. Acta 585: 405-415.

Entry 1.15

THROMBIN

Summary

EC Number: 3.4.21.5

Earlier names: The precursor is prothrombin, or coagulation factor II; the enzyme has been called fibrinogenase, and is factor II_a.

Distribution: Plasma of many mammals (as the precursor, prothrombin).

Source: Usually bovine plasma.

Action: Cleaves -Arg-Gly- bonds in the A(α) and B(β) chains of fibrinogen, liberating N-terminal fibrinopeptides to generate fibrin. Also acts on factors V, VII, VIII and XIII, always cleaving arginyl bonds. Some low mol.wt. substrates of trypsin are susceptible to thrombin.

Requirements: pH optimum 6.4 for clotting, but 9.0 for proteolysis.

Substrate, usual: Fibrinogen, Tos-Arg-OMe, Bz-Arg-NPhNO$_2$.

Substrate, special: Tos-Gly-Pro-Arg-NPhNO$_2$, Bz-Phe-Val-Arg-NPhNO$_2$ (S-2160), D-Phe-Pip-Arg-NPhNO$_2$ (S-2238), Tos-Gly-Pro-Arg-NPhNO$_2$ (Chromozym TH) (all relatively insensitive to factor X_a). Boc-Val-Pro-Arg-NMec for fluorimetric assays. Titration with Gbz-OPhNO$_2$.

Inhibitors: Dip-F, Pms-F, Tos-Lys-CH$_2$Cl, D-Phe-Pro-Arg-CH$_2$Cl, benzamidine, Dns-Arg-N-(3-ethyl-1,5-pentanediyl)amide (K$_i$ 10^{-7} M) [1], proflavine. Antithrombin III-heparin, α_1-proteinase inhibitor, α_2-macroglobulin, hirudin. (Not soybean trypsin inhibitor, aprotinin, or ovomucoid).

Molecular properties: Bovine prothrombin is a single-chain glycoprotein (11% carbohydrate), mol.wt. about 70,000, pI 4.2, containing Gla residues, and therefor vitamin K-dependent for biosynthesis.

Comment: Proteolytic activation is catalyzed by factor X_a in the presence of factor V_a (once known as factor VI), Ca^{2+} and phospholipid. The first stage is cleavage of -Arg-Thr-, which releases a catalytically inactive glycopeptide, mol.wt. 33,500, and a sigle chain polypeptide called intermediate II. Intermediate II is activated by further cleavage at -Arg-Ile- to form thrombin. Bovine thrombin has chains of 5700 and 32,000 mol.wt., the larger containing the catalytic residues; pI is

5.5. Activation by thrombin itself follows a slightly different course. Human prothrombin (mol.wt. 68,700) is activated similarly, to α-thrombin, then autolysis yields the β- and γ-forms. pI of human thrombin is 7.0-7.6.

References

[1] Nesheim et al. Biochemistry 18: 996-1003, 1979.

Bibliography

1954

Sherry, S. & Troll, W. The action of thrombin on synthetic substrates. J. Biol. Chem. 208: 95-105.

1956

Gladner, J. A. & Laki, K. The inhibition of thrombin by diisopropyl phosphofluoridate. Arch. Biochem. Biophys. 62: 501-503.

Miller, K. D. & van Vunakis, H. The effect of diisopropyl fluorophosphate on the proteinase and esterase activities of thrombin and on prothrombin and its activators. J. Biol. Chem. 223: 227-237.

1958

Laki, K., Gladner, J., Folk, J. E. & Kominz, D. R. The mode of action of thrombin. Thromb. Diath. Haemorrh. 2: 205-217.

1959

Hummel, B. C. W. A modified spectrophotometric determination of chymotrypsin, trypsin, and thrombin. Can. J. Biochem. 37: 1393-1399.

Ronwin, E. The active center of thrombin and trypsin. Biochim. Biophys. Acta 33: 326-332.

1961

Buluz, K., Januszko, T. & Olbromski, J. Conversion of fibrin to desmofibrin. Nature 191: 1093-1094.

1963

Rapaport, S. I., Schiffman, S., Patch, M. J. & Ames, S. B. The importance of activation of antihemophilic globulin and proaccellerin by traces of thrombin in the generation of intrinsic prothrombinase activity. Blood 21: 221-236.

1964

Lorand, L. & Konishi, K. Activation of the fibrin stabilizing factor of plasma by thrombin. Arch. Biochem. Biophys. 105: 58-67.

1965

Biggs, R., Macfarlane, R. G., Denson, K. W. E. & Ash, B. J. Thrombin and the interaction of factors VIII and IX. Br. J.

Haematol. 11: 276-295.

1966

Sherry, S., Alkjaersig, N. & Fletcher, A. P.   Comparative activity
of thrombin on substituted arginine and lysine esters.   Am. J.
Physiol. 209: 577-583.

1967

Cole, E. R., Koppel, J. L. & Olwin, J. H.   Multiple specificity of
thrombin for synthetic substrates.   Nature 213: 405-406.

1968

Esnouf, M. P. & Macfarlane, R. G.   Enzymology and the blood
clotting mechanism.   Advan. Enzymol. 30: 255-315.
Markwardt, F., Landmann, H. & Walsmann, P.   Comparative studies on
the inhibition of trypsin, plasmin and thrombin by derivatives
of benzylamine and benzamidine.   Eur. J. Biochem. 6: 502-506.

1969

Chase, T. & Shaw, E.   Comparison of the esterase activities of
trypsin, plasmin, and thrombin on guanidinobenzoate esters.
Titration of the enzymes.   Biochemistry 8: 2212-2224.
Feeney, R. E., Means, G. E. & Bigler, J. C.   Inhibition of human
trypsin, plasmin, and thrombin by naturally occurring inhibitors
of proteolytic enzymes.   J. Biol. Chem. 244: 1957-1960.

1970

Batt, C. W., Mikulka, T. W., Mann, K. G., Guarracino, C. L.,
Altiere, R. J., Graham, R. G., Quigley, J. P., Wolf, J. W. &
Zafonte, C. W.   The purification and properties of bovine
thrombin.   J. Biol. Chem. 245: 4857-4862.
Hartley, B. S.   Homologies in serine proteinases.   Philos. Trans.
Roy. Soc. London Ser. B 257: 77-87.
Rosenberg, R. D. & Waugh, D. F.   Multiple bovine thrombin
components.   J. Biol. Chem. 245: 5049-5056.

1971

Lundblad, R. L.   A rapid method for the purification of bovine
thrombin and the inhibition of the purified enzyme with
phenylmethylsulfonyl fluoride.   Biochemistry 10: 2501-2506.
Magnusson, S.   Thrombin and prothrombin.   In: The Enzymes (Boyer,
P. D. ed.), 3rd edn., vol.3, pp. 277-321, Academic Press, New
York.
Malhotra, O. P. & Carter, J. R.   Isolation and purification of
prothrombin from dicumarolized steers.   J. Biol. Chem. 246:
2665-2671.
Muramatu, M. & Fujii, S.   Inhibitory effects of ω-amino acid
esters on trypsin, plasmin, plasma kallikrein and thrombin.
Biochim. Biophys. Acta 242: 203-208.
Thompson, A. R. & Davie, E. W.   Affinity chromatography of
thrombin.   Biochim. Biophys. Acta 250: 210-215.

1972

De Lange, J. A. & Hemker, H. C. Inhibition of blood coagulation factors by serine esterase inhibitors. FEBS Lett. 24: 265-268.

Geratz, J. D. Synthetic low molecular weight inhibitors of thrombin. Fol. Haematol. 98: 455-470.

Kandall, C., Akinbami, T. K. & Colman, R. W. Determinants of prothrombinase activity and modification of prothrombin conversion by thrombin-treated factor V. Br. J. Haematol. 23: 655-668.

Maki, M. & Sasaki, K. A new broad spectrum inhibitor of esteroproteases, 4-(2-carboxyethyl)phenyl-trans-4-aminomethyl cyclohexane carboxylate hydrochloride (DV 1006). Tohoku J. Exp. Med. 106: 233-248.

McCoy, L. E. & Seegers, W. H. Preparation of bovine prothrombin complex and purified prothrombin. Thromb. Res. 1: 461-472.

Muramatu, M. & Fujii, S. Inhibitory effects of ω-guanidino acid esters on trypsin, plasmin, plasma kallikrein and thrombin. Biochim. Biophys. Acta 268: 221-224.

Nelsestuen, G. L. & Suttie, J. W. Mode of action of vitamin K. Calcium binding properties of bovine prothrombin. Biochemistry 11: 4961-4964.

Reekers, P. & Hemker, H. C. Purification of blood coagulation factors II, VII, IX and X from bovine citrated plasma. Haemostasis 1: 2-22.

Roberts, P. S. & Fleming, P. B. The effects of salts and of ionic strength on the hydrolysis of TAME (p-toluenesulfonyl-L-arginine methyl ester) by thrombin and thrombokinase. Thromb. Diath. Haemorrh. 27: 573-583.

Schmer, G. The purification of bovine thrombin by affinity chromatography on benzamidine-agarose. Hoppe-Seyler's Z. Physiol. Chem. 353: 810-814.

Stenn, K. S. & Blout, E. R. Mechanism of bovine prothrombin activation by an insoluble preparation of bovine factor Xa (thrombokinase). Biochemistry 11: 4502-4515.

Svendsen, L., Blomback, B., Blomback, M. & Olsson, P. Synthetic chromogenic substrates for determination of trypsin, thrombin and thrombin-like enzymes. Thromb. Res. 1: 267-278.

Weinstein, M. J. & Doolittle, R. F. Differential specificities of thrombin, plasmin and trypsin with regard to synthetic and natural substrates and inhibitors. Biochim. Biophys. Acta 258: 577-590.

1973

Bajaj, S. P. & Mann, K. G. Simultaneous purification of bovine prothrombin and factor X. Activation of prothrombin by trypsin-activated factor X. J. Biol. Chem. 248: 7729-7741.

Elmore, D. T. The enzymatic properties of thrombin and factor Xa. Biochem. Soc. Trans. 1: 1191-1194.

Esnouf, M. P., Lloyd, P. H. & Jesty, J. A method for simultaneous isolation of factor X and prothrombin from bovine plasma.

Biochem. J. 131: 781-789.

Exner, T. & Koppel, J. L.   Cholate enhancement of interaction between thrombin and tosyl lysine chloromethyl ketone. Biochim. Biophys. Acta 329: 233-240.

Fasco, M. J. & Fenton, J. W.   Specificity of thrombin. 1. Esterolytic properties of thrombin, plasmin, trypsin and chymotrypsin with N_β-substituted guanidino derivatives of p-nitrophenyl-p'-guanidinobenzoate. Arch. Biochem. Biophys. 159: 802-812.

Hixson, H. F. & Nishikawa, A. H.   Affinity chromatography: purification of bovine trypsin and thrombin. Arch. Biochem. Biophys. 154: 501-509.

Jameson, G. W., Roberts, D. V., Adams, R. W., Kyle, W. S. A. & Elmore, D. T.   Determination of the operational molarity of solutions of bovine α-chymotrypsin, trypsin, thrombin, and factor Xa by spectrofluorimetric titration. Biochem. J. 131: 107-117.

Jesty, J. & Esnouf, M. P.   The preparation of activated factor X and its action on prothrombin. Biochem. J. 131: 791-797.

Kisiel, W. & Hanahan, D. J.   The action of factor Xa, thrombin and trypsin on human factor II. Biochim. Biophys. Acta 329: 221-232.

Lanchantin, G. F., Friedmann, J. A. & Hart, D. W.   Two forms of human thrombin. J. Biol. Chem. 248: 5956-5966.

Lundblad, R. L.   Observations on the hydrolysis of p-nitrophenyl acylates by purified bovine thrombin. Thromb. Diath. Haemorrh. 30: 248-254.

Mikuni, Y., Iwanaga, S. & Konishi, K.   A peptide released from plasma fibrin stabilizing factor in the conversion to the active enzyme by thrombin. Biochem. Biophys. Res. Commun. 54: 1393-1402.

Rosenberg, R. & Damus, P.   The purification and mechanism of action of human antithrombin-heparin cofactor. J. Biol. Chem. 248: 6490-6505.

Stenflo, J. & Ganrot, P.-O.   Binding of Ca^{2+} to normal and dicoumarol-induced prothrombin. Biochem. Biophys. Res. Commun. 50: 98-104.

Triantaphyllopoulos, D. C. & Chandra, S.   Digestion of fibrin by thrombin. Biochim. Biophys. Acta 328: 229-232.

1974

Brecher, A. S., Howe, W. F., Burkholder, J. S. & Burkholder, D. E. An evaluation of the utility of benzoyl-DL-arginine-β-naphthylamide for the assay of thrombin activity in blood. Thromb. Res. 5: 157-171.

Esmon, C. T. & Jackson, C. M.   The conversion of prothrombin to thrombin. III. The factor Xa-catalyzed activation of prothrombin. J. Biol. Chem. 249: 7782-7790.

Esmon, C. T. & Jackson, C. M.   The conversion of prothrombin to thrombin. IV. The function of the fragment 2 region during activation in the presence of factor V. J. Biol. Chem. 249:

7791-7797.

Esmon, C. T., Owen, W. G. & Jackson, C. M. A plausible mechanism for prothrombin activation by factor X_a, factor V_a, phospholipid and calcium ions. J. Biol. Chem. 249: 8045-8047.

Esmon, C. T., Owen, W. G. & Jackson, C. M. The conversion of prothrombin to thrombin. II. Differentiation between thrombin and factor X_a-catalyzed proteolysis. J. Biol. Chem. 249: 606-611.

Fujikawa, K., Coan, M. H., Enfield, D. L., Titani, K., Ericsson, L. H. & Davie, E. W. A comparison of bovine prothrombin, factor IX (Christmas factor), and factor X (Stuart factor). Proc. Natl. Acad. Sci. U.S.A. 71: 427-430.

Kisiel, W. & Hanahan, D. J. Proteolysis of human factor II by factor Xa in the presence of hirudin. Biochem. Biophys. Res. Commun. 59: 570-577.

Morita, T., Nishibe, H., Iwanaga, S. & Suzuki, T. Studies on the activation of bovine prothrombin. Isolation and characterization of the fragments released from the prothrombin by activated factor X. J. Biochem. 76: 1031-1048.

Owen, W. G., Esmon, C. T. & Jackson, C. M. The conversion of prothrombin to thrombin. I. Characterization of the reaction products formed during the activation of bovine prothrombin. J. Biol. Chem. 249: 594-605.

Reuterby, J., Walz, D. A., McCoy, L. E. & Seegers, W. H. Amino acid sequence of O fragment of bovine prothrombin. Thromb. Res. 4: 885-890.

Seegers, W. H., McCoy, L. E. & Sakuragawa, N. Mechanisms of thrombin formation. Thromb. Diath. Haemorrh. Suppl. 57: 255-272.

Stenflo, J. Vitamin K and the biosynthesis of prothrombin. IV. Isolation of peptides containing prosthetic groups from normal prothrombin and the corresponding peptides from dicoumarol-induced prothrombin. J. Biol. Chem. 249: 5527-5535.

Stenflo, J., Fernlund, P., Egan, W. & Roepstorff, P. Vitamin K dependent modifications of glutamic acid residues in prothrombin. Proc. Natl. Acad. Sci. U.S.A. 71: 2730-2733.

Takagi, T. & Doolittle, R. F. Amino acid sequence studies on factor XIII and the peptide released during its activation by thrombin. Biochemistry 13: 750-756.

Walz, D. A. & Seegers, W. H. Amino acid sequence of human thrombin A chain. Biochem. Biophys. Res. Commun. 60: 717-722.

Walz, D. A., Seegers, W. H., Reuterby, J. & McCoy, L. E. Proteolytic specificity of thrombin. Thromb. Res. 4: 713-717.

1975

Davie, E. W. & Fujikawa, K. Basic mechanisms in blood coagulation. Annu. Rev. Biochem. 44: 799-829.

Davie, E. W., Fujikawa, K., Legaz, M. E. & Kato, H. Role of proteases in blood coagulation. In: Proteases in Biological Control (Reich, E., Rifkin, D. B. & Shaw, E. eds), pp. 65-77, Cold Spring Harbor Laboratory, New York.

Downing, M. R., Butkowski, R. J., Clark, M. M. & Mann, K. G. Human prothrombin activation. J. Biol. Chem. 250: 8897–8906.

Geratz, J. D., Cheng, M. C.-F. & Tidwell, R. R. New aromatic diamidines with central α-oxyalkane or α,ω-dioxyalkane chains. Structure–activity relationships for the inhibition of trypsin, pancreatic kallikrein, and thrombin and for the inhibition of the overall coagulation process. J. Med. Chem. 18: 477–481.

Legaz, M. E., Weinstein, M. J., Heldebrandt, C. M. & Davie, E. W. Isolation, subunit structure and proteolytic modification of bovine factor VIII. Ann N. Y. Acad. Sci. 240: 43–61.

Loeffler, L. J., Mar, E.-C., Geratz, J. D. & Fox, L. B. Synthesis of isosteres of p-amidinophenylpyruvic acid. Inhibitors of trypsin, thrombin and pancreatic kallikrein. J. Med. Chem. 18: 287–292.

Magnusson, S., Peterson, T. E., Sottrup–Jensen, L. & Claeys, H. Complete primary structure of prothrombin: isolation, structure and reactivity of ten carboxylated glutamic acid residues and regulation of prothrombin activation by thrombin. In: Proteases and Biological Control (Reich, E., Rifkin, D. B. & Shaw, E. eds), pp. 123–149, Cold Spring Harbor Laboratory, New York.

Radcliffe, R. & Nemerson, Y. Activation and control of factor VII by activated factor X and thrombin. J. Biol. Chem. 250: 388–395.

Sottrup–Jensen, L., Zajdel, M., Claeys, H., Peterson, T. E. & Magnusson, S. Amino-acid sequence of activation cleavage site in plasminogen: homology with "pro" part of prothrombin. Proc. Natl. Acad. Sci. U.S.A. 72: 2577–2581.

1976

Axelsson, G., Korsan-Bengtsen, K. & Waldenström, J. Prothrombin determination by means of a chromogenic peptide substrate. Thromb. Haemostas. 36: 517–524.

Bajaj, S. P., Harmony, J. A. K., Martinez-Carrion, M. & Castellino, F. J. Human plasma lipoproteins as accelerators of prothrombin activation. J. Biol. Chem. 251: 5233–5236.

Claeys, H., Sottrup–Jensen, L., Zajdel, M., Petersen, T. E. & Magnusson, S. Multiple gene duplication in the evolution of plasminogen. Five regions of sequence homology with the two internally homologous structures in prothrombin. FEBS Lett. 61: 20–24.

Downing, M. R., Taylor, J. C. & Mann, K. G. Interaction of human α_1-antitrypsin and human α-thrombin. Blood 48: 975 only.

Filip, D. J., Eckstein, J. D. & Veltkamp, J. J. Hereditary antithrombin III deficiency and thromboembolic disease. Am. J. Hematol. 2: 343–349.

Geratz, J. D., Cheng, M. C.-F. & Tidwell, R. R. Novel bis(benzamidine) compounds with an aromatic central link. Inhibitors of thrombin, pancreatic kallikrein, trypsin and complement. J. Med. Chem. 19: 634–639.

Lobo, A. P., Wos, J. D., Yu, S. M. & Lawson, W. B. Active site studies of human thrombin and bovine trypsin: peptide substrates.

Arch. Biochem. Biophys. 177: 235-244.
Lundblad, R. L., Kingdon, H. S. & Mann, K. G. Thrombin. Methods
Enzymol. 45: 156-176.
Magnusson, S., Sottrup-Jensen, L., Peterson, T. E., Dudek-
Wojciechowska, G. & Claeys, H. Homologous "kringle" structures
common to plasminogen and prothrombin. Substrate specificity of
enzymes activating prothrombin and plasminogen. In: Proteolysis
and Physiological Regulation (Brew, K. & Ribbons, D. W. eds),
pp. 203-238, Academic Press, New York.
Mann, K. G. Prothrombin. Methods Enzymol. 45: 123-256.
Matheson, N. R. & Travis, J. Inactivation of human thrombin in
the presence of human α_1-proteinase inhibitor. Biochem. J. 159:
495-502.
Smith, C. M. & Hanahan, D. J. The activation of factor V by
factor Xa or α-chymotrypsin and comparison with thrombin and
RVV-V action. Biochemistry 15: 1830-1838.
Tidwell, R. R., Fox, L. L. & Geratz, J. D. Aromatic tris-amidines.
A new class of highly active inhibitors of trypsin-like
proteases. Biochim. Biophys. Acta 445: 729-738.

1977

Aronson, D. L., Stevan, L., Ball, A. P., Franza, B. R. &
Finlayson, J. S. Generation of the combined prothrombin
activation peptide (F1.2) during the clotting of blood and
plasma. J. Clin. Invest. 60: 1410-1418.
Bang, N. U. & Mattler, L. E. Thrombin sensitivity and specificity
of three chromogenic peptide substrates. In: Chemistry and
Biology of Thrombin (Lundblad, R. L., Fenton, J. W., & Mann, K.
G. eds), pp. 305-310, Ann Arbor Science Publishers, Ann Arbor.
Baskova, I. P., Memon, M. S. & Cherkesova, D. U. Use of
N-benzoyl-phenylalanyl-valyl-arginine p-nitroanilide and hirudin
for the detection of enzyme activity in the acylation of
prothrombin with citraconic anhydride. Biochemistry U.S.S.R.
(Engl. Transl.) 42: 1161-1164.
Blömback, B., Hessel, B., Hogg, D. & Claesson, G. Substrate
specificity of thrombin on proteins and synthetic substrates.
In: Chemistry and Biology of Thrombin (Lundblad, R. L., Fenton,
J. W., & Mann, K. G. eds), pp. 275-290, Ann Arbor Science
Publishers, Ann Arbor.
Butkowski, R. J., Elion, J., Downing, M. R. & Mann, K. G. Primary
structure of human prethrombin 2 and α-thrombin. J. Biol. Chem.
252: 4942-4957.
Claeson, G., Aurell, L., Karlsson, G. & Friberger, P. Substrate
structure and activity relationship. In: New Methods for the
Analysis of Coagulation using Chromogenic Substrates (Witt, I.,
ed.), pp. 37-54, Walter de Gruyter, Berlin.
Di Scipio, R. G., Hermodson, M. A., Yates, S. G. & Davie, E. W. A
comparison of human prothrombin, factor IX (Christmas factor),
factor X (Stuart factor), and protein S. Biochemistry 16: 698-
706.
Edy, J. & Collen, D. The interaction in human plasma of

antiplasmin, the fast-reacting plasmin inhibitor, with plasmin, thrombin, trypsin and chymotrypsin. Biochim. Biophys. Acta 484: 423-432.

Elion, J., Downing, M. R., Butkowski, R. J. & Mann, K. G. Structure of human thrombin: comparison with other serine proteases. In: Chemistry and Biology of Thrombin (Lundblad, R. L., Fenton, J. W., & Mann, K. G. eds), pp. 97-111, Ann Arbor Science Publishers, Ann Arbor.

Fenton, J. W., Fasco, M. J., Stackrow, A. B., Aronson, D. L., Young, A. M. & Finlayson, J. S. Human thrombins. Production, evaluation and properties of α-thrombin. J. Biol. Chem. 252: 3587-3598.

Gaffney, P. J., Lord, K., Brasher, M. & Kirkwood, T. B. L. Problems in the assay of thrombin using synthetic peptides as substrates. Thromb. Res. 10: 549-556.

Hageman, T. C. & Scheraga, H. A. Mechanism of action of thrombin on fibrinogen. Reaction of the N-terminal CNBr fragment from the Bβ chain of bovine fibrinogen with bovine thrombin. Arch. Biochem. Biophys. 179: 506-517.

Hatton, M. W. C. & Regoeczi, E. The inactivation of thrombin and plasmin by antithrombin III in the presence of Sepharose-heparin. Thromb. Res. 10: 645-660.

Kettner, C. & Shaw, E. The selective inactivation of thrombin by peptides of arginine chloromethyl ketone. In: Chemistry and Biology of Thrombin (Lundblad, R. L., Fenton, J. W. & Mann, K. G. eds), pp. 129-143, Ann Arbor Science Publishers, Ann Arbor, Michigan.

Lundblad, R. L., Fenton, J. W., II. & Mann, K. G. (eds) Chemistry and Biology of Thrombin, Ann Arbor Science Publishers, Ann Arbor.

Mattler, L. E. & Bang, N. U. Serine protease specificity for peptide chromogenic substrates. Thromb. Haemostas. 38: 776-792.

Morita, T., Kato, H., Iwanaga, S., Takada, K., Kimura, T. & Sakakibara, S. New fluorogenic substrates for α-thrombin, factor X, kallikreins and urokinase. J. Biochem. 82: 1495-1498.

Roth, M. & Haarsma, M. Enzymatic determination of thrombin and thrombin inhibitors. In: New Methods for the Analysis of Coagulation using Chromogenic Substrates (Witt, I., ed.), pp. 91-104, Walter de Gruyter, Berlin.

Scheraga, H. A. Active site mapping of thrombin. In: Chemistry and Biology of Thrombin (Lundblad, R. L., Fenton, J. W. & Mann, K. G. eds), pp. 145-158, Ann Arbor Science Publishers, Ann Arbor.

Scully, M. F. & Kakkar, V. V. Methods for semi micro or automated determination of thrombin, antithrombin, and heparin cofactor using the substrate H-D-Phe-Pip-Arg-p-nitroanilide.2HCl. Clin. Chim. Acta 79: 595-602.

Tamura, Y., Hirado, M., Okamura, K., Minato, Y. & Fujii, S. Synthetic inhibitors of trypsin, plasmin, thrombin, C_1r and C_1 esterase. Biochim. Biophys. Acta 484: 417-422.

Thompson, A. R., Enfield, D. L., Ericsson, L. H., Legaz, M. E. &
Fenton, J. W. Human thrombin: partial primary structure. Arch.
Biochem. Biophys. 178: 356–367.

Van Nispen, J. W., Hageman, T. C. & Scheraga, H. A. Mechanism of
action of thrombin on fibrinogen. The reaction of thrombin with
fibrinogen–like peptides containing 11, 14 and 16 residues.
Arch. Biochem. Biophys. 182: 227–243.

Vehar, G. A. & Davie, E. W. Formation of a serine enzyme in the
presence of bovine factor VIII (antihemophilic factor) and
thrombin. Science 197: 374–376.

Walsmann, P. Inhibition of serine proteinases by benzamidine
derivatives. Acta Biol. Med. Ger. 36: 1931–1937.

Wylie, A. R. G., Lonsdale-Eccles, J. D., Blumson, N. L. & Elmore,
D. T. The effect of the proteinase from rat peritoneal mast
cells on prothrombin and Factor X. Biochem. Soc. Trans. 5:
1449–1452.

1978

Bajusz, S., Barabás, E., Tolnay, P., Széll, E. & Bagdy, D.
Inhibition of thrombin and trypsin by tripeptide aldehydes.
Int. J. Pep. Prot. Res. 12: 217–221.

Downing, M. R., Bloom, J. W. & Mann, K. G. Comparison of the
inhibition of thrombin by three plasma protease inhibitors.
Biochemistry 17: 2649–2653.

Geratz, J. D. & Tidwell, R. R. Current concepts on action of
synthetic thrombin inhibitors. Haemostasis 7: 170–176.

Jackson, C. M. The biochemistry of prothrombin activation. Br.
J. Haematol. 39: 1–8.

Kirchhof, B. R. J., Vermeer, C. & Hemker, H. C. The determination
of prothrombin using synthetic chromogenic substrates; choice of
a suitable activator. Thromb. Res. 13: 219–232.

Klausner, Y. S., Rigbi, M., Ticho, T., De Jong, P. J., Neginsky,
E. J. & Rindt, Y. The interaction of α-N-(p-toluenesulphonyl)-
p-guanidino-L-phenylalanine methyl ester with thrombin and
trypsin. Biochem. J. 169: 157–167.

Machovich, R. & Arányi, P. Effect of heparin on thrombin
inactivation by antithrombin-III. Biochem. J. 173: 869–875.

Mosher, D. F. & Vaheri, A. Thrombin stimulates the production and
release of a major surface-associated glycoprotein (fibronectin)
in cultures of human fibroblasts. Exp. Cell Res. 112: 323–334.

Tam, S. W. & Detwiler, T. C. Binding of thrombin to human
platelet plasma membranes. Biochim. Biophys. Acta 543: 194–201.

Walker, F. J. & Esmon, C. T. The molecular mechanisms of heparin
action. III. The anticoagulant properties of
polyanetholesulfonate. Biochem. Biophys. Res. Commun. 83: 1339–
1346.

Workman, E. F., Jr. & Lundblad, R. L. On the preparation of
bovine α-thrombin. Thromb. Haemost. 39: 193–200.

Workman, E. F., Jr. & Lundblad, R. L. The effect of monovalent
cations on the catalytic activity of thrombin. Arch. Biochem.
Biophys. 185: 544–548.

1979

Borsodi, A. & Machovich, R.   Inhibition of esterase and amidase
   activities of α- and β-thrombin in the presence of antithrombin
   III and heparin. Biochim. Biophys. Acta 566: 385–389.
Chang, T., Feinman, R. D., Landis, B. H. & Fenton, J. W., II.
   Antithrombin reactions with α- and γ-thrombins. Biochemistry
   18: 113–119.
Davie, E. W., Fujikawa, K., Kurachi, K. & Kisiel, W.   The role of
   serine proteases in the blood coagulation cascade. Adv.
   Enzymol. 48: 277–318.
Glenn, K. C. & Cunningham, D. D.   Thrombin-stimulated cell
   division involves proteolysis of its cell surface receptor.
   Nature 278: 711–714.
Griffith, M. J.   Covalent modification of human α-thrombin with
   pyridoxal 5'-phosphate. Effect of phosphopyridoxylation on the
   interaction of thrombin with heparin. J. Biol. Chem. 254: 3401–
   3406.
Griffith, M. J., Kingdon, H. S. & Lundblad, R. L.   Inhibition of
   the effectiveness of heparin in accelerating the antithrombin
   III/thrombin reaction by peptidylarginine chloromethyl ketone
   inactivated thrombin. Fed. Proc. Fed. Am Soc. Exp. Biol. 38:
   793 only.
Griffith, M. J., Kingdon, H. S. & Lundblad, R. L.   Inhibition of
   the heparin-antithrombin III/thrombin reaction by active site
   blocked-thrombin. Biochem. Biophys. Res. Commun. 87: 686–692.
Jesty, J.   Dissociation of complexes and their derivatives formed
   during inhibition of bovine thrombin and activated Factor X by
   antithrombin III. J. Biol. Chem. 254: 1044–1049.
Karpatkin, M. & Karpatkin, S.   Humoral factor that specifically
   regulates prothrombin (factor II) levels (coagulopoietin-II).
   Proc. Natl. Adac. Sci. U.S.A. 76: 491–493.
Kettner, C. & Shaw, E.   D-Phe-Pro-ArgCH$_2$Cl – a selective affinity
   label for thrombin. Thromb. Res. 14: 969–973.
Lundblad, R. L., Noyes, C. M., Mann, K. G. & Kingdon, H. S.   The
   covalent differences between bovine α- and β-thrombin. A
   structural explanation for the changes in catalytic activity.
   J. Biol. Chem. 254: 8524–8528.
Mizuochi, T., Yamashita, K., Fujikawa, K., Kisiel, W. & Kobata, A.
   The carbohydrate of bovine prothrombin. J. Biol. Chem. 254:
   6419–6425.
Nesheim, M. E. & Mann, K. G.   Thrombin catalysed activation of
   single chain bovine Factor V. J. Biol. Chem. 254: 1326–1334.
Nesheim, M. E., Prendergast, F. G. & Mann, K. G.   Interactions of
   a fluorescent active-site-directed inhibitor of thrombin:
   dansylarginine N-(3-ethyl-1,5-pentanediyl)amide. Biochemistry
   18: 996–1003.
Orthner, C. L. & Kosow, D. P.   Monovalent cation activation of
   human α-thrombin. Fed. Proc. Fed. Am. Soc. Exp. Biol. 38: 793
   only.
Paulssen, M. M. P., Kolhorn, A., Rothuizen, J. & Planje, M. C.   An

automated amidolytic assay of thrombin generation: an alternative for the prothrombin-time test. Clin. Chim. Acta 92: 465-468.

Saito, H., Goldsmith, G. H., Moroi, M. & Aoki, N. Inhibitory spectrum of alpha-2-plasmin inhibitor. Proc. Natl. Acad. Sci. U.S.A. 76: 2013-2017.

Silverberg, S. A. Proteolysis of prothrombin by thrombin. Determination of kinetic parameters and demonstration and characterization of an unusual inhibition by Ca^{2+} ions. J. Biol. Chem. 254: 88-94.

Valenty, V. B., Wos, J. D., Lobo, A. P. & Lawson, W. B. 5-Acyloxyazoles as serine protease inhibitors. Rapid inactivation of thrombin in contrast to plasmin. Biochem. Biophys. Res. Commun. 88: 1375-1381.

Walker, F. J. & Esmon, C. T. Factors that influence the rate of antithrombin III inhibition of thrombin and factor Xa. Fed. Proc. Fed. Am. Soc. Exp. Biol. 38: 809 only.

Walker, F. J. & Esmon, C. T. The effect of prothrombin fragment 2 on the inhibition of thrombin by antithrombin III. J. Biol. Chem. 254: 5618-5622.

Entry 1.16

COAGULATION FACTOR VII$_a$

Summary

EC Number: 3.4.21.21

Earlier names: Precursor: proconvertin. Enzyme: convertin, serum prothrombin conversion accelerator (SPCA).

Distribution: Plasma, as the precursor.

Source: Human or bovine plasma.

Action: Cleaves an -Arg-Ile- bond in factor X to form factor X$_a$ in the extrinsic route of coagulation; no other protein substrate is known. Hydrolyzes several blocked arginine esters [1].

Requirements: Neutral pH, tissue factor.

Substrate, usual: Factor X, in assay by correction of clotting time of factor VII-deficient plasma.

Inhibitors: Dip-F, Pms-F, benzamidine. (Not plasma inhibitors.)

Molecular properties: Precursor (bovine) is a single chain glycoprotein, 46,000 mol.wt., pI 5.6. It contains Gla residues, so biosynthesis is vitamin K-dependent.
    Factor VII is activated by factor X$_a$, thrombin, kallikrein, factor XII$_a$ or factor XI$_a$, through cleavage of an -Arg-Ile- bond, to produce factor VII$_a$ containing chains of 27,000 and 18,000 mol.wt. Further proteolysis can inactivate.

Comment: Factor VII is a component of the extrinsic route of coagulation. The uncleaved factor VII has distinct catalytic activity (although less than factor VII$_a$) [1], and reacts with Dip-F; since precursors of other serine proteinases in plasma do not react with Dip-F, this reagent can be used to render bovine plasma specifically deficient in this factor. Factor VII$_a$ is the only proteinase in the coagulation system that is not inhibited by antithrombin III.

References
[1] Zur & Nemerson J. Biol. Chem. 253: 2203-2209, 1978.

Bibliography

1951

Koller, F., Loeliger, A. & Duckert, F. Experiments on a new clotting factor (factor VII). Acta Haematol. 6: 1-18.

1961

Denson, K. W. E. The specific assay of Prower-Stuart factor and
factor VII. Acta Haematol. 25: 105-120.

1966

Nemerson, Y. The reaction between bovine brain tissue factor and
factors VII and X. Biochemistry 5: 601-608.

1970

Gladhaug, Å. & Prydz, H. Purification of the coagulation factors
VII and X from human serum. Some properties of factor VII.
Biochim. Biophys. Acta 215: 105-111.

1972

De Lange, J. A. & Hemker, H. C. Inhibition of blood coagulation
factors by serine esterase inhibitors. FEBS Lett. 24: 265-268.
Nemerson, Y. & Esnouf, M. P. Activation of a proteolytic system
by a membrane lipoprotein: mechanism of action of tissue factor.
Proc. Natl. Acad. Sci. U.S.A. 70: 310-314.
Reekers, P. & Hemker, H. C. Purification of blood coagulation
factors II, VII, IX and X from bovine citrated plasma.
Haemostasis 1: 2-22.

1973

Nemerson, Y. & Esnouf, M. P. Activation of a proteolytic system
by a membrane lipoprotein: mechanism of action of tissue factor.
Proc. Natl. Acad. Sci. U.S.A. 70: 310-314.
Stormorken, H., Gjönnaess, H. & Laake, K. Interrelations between
the clotting and kinin systems. Haemostasis 2: 245-252.

1974

Fujikawa, K., Legaz, M. E., Kato, H. & Davie, E. W. The mechanism
of activation of bovine factor IX (Christmas factor) by bovine
factor XIa (activated plasma thromboplastin antecedent).
Biochemistry 13: 4508-4516.
Jesty, J. & Nemerson, Y. Purification of factor VII from bovine
plasma. Reaction with tissue factor and activation of factor X.
J. Biol. Chem. 249: 509-515.
Laake, K. & Ellingsen, R. Purification and some characteristics
of factor VII in human citrated plasma, glass-activated serum,
and cold-activated plasma. Thromb. Res. 5: 539-556.
Laake, K. & Osterud, B. Activation of purified plasma factor VII
by human plasmin, plasma kallikrein, and activated components of
the human intrinsic blood coagulation system. Thromb. Res. 5:
759-772.
Laake, K., Venneröd, A. M., Haugen, G. & Gjönnaess, H. Cold-
promoted activation of factor VII in human plasma: studies on
the associated acyl-arginine esterase activity. Thromb. Res. 4:
769-785.

1975

Davie, E. W. & Fujikawa, K. Basic mechanisms in blood coagulation. Annu. Rev. Biochem. 44: 799-829.

Kisiel, W. & Davie, E. W. Isolation and characterization of bovine factor VII. Biochemistry 14: 4928-4934.

Radcliffe, R. & Nemerson, Y. Activation and control of factor VII by activated factor X and thrombin. J. Biol. Chem. 250: 388-395.

Saito, H. & Ratnoff, O. D. Alteration of factor VII activity by activated Fletcher factor (a plasma kallikrein): a potential link between the intrinsic and extrinsic blood-clotting systems. J. Lab. Clin. Med. 85: 405-415.

1976

Østerud, B., Miller-Andersson, M., Abildgaard, U. & Prydz, H. The effect of antithrombin III on the activity of the coagulation factors VII, IX, and X. Thromb. Haemost. 35: 295-304.

Radcliff, R. & Nemerson, Y. Mechanism of activation of bovine factor VII. Products of cleavage by factor X$_a$. J. Biol. Chem. 251: 4797-4802.

Radcliffe, R. & Nemerson, Y. Bovine factor VII. Methods Enzymol. 45: 49-56.

Silverberg, S. A., Ostapchuk, P. & Nemerson, Y. The enzymatic activity of factor VII: a novel assay of a coagulation factor. Blood 48: 976 only.

1977

Kisiel, W., Fujikawa, K. & Davie, E. W. Activation of bovine factor VII (proconvertin) by factor XIIa (activated Hageman factor). Biochemistry 16: 4189-4194.

Østerud, B. & Rapaport, S. I. Activation of factor IX by the reaction product of tissue factor and factor VII: additional pathway for initiating blood coagulation. Proc. Natl. Acad. Sci. U.S.A. 74: 5260-5264.

Radcliffe, R., Bagdasarian, A., Colman, R. & Nemerson, Y. Activation of bovine factor VII by Hageman factor fragments. Blood 50: 611-617.

Silverberg, S. A., Nemerson, Y. & Zur, M. Kinetics of the activation of bovine coagulation factor X by components of the extrinsic pathway. Kinetic behavior of two-chain factor VII in the presence and absence of tissue factor. J. Biol. Chem. 252: 8481-8488.

1978

Jesty, J. The inhibition of activated bovine coagulation factors X and VII by antithrombin III. Arch. Biochem. Biophys. 185: 165-173.

Seligsohn, U., Østerud, B., Griffin, J. H. & Rapaport, S. I. Evidence for the participation of both activated factor XII and activated factor IX in cold-promoted activation of factor VII. Thromb. Res. 13: 1049-1056.

Zur, M. & Nemerson, Y.　The esterase activity of coagulation factor VII. Evidence for intrinsic activity of the zymogen.　J. Biol. Chem. 253: 2203–2209.

1979

Davie, E. W., Fujikawa, K., Kurachi, K. & Kisiel, W.　The role of serine proteases in the blood coagulation cascade.　Adv. Enzymol. 48: 277–318.

Flengsrud, R.　Purification and some characteristics of the human coagulation factor VII.　Eur. J. Biochem. 98: 455–464.

Seligsohn, U., Østerud, B., Brown, S. F., Griffin, J. H. & Rapaport, S. I.　Activation of human factor VII in plasma and in purified systems. Roles of activated factor IX, kallikrein, and activated factor XII.　J. Clin. Invest. 64: 1056–1065.

Entry 1.17

PROTEIN C

Summary

EC Number: (3.4.21.-)

Earlier names: -

Distribution: Plasma.

Source: Plasma.

Action: Destroys coagulation factor V_a, thus slowing activation of
prothrombin by factor X_a, which increases the kaolin-cephalin
clotting time. No other substrates are known for this enzyme.

Requirements: Neutral pH, negatively charged phospholipid, Ca^{2+}.

Substrate, usual: Factor V_a, in a clotting-time assay.

Inhibitors: Dip-F, Pms-F.

Molecular properties: The zymogen is a 62,000 mol.wt. glycoprotein
(23% carbohydrate), comprising a heavy (41,000) and a light
(21,000) chain; it contains Gla residues, so biosynthesis is
vitamin K-dependent.
    The heavy chain of bovine protein C is cleaved between
residues 14 and 15, -Arg-Ile-, during activation by α-thrombin,
Russell's viper venom factor X inactivator, or trypsin, with
loss of the N-terminal peptide. The heavy chain then contains a
Ser residue reactive with Dip-F in a sequence homologous with
those of other plasma serine proteinases.

Comment: Protein C may serve as a regulator of coagulation. The
bovine and human proteins are closely similar [1].

References
[1] Kisiel J. Clin. Invest. 64: 761-769, 1979.

Bibliography

1976

Bucher, D., Nebelin, E., Thomsen, J. & Stenflo, J. Identification
of γ-carboxyglutamic acid residues in bovine factors IX and X,
and in a new vitamin K-dependent protein. FEBS Lett. 68: 293-
296.
Esmon, C. T., Stenflo, J., Suttie, J. W. & Jackson, C. M. A new
vitamin K-dependent protein. A phospholipid-binding zymogen of a

serine esterase. J. Biol. Chem. 251: 3052-3056.

Kisiel, W., Ericsson, L. H. & Davie, E. W. Proteolytic activation of protein C from bovine plasma. Biochemistry 15: 4893-4900.

Seegers, W. H., Novoa, E., Henry, R. L. & Hassouna, H. I. Relationship of "new" vitamin K-dependent protein C and "old" autoprothrombin II-A. Thromb. Res. 8: 543-552.

Stenflo, J. A new vitamin K-dependent protein. Purification from bovine plasma and preliminary characterization. J. Biol. Chem. 251: 355-363.

1977

Kisiel, W., Canfield, W. M., Ericsson, L. H. & Davie, E. W. Anticoagulant properties of bovine plasma protein C following activation by thrombin. Biochemistry 16: 5824-5831.

1978

Fernland, P., Stenflo, J. & Tufvesson, A. Bovine protein C: amino acid sequence of the light chain. Proc. Natl. Acad. Sci. U.S.A. 75: 5889-5892.

1979

Kisiel, W. Human plasma protein-C - isolation, characterization, and mechanism of activation by alpha-thrombin. J. Clin. Invest. 64: 761-769.

Walker, F. J., Sexton, P. W. & Esmon, C. T. The inhibition of blood coagulation by activated protein C through the selective inactivation of activated factor V. Biochim. Biophys. Acta 571: 333-342.

Entry 1.18

PROTEIN S

Summary

EC Number: Possibly 3.4.21.- ?

Earlier names: -

Distribution: Bovine and human plasma.

Source: Plasma.

Action: Unknown.

Requirements: -

Substrate, usual: -

Inhibitors: -

Molecular properties: Mol.wt. 64,200 (bovine) or 69,000 (human), single chain glycoprotein (7% carbohydrate) containing 10 Gla residues (so biosynthesis is vitamin K-dependent).

Comment: Although not known to be a proteinase, this protein is homologous with thrombin and the other vitamin K-dependent coagulation factors.

Bibliography

1977

Di Scipio, R. G., Hermodson, M. A., Yates, S. G. & Davie, E. W. A comparison of human prothrombin, factor IX (Christmas factor), factor X (Stuart factor), and protein S. Biochemistry 16: 698-706.

1979

Di Scipio, R. G. & Davie, E. W. Characterization of protein S, a γ-carboxyglutamic acid containing protein from bovine and human plasma. Biochemistry 18: 899-904.

Stenflo, J. & Jönsson, M. Protein S, a new vitamin K-dependent protein from bovine plasma. FEBS Lett. 101: 377-381.

Entry 1.19

PLASMA FACTOR PF/Dil

Summary

EC Number: (3.4.21.-)

Earlier names: Permeability globulin B, permeability factor of
dilution. See Comments.

Distribution: Plasma of guinea pig, rat, rabbit and man, as the
inactive precursor.

Source: Plasma.

Action: Increases vascular permeability _in vivo_. Generates kinins
in unheated plasma, perhaps by activating a kininogenase.
Hydrolyzes Tos-Arg-OEt, Tos-Arg-OMe.

Requirements: pH optimum about 9 for esterase activity.

Substrate, usual: (Assayed by a capillary permeability test in
skin.)

Inhibitors: Dip-F. $\overline{\text{CI}}$ inhibitor, soybean trypsin inhibitor. (Not
ovomucoid.)

Molecular properties: Sedimentation 6S.

Comment: The factor is apparently generated from an inactive
precursor, pro-PF, classically following dilution of plasma in a
glass vessel.
   There is evidence that PF/Dil may be identical with an
activated form of factor XII (#1.09) [1,2].

References
[1] Johnson et al. J. Immunol. 113: 103-109, 1974.
[2] Oh-ishi & Webster Biochem. Pharmacol. 24: 591-598, 1975.

Bibliography

1953

Mackay, M. E., Miles, A. A., Shachter, C. B. E. & Wilhelm, D. L.
Susceptibility of the guinea pig to pharmacological factors from
its own serum. Nature 172: 714-716.

1955

Wilhelm, D. L., Miles, A. A. & Mackay, M. E. Enzyme-like
globulins from serum reproducing the vascular phenomena of

inflammation. II. Isolation and properties of the permeability factor and its inhibitor. Br. J. Exp. Path. 36: 82-104.

1958

Mill, P. J., Elder, J. M., Miles, A. A. & Wilhelm, D. L. Enzyme-like globulins from serum reproducing the vascular phenomena of inflammation. VI. Isolation and properties of permeability factor and its inhibitor in human plasma. Br. J. Exp. Path. 39: 343-355.

1959

Becker, E. L., Wilhelm, D. L. & Miles, A. A. Enzymic nature of the serum globulin permeability factor. Nature 183: 1264-1265.

1962

Mason, B. & Miles, A. A. Globulin permability factors without kininogenase activity. Nature 196: 587-588.

1964

Kagen, L. J. Some biochemical and physical properties of the human permeability globulins. Br. J. Exp. Path. 45: 604-611.

Ratnoff, O. D. & Miles, A. A. The induction of permeability-increasing activity in human plasma by activated Hageman factor. Br. J. Exp. Path. 45: 328-345.

1969

Ratnoff, O. D., Pensky, J., Ogston, D. & Naff, G. B. The inhibition of plasmin, plasma kallikrein, plasma permeability factor, and the C'1r subcomponent of the first component of complement by serum C'1 esterase inhibitor. J. Exp. Med. 129: 315-331.

1973

Oh-ishi, S. PF/Dil (permeability factor dilute): its relationship to Hageman factor and the kallikrein-kinin system. Fed. Proc. Fed. Am. Soc. Exp. Biol. 32: 846 only.

Wuepper, K. D. Prekallikrein deficiency in man. J. Exp. Med. 138: 1345-1355.

1974

Johnston, A. R., Cochrane, C. G. & Revak, S. D. The relationship between PF/Dil and activated human Hageman factor. J. Immunol. 113: 103-109.

1975

Oh-ishi, S. & Webster, M. E. Vascular permeability factors (PF/Nat and PF/Dil) - their relationship to Hageman factor and the kallikrein-kinin system. Biochem. Pharmacol. 24: 591-598.

1976

Oh-ishi, S. & Webster, M. E. Formation of prekallikrein activator and TAME esterase activity by dilution of human plasma. In: Chemistry and Biology of the Kallikrein-Kinin System in Health

and Disease, DHEW Publication No. (NIH) 76-791 (Pisano, J. J. &
Austen, K. F. eds), pp. 61-64, U.S. Government Printing Office,
Washington.
Wuepper, K. D.   Plasma prekallikrein: its characterization,
mechanism of action, and inherited deficiency in man.   In:
Chemistry and Biology of the Kallikrein-Kinin System in Health
and Disease, DHEW Publication No. (NIH) 76-791 (Pisano, J. J. &
Austen, K. F. eds), pp. 37-51, U.S. Government Printing Office,
Washington.

Entry 1.20

PLASMA PLASMINOGEN ACTIVATOR

Summary

EC Number: (3.4.21.-)

Earlier names: -

Distribution: Plasma, as the latent plasminogen proactivator.

Source: Plasma.

Action: Activates plasminogen to plasmin.

Requirements: Neutral pH.

Substrate, usual: Plasminogen.

Inhibitors: Dip-F. ($\overline{CI}$ inhibitor variable, not α_1-proteinase inhibitor.)

Molecular properties: Precursor mol.wt. 95,000-165,000 (so much higher than urokinase, #1.24 or tissue activators). pI 8.9. Activated by coagulation factor XII_a (#1.09).

Comment: Much of the activity in plasma shows properties closely resembling those of prekallikrein (#1.10), and the remainder has not been separated from factor XI (#1.11), so the existence of this enzyme as a separate entity remains to be proved.

Bibliography

1953

Müllertz, S. A plasminogen activator in spontaneously active human blood. Proc. Soc. Exp. Biol. Med. 82: 291-295.

1968

Kucinski, C. S., Fletcher, A. P. & Sherry, S. Effect of urokinase antiserum on plasminogen activators: demonstration of immunological dissimilarity between plasma plasminogen activator and urokinase. J. Clin. Invest. 47: 1238-1253.

1972

Kaplan, A. P. & Austen, K. F. The fibrinolytic pathway of human plasma. Isolation and characterization of the plasminogen proactivator. J. Exp. Med. 136: 1378-1393.

1973

Tomikawa, M. & Abiko, Y.   The plasminogen-plasmin system. Part
VIII. On the mechanism of chemical induction of fibrinolysis.
Isolation of a latent plasminogen activator from human plasma.
Thromb. Diath. Haemorrh. 29: 50-62.

1974

Astrup, T. & Rosa, A. T.   A plasminogen proactivator-activator
system in human blood effective in the absence of Hageman
factor. Thromb. Res. 4: 609-613.

Laake, K. & Vennerod, A. M.   Factor XII induced fibrinolysis:
studies on the separation of prekallikrein, plasminogen
proactivator, and factor XI in human plasma. Thromb. Res. 4:
285-302.

1975

Colman, R. W., Bagdasarian, A., Talamo, R. C., Seavey, M., Scott,
C. F. & Kaplan, A.   Williams trait: combined deficiency of
plasma plasminogen proactivator, kininogen and a new procoagulant
factor. Fed. Proc. Fed. Am. Soc. Exp. Biol. 34: 859 only.

1976

Kaplan, A. P., Meier, H. L., Yecies, L. D. & Heck, L. W.   Hageman
factor and its substrates: the role of factor XI, prekallikrein,
and plasminogen proactivator in coagulation, fibrinolysis, and
kinin generation.   In: Chemistry and Biology of the Kallikrein-
Kinin System in Health and Disease, DHEW Publication No. (NIH)
76-791 (Pisano, J. J. & Austen, K. F. eds), pp. 237-254, U.S.
Government Printing Office, Washington.

1977

Christman, J. K., Silverstein, S. C. & Acs, G.   Plasminogen
activators.   In: Proteinases in Mammalian Cells and Tissues
(Barrett, A. J. ed.), pp. 91-149, North-Holland Publishing Co.,
Amsterdam.

1978

Yecies, L., Meier, H., Mandle, R., Jr. & Kaplan, A. P.   The
Hageman factor-dependent fibrinolytic pathway.   In: Progress in
Chemical Fibrinolysis and Thrombolysis (Davidson, J. F., Rowan,
R. M., Samama, M. M. & Desnoyers, P. C. eds), vol. 3, pp. 121-
139, Raven Press, New York.

Entry 1.21

PLASMIN

Summary

EC Number: 3.4.21.7

Earlier names: The precursor zymogen, plasminogen, has been called
profibrinolysin. Enzyme: fibrinase, fibrinolysin.

Distribution: Plasma, as plasminogen (about 0.2 mg/ml in human
plasma).

Source: Plasma, especially of rabbit and man.

Action: Degrades fibrin, casein and other proteins. Preferentially
cleaves –Lys-X- and -Arg-X- bonds, but is more selective than
trypsin.

Requirements: Neutral pH.

Substrate, usual: Fibrin, Tos-Arg-OMe, Z-Lys-OPhNO$_2$, Bz-Arg-NPhNO$_2$.

Substrate, special: D-Val-Leu-Lys-NPhNO$_2$ (S-2251), Tos-Gly-Pro-Lys-
NPhNO$_2$ (Chromozym PL). MeOSuc-Ala-Phe-Lys-NMec allows excellent
discrimination from plasminogen activators and thrombin [3].

Inhibitors: Dip-F, Pms-F, Tos-Lys-CH$_2$Cl, leupeptin. Aprotinin,
soybean inhibitor, α_2-plasmin inhibitor, α_2-macroglobulin,
antithrombin III-heparin. Lysine and analogues cause allosteric
activation at low concentration, and competitive inhibition at
higher concentrations (K_i about 0.3 mM).

Molecular properties: Plasminogen has A280,1% about 20. The form
with N-terminal Glu- (mol.wt. 93,000) is readily converted to
forms with terminal Lys- or Val- (mol.wt. 86,000). Two forms of
plasminogen (I and II) differing in glycosylation can be
separated in the basis of their affinity for lysine-Sepharose
[1]. Multiple forms (pI 6-9) can be resolved by isoelectric
focusing [2].
    Proteolytic activation of plasminogen yields plasmin with two
chains (A, 60,000 and B, 25,000) disulfide-linked, plus a free
peptide (7,000). The B (light) chain contains the catalytic Ser
residue.

Comment: Zymogen and enzyme have binding sites for lysine and
6-aminohexanoate: insolubilized lysine is used in affinity
chromatography, and 6-aminohexanoate causes a conformational
change and inhibits activation of plasminogen.

References
[1] Hayes & Castellino J. Biol. Chem. 254: 8768-8771, 1979.
[2] Summaria et al. J. Biol. Chem. 247: 4691-4702, 1972.
[3] Pierzchala et al. Biochem. J. 183: 555-559, 1979.

Bibliography

1947
Astrup, T. & Permin, P. M. Fibrinolysis in the animal organism.
Nature 159: 681-682.

1957
Sherry, S. & Alkjaersig, N. Biochemical, experimental, and
clinical studies of proteolytic enzymes: with particular
reference to the fibrinolytic enzyme of human plasma. Ann. N.
Y. Acad. Sci. 68: 52-66.

1958
Roberts, P. S. Measurement of the rate of plasmin action on
synthetic substrates. J. Biol. Chem. 232: 285-291.

1959
Alkjaersig, N., Fletcher, A. P. & Sherry, S. ϵ-Aminocaproic acid:
an inhibitor of plasminogen activation. J. Biol. Chem. 234:
832-837.

1960
Sgouris, J. T., Inman, J. K., McCall, K. B., Hyndman, L. A. &
Anderson, H. D. The preparation of human fibrinolysin
(plasmin). Vox Sang. 5: 357-376.

1963
Nagamatsu, A., Okuma, T., Watanabe, M. & Yamamura, Y. The
inhibition of plasmin by some amino acid derivatives. J.
Biochem. 54: 491-496.

1964
Alkjaersig, N. The purification and properties of human
plasminogen. Biochem. J. 93: 171-182.
Bickford, A. F., Taylor, F. B. & Sheena, R. Inhibition of the
fibrinogen-plasmin reaction by ω-aminocarboxylic acids and
alkylamines. Biochim. Biophys. Acta 92: 328-333.

1965
Hummel, B. C. W., Schor, J. M., Buck, F. F., Boggiano, E. &
DeRenzo, E. C. Quantitative enzymic assay of human plasminogen
and plasmin with azocasein as substrate. Anal. Biochem. 11:
532-547.
Robbins, K. C., Summaria, L., Elwyn, D. & Barlow, G. H. Further
studies on the purification and characterization of human
plasminogen and plasmin. J. Biol. Chem. 240: 541-550.

1967

Ling, C.-M., Summaria, L. & Robbins, K. C. Isolation and characterization of bovine plasminogen activator from a human plasminogen-streptokinase mixture. J. Biol. Chem. 242: 1419-1425.

Robbins, K. C., Summaria, L., Hsieh, B. & Shah, R. J. The peptide chains of human plasmin. Mechanism of activation of human plasminogen to plasmin. J. Biol. Chem. 242: 2333-2342.

Summaria, L., Hsieh, B. & Robbins, K. C. The specific mechanism of activation of human plasminogen to plasmin. J. Biol. Chem. 242: 4279-4283.

Summaria, L., Hsieh, B., Groskopf, W. R., Robbins, K. C. & Barlow, G. H. The isolation and characterization of the S-carboxymethyl β (light) chain derivatives of human plasmin. The localization of the active site on the β (light) chain. J. Biol. Chem. 242: 5046-5052.

1968

Markwardt, F., Landmann, H. & Walsmann, P. Comparative studies on the inhibition of trypsin, plasmin and thrombin by derivatives of benzylamine and benzamidine. Eur. J. Biochem. 6: 502-506.

Okamoto, S., Oshiba, S., Mihara, H. & Okamoto, U. Synthetic inhibitors of fibrinolysis: in vitro and in vivo mode of action. Ann. N. Y. Acad. Sci. 146: 414-429.

1969

Chase, T. & Shaw, E. Comparison of the esterase activities of trypsin, plasmin, and thrombin on guanidinobenzoate esters. Titration of the enzymes. Biochemistry 8: 2212-2224.

Feeney, R. E., Means, G. E. & Bigler, J. C. Inhibition of human trypsin, plasmin, and thrombin by naturally occurring inhibitors of proteolytic enzymes. J. Biol. Chem. 244: 1957-1960.

Groskopf, W. R., Hsieh, B., Summaria, L. & Robbins, K. C. Studies on the active center of human plasmin. The serine and histidine residues. J. Biol. Chem. 244: 359-365.

Ratnoff, O. D., Pensky, J., Ogston, D. & Naff, G. B. The inhibition of plasmin, plasma kallikrein, plasma permeability factor, and the C'lr subcomponent of the first component of complement by serum C'l esterase inhibitor. J. Exp. Med. 129: 315-331.

1970

Deutsch, D. G. & Mertz, E. T. Plasminogen: purification from human plasma by affinity chromatography. Science 170: 1095-1096.

Robbins, K. C. & Summaria, L. Human plasminogen and plasmin. Methods Enzymol. 19: 184-199.

Wallén, P. & Wiman, B. Characterization of human plasminogen. I. On the relationship between different molecular forms of plasminogen demonstrated in plasma and found in purified preparations. Biochim. Biophys. Acta 221: 20-30.

1971

Brockway, J. & Castellino, F.   The mechanism of the inhibition of
plasmin activity by ε-aminocaproic acid.   J. Biol. Chem. 246:
4641-4647.

Kaplan, A. P. & Austen, K. F.   A prealbumin activator of
prekallikrein. II. Derivation of activators of prekallikrein
from active Hageman factor by digestion with plasmin.   J. Exp.
Med. 133: 696-712.

Muramatu, M. & Fujii, S.   Inhibitory effects of ω-amino acid
esters on trypsin, plasmin, plasma kallikrein and thrombin.
Biochim. Biophys. Acta 242: 203-208.

Rickli, E. E. & Cuendet, P. A.   Isolation of plasmin-free human
plasminogen with N-terminal glutamic acid.   Biochim. Biophys.
Acta 250: 447-451.

1972

Brockway, W. J. & Castellino, F. J.   Measurement of the binding of
antifibrinolytic amino acids to various plasminogens.   Arch.
Biochem. Biophys. 151: 194-199.

Maki, M. & Sasaki, K.   A new broad spectrum inhibitor of
esteroproteases, 4-(2-carboxyethyl)phenyl-trans-4-aminomethyl
cyclohexane carboxylate hydrochloride (DV 1006).   Tohoku J. Exp.
Med. 106: 233-248.

Muramatu, M. & Fujii, S.   Inhibitory effects of ω-guanidino acid
esters on trypsin, plasmin, plasma kallikrein and thrombin.
Biochim. Biophys. Acta 268: 221-224.

Reddy, K. N. N. & Marcus, G.   Mechanism of activation of human
plasminogen by streptokinase. Presence of an active center in
streptokinase-plasminogen complex.   J. Biol. Chem. 247: 1683-
1691.

Sodetz, J. M., Brockway, W. J. & Castellino, F. J.   Multiplicity
of rabbit plasminogen. Physical characterisation.   Biochemistry
11: 4451-4458.

Summaria, L., Arzadon, L., Bernabe, P. & Robbins, K. C.   Studies
on the isolation of the multiple molecular forms of human
plasminogen and plamin by isoelectric focusing methods.   J.
Biol. Chem. 247: 4691-4702.

Wallén, P. & Wiman, B.   Characterisation of human plasminogen. II.
Separation and partial characterization of different molecular
forms of human plasminogen.   Biochim. Biophys. Acta 257: 122-
134.

Weinstein, M. J. & Doolittle, R. F.   Differential specificities of
thrombin, plasmin and trypsin with regard to synthetic and
natural substrates and inhibitors.   Biochim. Biophys. Acta 258:
577-590.

Zolton, R. P. & Mertz, E. T.   Studies on plasminogen. X. Isolation
of plasminogen by affinity chromatography using Sepharose-
butesin.   Can. J. Biochem. 50: 529-537.

1973

Geratz, J. D.   Structure-activity relationships for the inhibition of plasmin and plasminogen activation by aromatic diamidines and a study of the effect of plasma proteins on the inhibition process. Thromb. Diath. Haemorrh. 24: 154-167.

Reddy, K. N. N. & Marcus, G. Further evidence of an active center in streptokinase-plasminogen complex; interaction with pancreatic trypsin inhibitor. Biochem. Biophys. Res. Commun. 51: 672-679.

Rickli, E. E. & Otavsky, W. I. Release of an N-terminal peptide from human plasminogen during activation with urokinase. Biochim. Biophys. Acta 295: 381-384.

Robbins, K. C., Bernabe, P., Arzadon, L. & Summaria, L. NH_2-terminal sequences of mammalian plasminogens and plasmin S-carboxymethyl heavy (A) and light (B) chain derivatives. A re-evaluation of the mechanism of activation of plasminogen. J. Biol. Chem. 248: 7242-7246.

Sjöholm, I., Wiman, B. & Wallén, P. Studies on the conformational changes of plasminogen induced during activation to plasmin and by 6-aminohexanoic acid. Eur. J. Biochem. 39: 471-479.

Wiman, B. Primary structure of peptides released during activation of human plasminogen by urokinase. Eur. J. Biochem. 39: 1-9.

Wiman, B. & Wallén, P. Activation of human plasminogen by an insoluble derivative of urokinase. Structural changes of plasminogen in the course of activation to plasmin and demonstration of a possible intermediate compound. Eur. J. Biochem. 36: 25-31.

                                1974

Christensen, U. & Müllertz, S. Mechanism of reaction of human plasmin with α-N-benzoyl-L-arginine-p-nitroanilide. Biochim. Biophys. Acta 334: 187-198.

Claeys, H. & Vermylen, J. Physico-chemical and proenzyme properties of NH_2-terminal glutamic acid and NH_2-terminal lysine human plasminogen. Biochim. Biophys. Acta 342: 351-359.

Hatton, M. W. C. & Regoeczi, E. Some observations on the affinity chromatography of rabbit plasminogen. Biochim. Biophys. Acta 359: 55-65.

Highsmith, R. F. & Rosenberg, R. D. The inhibition of human plasmin by human antithrombin-heparin cofactor. J. Biol. Chem. 249: 4335-4338.

Laake, K. & Vennerod, A. M. Factor XII induced fibrinolysis: studies on the separation of prekallikrein, plasminogen proactivator, and factor XI in human plasma. Thromb. Res. 4: 285-302.

Müllertz, S. Different molecular forms of plasminogen and plasmin produced by urokinase in human plasma and their relation to protease inhibitors and lysis of fibrinogen and fibrin. Biochem. J. 143: 273-283.

Schick, L. A. & Castellino, F. J. Direct evidence for the generation of an active site in the plasminogen moiety of the streptokinase-human plasminogen activator complex. Biochem.

Biophys. Res. Commun. 57: 47-54.

1975

Christensen, U. pH effects in plasmin-catalysed hydrolysis of α-N-benzoyl-L-arginine compounds. Biochim. Biophys. Acta 397: 459-467.

Crawford, G. P. M. & Ogston, D. The action of antithrombin III on plasmin and activators of plasminogen. Biochim. Biophys. Acta 391: 189-192.

Davie, E. W. & Fujikawa, K. Basic mechanisms in blood coagulation. Annu. Rev. Biochem. 44: 799-829.

Hatton, M. W. C. & Regoeczi, E. The effect of heparin on the affinity chromatography of plasminogen. Demonstration of heparin-plasminogen interaction. Biochim. Biophys. Acta 386: 451-460.

Hatton, M. W. C. & Regoeczi, E. The relevance of the structure of lysine bound to Sepharose for the affinity of rabbit plasminogen. Biochim. Biophys. Acta 379: 504-511.

Ossowski, L., Quigley, J. P. & Reich, E. Plasminogen, a necessary factor for cell migration in vitro. In: Proteases and Biological Control (Reich, E., Rifkin, D. B. & Shaw, E. eds), pp. 901-913, Cold Spring Harbor Laboratory, New York.

Rickli, E. E. & Otavsky, W. I. A new method of isolation and some properties of the heavy chain of human plasmin. Eur. J. Biochem. 59: 441-447.

Robbins, K. C., Boreisha, J. G., Arzadon, L. & Summaria, L. Physical and chemical properties of the NH_2-terminal glutamic acid and lysine forms of human plasminogen and their derived plasmins with an NH_2-terminal lysine heavy (A) chain. Comparative data. J. Biol. Chem. 250: 4044-4047.

Robbins, K. C., Summaria, L. & Barlow, G. H. Activation of plasminogen. In: Proteases and Biological Control (Reich, E., Rifkin, D. B. & Shaw, E. eds), pp. 305-310, Cold Spring Harbor Laboratory, New York.

Silverstein, R. M. The determination of human plasminogen using Nα-CBZ-L-lysine p-nitrophenyl ester as substrate. Anal. Biochem. 65: 500-506.

Sodetz, J. M. & Castellino, F. J. A comparison of mechanisms of activation of rabbit plasminogen isozymes by urokinase. In: Proteases and Biological Control (Reich, E., Rifkin, D. B. & Shaw, E. eds), pp. 311-324, Cold Spring Harbor Laboratory, New York.

Sodetz, J. M. & Castellino, F. J. The mechanism of activation of rabbit plasminogen by urokinase. J. Biol. Chem. 250: 3041-3049.

Sottrup-Jensen, L., Zajdel, M., Claeys, H., Peterson, T. E. & Magnusson, S. Amino-acid sequence of activation cleavage site in plasminogen: homology with "pro" part of prothrombin. Proc. Natl. Acad. Sci. U.S.A. 72: 2577-2581.

Summaria, L., Arzadon, L., Bernabe, P. & Robbins, K. C. The activation of plasminogen to plasmin by urokinase in the presence of the plasmin inhibitor trasylol. The preparation of plasmin with the same NH_2-terminal heavy (A) chain sequence as the parent zymogen. J. Biol. Chem. 250: 3988-3995.

Thorsen, S. Differences in the binding to fibrin of native plasminogen and plasminogen modified by proteolytic degradation. Influence of ω-aminocarboxylic acids. Biochim. Biophys. Acta 393: 55-65.

Wallén, P. & Wiman, B. On the generation of intermediate plasminogen and its significance for activation. In: Proteases and Biological Control (Reich, E., Rifkin, D. B. & Shaw, E. eds), pp. 291-303, Cold Spring Harbor Laboratory, New York.

Wiman, B. & Wallén, P. Amino-acid sequence of the cyanogen bromide fragment from human plasminogen that forms the linkage between the plasmin chains. Eur. J. Biochem. 58: 539-547.

Wiman, B. & Wallén, P. On the primary structure of human plasminogen and plasmin. Purification and characterization of cyanogen-bromide fragments. Eur. J. Biochem. 57: 387-394.

Wiman, B.& Wallén, P. Structural relationships between "glutamic acid" and "lysine" forms of human plasminogen and their interaction with the NH_2-teminal activation peptide as studies by affinity chromatography. Eur. J. Biochem. 50: 489-494.

1976

Castellino, F. J. & Sodetz, J. M. Rabbit plasminogen and plasmin isozymes. Methods Enzymol. 45: 273-286.

Claeys, H., Sottrup-Jensen, L., Zajdel, M., Petersen, T. E. & Magnusson, S. Multiple gene duplication in the evolution of plasminogen. Five regions of sequence homology with the two internally homologous structures in prothrombin. FEBS Lett. 61: 20-24.

Kaplan, A. P., Meier, H. L., Yecies, L. D. & Heck, L. W. Hageman factor and its substrates: the role of factor XI, prekallikrein, and plasminogen proactivator in coagulation, fibrinolysis, and kinin generation. In: Chemistry and Biology of the Kallikrein-Kinin System in Health and Disease, DHEW Publication No. (NIH) 76-791 (Pisano, J. J. & Austen, K. F. eds), pp. 237-254, U.S. Government Printing Office, Washington.

Magee, S. C., Geren, C. R. & Ebner, K. E. Plasmin and the conversion of the molecular forms of bovine milk galactosyltransferase. Biochim. Biophys. Acta 420: 187-194.

Magnusson, S., Sottrup-Jensen, L., Peterson, T. E., Dudek-Wojciechowska, G. & Claeys, H. Homologous "kringle" structures common to plasminogen and prothrombin. Substrate specificity of enzymes activating prothrombin and plasminogen. In: Proteolysis and Physiological Regulation (Brew, K. & Ribbons, D. W. eds), pp. 203-238, Academic Press, New York.

Rickli, E. E., Lergier, W. & Gillessen, D. Investigations on the primary structure of human plasminogen. Further evidence for sequence homology. Biochim. Biophys. Acta 439: 47-50.

Robbins, K. C. & Summaria, L. Plasminogen and plasmin. Methods Enzymol. 45: 257-273.

Siefring, G. E., Jr. & Castellino, F. J. Interaction of

streptokinase with plasminogen. Isolation and characterization of a streptokinase degradation product. J. Biol. Chem. 251: 3913-3920.

Spragg, J. Immunological assays for components of the human plasma kinin-forming system. In: Chemistry and Biology of the Kallikrein-Kinin System in Health and Disease, DHEW Publication No. (NIH) 76-791 (Pisano, J. J. & Austen, K. F. eds), pp. 195-204, U.S. Government Printing Office, Washington.

Violand, B. N. & Castellino, F. J. Mechanism of the urokinase-catalyzed activation of human plasminogen. J. Biol. Chem. 251: 3906-3912.

1977

Bajaj, S. P. & Castellino, F. J. Activation of human plasminogen by equimolar levels of streptokinase. J. Biol. Chem. 252: 492-498.

Christensen, U. Kinetic studies of the urokinase-catalysed conversion of NH_2-terminal glutamic acid plasminogen to plasmin. Biochim. Biophys. Acta 481: 638-647.

Claeson, G., Aurell, L., Karlsson, G. & Friberger, P. Substrate structure and activity relationship. In: New Methods for the Analysis of Coagulation using Chromogenic Substrates (Witt, I., ed.), pp.37-54, Walter de Gruyter, Berlin.

Clavin, S. A., Bobbitt, J. L., Shuman, R. T. & Smithwick, E. L. Use of peptidyl-4-methoxy-2-naphthylamides to assay plasmin. Anal. Biochem. 80: 355-365.

Edy, J. & Collen, D. The interaction in human plasma of antiplasmin, the fast-reacting plasmin inhibitor, with plasmin, thrombin, trypsin and chymotrypsin. Biochim. Biophys. Acta 484: 423-432.

Eeckhout, Y. & Vaes, G. Further studies on the activation of procollagenase, the latent precursor of bone collagenase. Effects of lysosomal cathepsin B, plasmin and kallikrein, and spontaneous activation. Biochem. J. 166: 21-31.

Gonzales-Gronow, M., Violand, B. N. & Castellino, F. J. Purification and some properties of the Glu- and Lys-human plasmin heavy chains. J. Biol. Chem. 252: 2175-2177.

Harpel, P. C. Plasmin inhibitor interactions. The effectiveness of $α_2$-plasmin inhibitor in the presence of $α_2$-macroglobulin. J. Exp. Med. 146: 1033-1040.

Hatton, M. W. C. & Regoeczi, E. The inactivation of thrombin and plasmin by antithrombin III in the presence of Sepharose-heparin. Thromb. Res. 10: 645-660.

Highsmith, R. F., Weirich, C. J. & Burnett, C. J. Protease inhibitors and human plasmin: interaction in a whole plasma system. Biochem. Biophys. Res. Commun. 79: 648-656.

Mattler, L. E. & Bang, N. U. Serine protease specificity for peptide chromogenic substrates. Thromb. Haemost. 38: 776-792.

Nieuwenhuizen, W., Wijngaards, G. & Groeneveld, E. Fluorogenic peptide amide substrates for the estimation of plasminogen activators and plasmin. Anal. Biochem. 83: 143-148.

Summaria, L., Boreisha, I. G., Arzadon, L. & Robbins, K. C.
Activation of human Glu-plasminogen to Glu-plasmin by urokinase
in presence of plasmin inhibitors. Streptomyces leupeptin and
human plasma α_1-antitrypsin and antithrombin III (plus heparin).
J. Biol. Chem. 252: 3945-3951.

Walsmann, P. Inhibition of serine proteinases by benzamidine
derivatives. Acta Biol. Med. Ger. 36: 1931-1937.

Wiman, B. Primary structure of the B-chain of human plasmin.
Eur. J. Biochem. 76: 129-137.

Wiman, B. & Wallén, P. The specific interaction between
plasminogen and fibrin. A physiological role of the lysine
binding site in plasminogen. Thromb. Res. 10: 213-222.

1978

Christensen, U. Allosteric effects of some antifibrinolytic amino
acids on the catalytic activity of human plasmin. Biochim.
Biophys. Acta 526: 194-201.

Claeys, H. & Collen, D. Influence of the protein conformation on
the activation of human plasminogen. In: Progress in Chemical
Fibrinolysis and Thrombolysis (Davidson, J. F., Rowan, R. M.,
Samama, M. M. & Desnoyers, P. C. eds), vol. 3, pp. 185-190,
Raven Press, New York.

Friedman, R. B., Kwaan, H. C. & Szczecinski, M. Improved
sensitivity of chromogenic substrate assays for urokinase and
plasmin. Thromb. Res. 12: 37-46.

Goldsmith, G. H., Jr., Saito, H. & Ratnoff, O. D. Activation of
plasminogen by Hageman factor (Factor-XII) and Hageman factor
fragments. J. Clin. Invest. 62: 54-60.

Gonzalez-Gronow, M., Siefring, G. E., Jr. & Castellino, F. J.
Mechanism of activation of human plasminogen by the activator
complex, streptokinase plasmin. J. Biol. Chem. 253: 1090-1094.

Kakinuma, A., Sugino, H., Moriya, N. & Isono, M. Plasminostreptin,
a protein proteinase inhibitor produced by Streptomyces
antifibrinolyticus. I. Isolation and characterization. J. Biol.
Chem. 253: 1529-1537.

Kakinuma, A., Sugino, H., Moriya, N. & Isono, M. Plasminostreptin,
a protein proteinase inhibitor produced by Streptomyces
antifibrinolyticus. II. Determination of the reactive site for
proteinases. J. Biol. Chem. 253: 1538-1545.

Markus, G., DePasquale, J. L. & Wissler, F. C. Quantitative
determination of the binding of ϵ-aminocaproic acid to native
plasminogen. J. Biol. Chem. 253: 727-732.

Markus, G., Evers, J. L. & Hobika, G. H. Comparison of some
properties of native (Glu) and modified (Lys) human plasminogen.
J. Biol. Chem. 253: 733-739.

Pochron, S. P., Mitchell, G. A., Albareda, I., Huseby, R. M. &
Gargiulo, R. J. A fluorescent substrate assay for plasminogen.
Thromb. Res. 13: 733-739.

Roblin, R. O., Bell, T. E. & Young, P. L. Assessment of
plasminogen synthesis in vitro by mouse tumor cells using a
competition radioimmunoassay for mouse plasminogen. Biochim.

Biophys. Acta 543: 383-396.
Sottrup-Jensen, L., Claeys, H., Zajdel, M., Petersen, T. E. &
Magnusson, S. The primary structure of human plasminogen:
isolation of two lysine-binding fragments and one "mini-"
plasminogen (MW,38,000) by elastase-catalyzed-specific limited
proteolysis. In: Progress in Chemical Fibrinolysis and
Thrombolysis (Davidson, J. F., Rowan, R. M., Samama, M. M. &
Desnoyers, P. C. eds), vol. 3, pp. 191-209, Raven Press, New
York.
Sugino, H., Kakinuma, A. & Iwanaga, S. Plasminostreptin, a
protein proteinase inhibitor produced by Streptomyces
antifibrinolyticus. III. Elucidation of the primary structure.
J. Biol. Chem. 253: 1546-1555.
Wallén, P. Chemistry of plasminogen and plasminogen activation.
In: Progress in Chemical Fribrinolysis and Thrombolysis
(Davidson, J. F., Rowan, R. M., Samama, M. M. & Desnoyers, P. C.
eds), vol. 3, pp. 167-181, Raven Press, New York.
Wiman, B. & Collen, D. On the kinetics of the reaction between
human antiplasmin and plasmin. Eur. J. Biochem. 84: 573-578.
Yecies, L., Meier, H., Mandle, R., Jr. & Kaplan, A. P. The
Hageman factor-dependent fibrinolytic pathway. In: Progress in
Chemical Fibrinolysis and Thrombolysis (Davidson, J. F., Rowan,
R. M., Samama, M. M. & Desnoyers, P. C. eds), vol. 3, pp. 121-
139, Raven Press, New York.

1979

Castellino, F. J. A unique enzyme-protein substrate modifier
reaction: plasmin/streptokinase interaction. Trends Biochem.
Sci. 4: 1-5.
Christensen, U. & Ipsen, H.-H. Steady-state kinetics of plasmin-
and trypsin-catalysed hydrolysis of a number of tripeptide-p-
nitroanilides. Biochim. Biophys. Acta 569: 177-183.
Christensen, U., Sottrup-Jensen, L., Magnusson, S., Petersen, T.
E. & Clemmensen, I. Enzymic properties of the neo-plasmin-Val-
442 (miniplasmin). Biochim. Biophys. Acta 567: 472-481.
Eigel, W. N., Hofmann, C. J., Chibber, B. A. K., Tomich, J. M.,
Keenan, T. W., & Mertz, E. T. Plasmin-mediated proteolysis of
casein in bovine milk. Proc. Natl. Acad. Sci. U.S.A. 76: 2244-
2248.
Hayes, M. L. & Castellino, F. J. Carbohydrate of the human
plasminogen variants. I. Carbohydrate composition, glycopeptide
isolation, and characterization. J. Biol. Chem. 254: 8768-8771.
Hayes, M. L. & Castellino, F. J. Carbohydrate of the human
plasminogen variants. II. Structure of the asparagine-linked
oligosaccharide unit. J. Biol. Chem. 254: 8772-8776.
Hayes, M. L. & Castellino, F. J. Carbohydrate of the human
plasminogen variants. III. Structure of the O-glycosidically
linked oligosaccharide unit. J. Biol. Chem. 254: 8777-8780.
Markus, G., Priore, R. L. & Wissler, F. C. The binding of
tranexamic acid to native (Glu) and modified (Lys) human
plasminogen and its effect on conformation. J. Biol. Chem. 254:

1211–1216.

Paoni, N. F. & Castellino, F. J. A comparison of the urokinase and streptokinase activation properties of the native and lower molecular weight forms of sheep plasminogen. J. Biol. Chem. 254: 2064–2070.

Pierzchala, P. A., Dorn, C. P. & Zimmerman, M. A new fluorogenic substrate for plasmin. Biochem. J. 183: 555–559.

Reddy, K. N. N. Plasmin/streptokinase interaction. Trends Biochem. Sci. 4: N212 only.

Summaria, L., Boreisha, I., Wohl, R. C. & Robbins, K. C. Recombinant human Lys-plasmin and the Lys-plasmin-streptokinase complex. J. Biol. Chem. 254: 6811–6814.

Wiman, B., Lijnen, H. R. & Collen, D. On the specific interaction between the lysine-binding sites in plasmin and complementary sites in α_2-antiplasmin and in fibrinogen. Biochim. Biophys. Acta 579: 142–154.

Wohl, R. C., Summaria, L. & Robbins, K. Physiological activation of the human fibrinolytic system. Isolation and characterization of human plasminogen variants. Chicago I and Chicago II. J. Biol. Chem. 254: 9063–9069.

Entry 1.22

TISSUE PLASMINOGEN ACTIVATOR

Summary

EC Number: (3.4.21.-)

Earlier names: Angiokinase (for the activator from blood vessel walls).

Distribution: Vascular endothelium, squamous epithelium. Probably smaller amounts in other tissue elements.

Source: Heart, blood vessels, ovary.

Action: Converts plasminogen to plasmin. Has little general proteolytic activity, but hydrolyzes some blocked peptide naphthylamides.

Requirements: Neutral pH.

Substrate, usual: Plasminogen (then detect activity of plasmin, often on fibrin).

Substrate, special: Boc-Val-Gly-Arg-2-NNap [6].

Inhibitors: Dip-F.

Molecular properties: Extracted from tissue with 2 M KSCN or 5 M urea and precipitated at pH 2.5 without loss of activity, so clearly a very stable enzyme. Pig heart: 52,000 mol.wt. basic protein. Human vasculature: 67,000 mol.wt. single chain protein, pI 8.2, immunologically distinct from urokinase.

Comment: The tissue plasminogen activator is typically found to consist of a single polypeptide chain, unlike urokinase, and to have a slightly higher mol.wt. There are also immunological differences [1-3], and differences in specificity for substrates and inhibitors [4,5]. It is likely that there is overlap between this entry and that for cellular plasminogen activator (#1.23), however.

The name "tissue plasminogen activator" is not ideal: all the plasminogen activators obviously are produced by cells of one type or another, and there is as yet too little understanding of the differences between activators produced by different cell types, and under different conditions, for satisfactory classification.

References
[1] Aoki J. Biochem. 75: 731-741, 1974.

[2] Christman & Acs Nature 264: 681 only, 1976.
[3] Åstedt Thromb. Res. 14: 535-539, 1979.
[4] Nieuwenhuizen et al. Thromb. Res. 11: 87-89, 1977.
[5] Kettner & Shaw Biochim. Biophys. Acta 569: 31-40, 1979.
[6] Nieuwienhuizen et al. Anal. Biochem. 83: 143-148, 1977.

Bibliography

1947

Astrup, T. & Permin, P. M. Fibrinolysis in the animal organism.
Nature 159: 681-682.

1949

Tagnon, H. J. & Petermann, M. L. Activation of proplasmin by a
tissue fraction. Proc. Sci. Exp. Biol. Med. 70: 359-360.

1952

Astrup, T. & Stage, A. Isolation of a soluble fibrinolytic
activator from animal tissue. Nature 170: 929 only.

1953

Müllertz, S. A plasminogen activator in spontaneously active
human blood. Proc. Soc. Exp. Biol. Med. 82: 291-295.

1956

Albrechtsen, O. K. The fibrinolytic activity of the human
endometrium. Acta Endocrinol. 23: 207-218.
Astrup, T. & Sterndorff, I. The plasminogen activator in animal
tissue. Acta Physiol. Scand. 36: 250-255.

1957

Albrechtsen, O. K. The fibrinolytic activity of human tissues.
Br. J. Haematol. 3: 284-291.
Astrup, T. & Albrechtssen, O. K. Estimation of the plasminogen
activator and the trypsin inhibitor in animal and human tissues.
Scand. J. Clin. Lab. Invest. 9: 233-243.
Roberts, H. R. & Astrup, T. Content of tissue activator of
plasminogen in monkey tissues. Thromb. Diath. Haemorrh. 1: 376-
379.

1958

Moltke, P. Plasminogen activator in animal meninges. Proc. Soc.
Exp. Biol. Med. 98: 377-379.

1959

Todd, A. S. The histological localisation of fibrinolysin
activator. J. Path. Bact. 78: 281-283.

1963

Holemans, R. Enhancing the finrinolytic activity of the blood by
vasoactive drugs. Med. Exp. 9: 5-12.
Kwaan, H. C. & Astrup, T. Localization of fibrinolytic activity

in the eye. Arch. Pathol. 76: 595-601.

1964

Bachmann, F., Fletcher, A. P., Alkjaersig, N. & Sherry, S.
Partial purification and properties of the plasminogen activator
from pig heart. Biochemistry 3: 1578-1585.
Lack, C. H. Proteolytic activity and connective tissue. Br. Med.
Bull. 20: 217-222.
Warren, B. A. Fibrinolytic activity of vascular endothelium. Br.
Med. Bull. 20: 213-216.

1965

Astrup, T. & Kok, P. Preparation and purification of a plasminogen
activator from porcine tissue. Thromb. Diath. Haemorrh. 13: 587
only.
Sugiyama, Y. Plasminogen activator in the cytoplasmic granules
from different organs of mammals: studies on intracellular
localization, releasing, and molecular size of two kinds of the
activators. Kobe J. Med. Sci. 11: 151-167.

1966

Astrup, T. Tissue activators of plasminogen. Fed. Proc. Fed. Am.
Soc. Exp. Biol. 25: 42-51.

1967

Kok, P. & Astrup, T. Some comparative studies on tissue activator
and urokinase. Thromb. Diath. Haemorrh. 18: 294-295.
Pandolfi, M. & Astrup, T. A histochemical study of the
fibrinolytic activity: cornea, conjunctiva and lacrimal gland.
Arch. Ophthalmol. 77: 258-264.

1968

Beard, E. L., Montuori, M. H., Danos, G. J., & Busuttil, R. W.
Plasminogen activator activity of rat lysosomes. Proc. Soc.
Exp. Biol. Med. 129: 804-808.
Lieberman, J. Substrate specificity of plasminogen activator in
the lungs of rabbits and rodents. Ionic enhancement of
plasminogen activator activity. Lab. Invest. 19: 466-472.
Nakahara, M. & Celander, D. R. Properties of microsomal activator
of profibrinolysin found in bovine and porcine heart muscle.
Thromb. Diath. Haemorrh. 19: 483-491.

1969

Beard, E. L., Carroll, G. F. & Danos, G. T. Release of plasminogen
activator from rat liver lysosomes induced by stress related
enzymes. Proc. Soc. Exp. Biol. Med. 131: 438-442.
Bernik, M. B. & Kwaan, H. C. Plasminogen activator activity in
cultures from human tissues. An immunological and histochemical
study. J. Clin. Invest. 48: 1740-1753.
Clarke, N. & Collins, M. Immediate fibrinolytic effect of
components from the interior of human cells. Br. J. Exp.
Pathol. 50: 153-157.

Constantini, R., Spöttl, F., Holzknecht, F. & Braunsteiner, H. On the role of the stable plasminogen activator of the venous wall in occlusion induced fibrinolytic state. Thromb. Diath. Haemorrh. 22: 544-551.
Kok, P. & Astrup, T. Isolation and purification of a tissue plasminogen activator and its comparison with urokinase. Biochemistry 8: 79-86.
von Kaulla, K. N. & Wasantapruek, S. Extraction of plasminogen activator from canine vascular segments in situ. Proc. Soc. Exp. Biol. Med. 132: 830-834.

1970

Astrup, T. & Kok, P. Assay and preparation of a tissue plasminogen activator. Methods Enzymol. 19: 821-834.
Glas, P. & Astrup, T. Thromboplastin and plasminogen activator in tissues of the rabbit. Am. J. Physiol. 219: 1140-1146.
Pandolfi, M. Persistence of fibrinolytic activity in fragments of human veins cultured in vitro. Thromb. Diath. Haemorrh. 24: 43-47.
Rickli, E. E. & Zaugg, H. Isolation and purification of highly enriched tissue plasminogen activator from pig heart. Thromb. Diath. Haemorrh. 23: 64-76.

1971

Åstedt, B. & Pandolfi, M. Ontogenisis of tissue plasminogen activator in the human. Thromb. Diath. Haemorrh. 25: 469-480.
Aoki, N. & von Kaulla, K. N. Dissimilarity of human vascular plasminogen activator and human urokinase. J. Lab. Clin. Med. 78: 354-362.
Aoki, N. & von Kaulla, K. N. The extraction of vascular plasminogen activator from human cadavers and a description of some of its properties. Am. J. Clin. Pathol. 55: 171-179.
Auerswald, W., Binder, B. & Doleschel, W. "Angiokinase" - molecular weights of proteins representing a perivascular plasminogen activator. Thromb. Diath. Haemorrh. 26: 411-413.
Beard, E. L. Retention of plasminogen activator and acid phosphatase in lysosomes of lipemic rats. Thromb. Diath. Haemorrh. 25: 134-141.

1972

Aoki, N. & von Kaula, K. N. Potentiation by synthetic fibrinolytic agents of vascular activator- and urokinase-induced lysis of human plasma clots as demonstrated by thrombelastography. Thromb. Diath. Haemorrh. 27: 246-251.
Constantini, R., Hilbe, G., Spöttl, F. & Holzknecht, F. The plasminogen activator content of the arterial wall in occlusive arterial diseases. Thromb. Diath. Haemorrh. 27: 649-654.
Thorsen, S., Glas-Greenwalt, P. & Astrup, T. Differences in the binding to fibrin of urokinase and tissue plasminogen activator. Thromb. Diath. Haemorrh. 28: 65-74.
Wünschmann-Henderson, B. & Astrup, T. Relation of fibrinolytic

activity in human oral epithelial cells to cellular maturation: the influence of smoking. J. Path. 108: 293-301.

1973

Ray, G. & Astrup, T. Effects of some polysaccharide sulfates and some phosphates on neutral proteases and plasminogen activators. Haemostasis 2: 269-276.

Thorsen, S. The inhibition of tissue plasminogen activator and urokinase-induced fibrinolysis by some natural proteinase inhibitors and by plasma and serum from normal and pregnant subjects. Scand. J. Clin. Lab. Invest. 31: 51-59.

Todd, A. S. Endothelium and fibrinolysis. Bibl. Anatom. 12: 98-105.

1974

Aoki, N. Preparation of plasminogen activator from vascular trees of human cadavers. Its comparison with urokinase. J. Biochem. 75: 731-741.

Bernik, M. B., White, W. F., Oller, E. P. & Kwaan, H. C. Immunological identity of plasminogen activator in human urine, heart, blood vessels, and tissue culture. J. Lab. Clin. Med. 84: 546-558.

Cartwright, T. Partial purification of a specific plasminogen activator from porcine parotid glands. Thromb. Diath. Haemorrh. 31: 403-414.

Hajikata, A., Fujimoto, K., Kitaguchi, H. & Okamoto, S. Some properties of the tissue plasminogen activator from pig heart. Thromb. Res. 4: 731-740.

Hegt, V. H. & Brakman, P. Histochemical study of an inhibitor of fibrinolysis in the human arterial wall. Nature 248: 75-76.

Rybák, M., Rybáková, B. & Koutský, J. Trypsin-like enzymes of human uterus wall and their relationship to plasminogen and kinin-system activator. Physiol. Bohemoslov. 23: 467-473.

Thorsen, S. & Astrup, T. Substrate composition and the effect of ε-aminocaproic acid on tissue plasminogen activator and urokinase-induced fibrinolysis. Thromb. Diath. Haemorrh. 32: 306-324.

Wunschmann-Henderson, B. & Astrup, T. Inhibition by hydrocortisone of plasminogen activator production in rat tongue organ cultures. Lab. Invest. 30: 427-433.

1975

Astrup, T. Cell induced fibrinolysis: a fundamental process. In: Proteases and Biological Control (Reich, E., Rifkin, D. B. & Shaw, E. eds), pp. 343-355, Cold Spring Harbor Laboratory.

Beers, W. H., Strickland, S. & Reich, E. Ovarian plasminogen activator: relationship to ovulation and hormonal regulation. Cell 6: 387-394.

Shaper, A. G., Marsh, N. A., Patel, I. & Kater, F. Response of fibrinolytic activity to venous occlusion. Br. Med. Bull. 3: 571-573.

1976

Christman, J. K. & Acs, G. Uniqueness of plasminogen activators. Nature 264: 681 only.

Hegt, V. H. Distribution and variation of fibrinolytic activity in the walls of human arteries and veins. Haemostasis 5: 355–372.

Ogston, D., Bennett, B. & Mackie, M. Properties of a partially purified preparation of a circulating plasminogen activator. Thromb. Res. 8: 275–284.

1977

Christman, J. K., Silverstein, S. C. & Acs, G. Plasminogen activators. In: Proteinases in Mammalian Cells and Tissues (Barrett, A. J. ed.), pp. 91–149, North–Holland Publishing Co., Amsterdam.

Cole, E. R. & Bachmann, F. W. Purification and properties of a plasminogen activator from pig heart. J. Biol. Chem. 252: 3729–3737.

Kluft, C., Jenks, R. L. & Astrup, T. Production of plasminogen activator activity in factor XII-deficient plasma. Thromb. Res. 10: 759–764.

Loskutoff, D. J. & Edgington, T. S. Synthesis of a fibrinolytic activator and inhibitor by endothelial cells. Proc. Natl. Acad. Sci. U.S.A. 74: 3903–3907.

Nieuwenhuizen, W., Wijngaards, G. & Groeneveld, E. Fluorogenic peptide amide substrates for the estimation of plasminogen activators and plasmin. Anal. Biochem. 83: 143–148.

Nieuwenhuizen, W., Wijngaards, G. & Groeneveld, E. Synthetic substrates and the discrimination between urokinase and tissue plasminogen activator activity. Thromb. Res. 11: 87–89.

1978

Blasini, R., Stemberger, A., Wriedt-Lübbe, I. & Blümel, G. Tissue proteases demonstrated by a histochemical method using chromogenic substrates. Thromb. Res. 13: 585–590.

Holmberg, L., Lecander, I., Persson, B. & Åstedt, B. An inhibitor from placenta specifically binds urokinase and inhibits plasminogen activator released from ovarian carcinoma in tissue culture. Biochim. Biophys. Acta 544: 128–137.

Nieuwenhuizen, W., Wijngaards, G. & Groeneveld, E. Fluorogenic substrates for sensitive and differential estimation of urokinase and tissue plasminogen activator. Haemostasis 7: 146–149.

Radcliffe, R. & Heinze, T. Isolation of plasminogen activator from human plasma by chromatography on lysine-Sepharose. Arch. Biochem. Biophys. 189: 185–194.

Wunschmannhenderson, B. & Astrup, T. Inhibition of plasminogen activator production in organ cultures by cycloheximide. Experientia 34: 1288–1289.

1979

Åstedt, B. No crossreaction between circulating plasminogen activator and urokinase. Thromb. Res. 14: 535–539.

Binder, B. R., Spragg, J. & Austen, K. F. Purification and characterization of human vascular plasminogen activator derived from blood vessel perfusates. J. Biol. Chem. 254: 1998–2003.

Svanberg, L. & Åstedt, B. Release of plasminogen activator from normal and neoplastic endometrium. Experientia 35: 818–819.

Entry 1.23

CELLULAR PLASMINOGEN ACTIVATOR

Summary

EC Number: (3.4.21.-)

Earlier names: -

Distribution: Known from many cell types, but virus transformed
cells have been the subject of particular interest. May be
membrane-bound. See Comment, however!

Source: Culture medium of virus-transformed cells.

Action: Converts plasminogen to plasmin. Has little general
proteolytic activity, but some lysine esterase action.

Requirements: Neutral pH.

Substrate, usual: Plasminogen with fibrin. Ac-Lys-OMe.

Substrate, special: Z-Gly-Gly-Arg-NMec [1].

Inhibitors: Dip-F, X-Gly-Arg-CH$_2$Cl [2], Gbz-OPhNO$_2$. Also
6-aminohexanoic acid indirectly through an effect on plasminogen.

Molecular properties: SV-40 transformed hamster cells: mol.wt.
50,000; pI 9.5.

Comment: This entry is unsatisfactory, because we do not yet know
how many separate plasminogen activators exist. The activators
dealt with in the reports cited may truly be of one distinct
type, or several, but most likely some resemble the tissue
activator best characterized from vessel walls (#1.22), whereas
others are the same as the renal activator (#1.24).

References
[1] Zimmerman et al. Proc. Natl. Acad. Sci. U.S.A. 75: 750-753,
1978.
[2] Colman et al. Biochim. Biophys. Acta 569: 41-51, 1979.

Bibliography

1962

Painter, R. H. & Charles, A. F. Characterization of a soluble
plasminogen activator from kidney cell cultures. Am. J.
Physiol. 202: 1125-1130.

1963

Künzer, W. & Haberhausen, A. Fibrinolytic activity of human erythrocytes. Nature 198: 396-397.

1969

Bernik, M. B. & Kwaan, H. C. Plasminogen activator activity in cultures from human tissues. An immunological and histochemical study. J. Clin. Invest. 48: 1740-1753.
Semar, M., Skoza, L. & Johnson, A.J. Partial purification and properties of a plasminogen activator from human erythrocytes. J. Clin. Invest. 48: 1777-1785.

1972

Wünschmann-Henderson, B., Horwitz, D. L. & Astrup, T. Release of plasminogen activator from viable leukocytes of man, baboon, dog and rabbit. Proc. Soc. Exp. Biol. Med. 141: 634-638.

1973

Bernik, M. B. Increased plasminogen activator (urokinase) in tissue culture after fibrin deposition. J. Clin. Invest. 52: 823-834.
Ossowski, L., Quigley, J. P., Kellerman, G. M. & Reich, E. Fibrinolysis associated with oncogenic transformation. Requirement of plasminogen for correlated changes in cellular morphology, colony formation in agar, and cell migration. J. Exp. Med. 138: 1056-1064.
Ossowski, L., Unkeless, J. C., Tobia, A., Quigley, J. P., Rifkin, D. B. & Reich, E. An enzymatic function associated with transformation of fibroblasts by oncogenic viruses. II. Mammalian fibroblast cultures transformed by DNA and RNA tumor viruses. J. Exp. Med. 137: 112-126.

1974

Bernik, M. B., White, W. F., Oller, E. P. & Kwaan, H. C. Immunological identity of plasminogen activator in human urine, heart, blood vessels, and tissue culture. J. Lab. Clin. Med. 84: 546-558.
Christman, J. K. & Acs, G. Purification and characterization of a cellular fibrinolytic factor associated with oncogenic transformation: the plasminogen activator from SV-40 transformed hamster cells. Biochim. Biophys. Acta 340: 339-347.
Kopitar, M., Stegnar, M., Accetto, B. & Lebez, D. Isolation and characterisation of plasminogen activator from pig leucocytes. Thromb. Diath. Haemorrh. 31: 72-85.
Ossowski, L., Quigley, J. P. & Reich, E. Fibrinolysis associated with oncogenic transformation. Morphological correlates. J. Biol. Chem. 249: 4312-4320.
Rifkin, D. B., Loeb, J. N., Moore, G. & Reich, E. Properties of plasminogen activators formed by neoplastic human cell cultures. J. Exp. Med. 139: 1317-1328.
Unkeless, J. C., Gordon, S. & Reich, E. Secretion of plasminogen

activator by stimulated macrophages. J. Exp. Med. 139: 834-850.
Wunschmann-Henderson, B. & Astrup, T. Inhibition by hydrocortisone
   of plasminogen activator production in rat tongue organ
   cultures. Lab. Invest. 30: 427-433.

1975

Beers, W. H., Strickland, S. & Reich, E. Ovarian plasminogen
   activator: relationship to ovulation and hormonal regulation.
   Cell 6: 387-394.
Christman, J. K., Acs, G., Silagi, S. & Silverstein, S. C.
   Plasminogen activator: biochemical characterization and
   correlation with tumorigenicity. In: Proteases and Biological
   Control (Reich, E., Rifkin, D. B. & Shaw, E. eds), pp. 827-839,
   Cold Spring Harbor Laboratory, New York.
Danø, K. & Reich, E. Inhibitors of plasminogen activation. In:
   Proteases and Biological Control (Reich, E., Rifkin, D. B. &
   Shaw, E., eds), pp. 357-366, Cold Spring Harbor Laboratory, New
   York.
Ossowski, L., Quigley, J. P. & Reich, E. Plasminogen, a necessary
   factor for cell migration in vitro. In: Proteases and
   Biological Control (Reich, E., Rifkin, D. B. & Shaw, E. eds),
   pp. 901-913, Cold Spring Harbor Laboratory, New York.
Reich, E. Plasminogen activator: secretion by neoplastic cells
   and macrophages. In: Proteases and Biological Control (Reich,
   E., Rifkin, D. B. & Shaw, E. eds), pp. 333-341, Cold Spring
   Harbor Laboratory, New York.
Rifkin, D. B., Beal, L. P. & Reich, E. Macromolecular determinants
   of plasminogen activator synthesis. In: Proteases and
   Biological Control (Reich, E., Rifkin, D. B. & Shaw, E. eds),
   pp. 841-847, Cold Spring Harbor Laboratory, New York.

1976

Åstedt, B. & Holmberg, L. Immunological identity of urokinase and
   ovarian carcinoma plasminogen activator released in tissue
   culture. Nature 261: 595-597.
Åstedt, B., Bladh, B., Holmberg, L. & Liedholm, P. Purification
   of plasminogen activator(s) from human seminal plasma.
   Experientia 32: 148-149.
Christman, J. K. & Acs, G. Uniqueness of plasminogen activators.
   Nature 264: 681 only.
Hamilton, J., Vassalli, J.-D. & Reich, E. Macrophage plasminogen
   activator: induction by asbestos is blocked by anti-inflammatory
   steroids. J. Exp. Med. 144: 1689-1694.
Quigley, J. P. Association of a protease (plasminogen activator)
   with a specific membrane fraction isolated from transformed
   cells. J. Cell Biol. 71: 472-486.

1977

Berger, H., Jr. Secretion of plasminogen activator by rheumatoid
   and nonrheumatoid synovial cells in culture. Arthritis Rheum.
   20: 1198-1205.

Chou, I.-N., O'Donnell, S. P., Black, P. H. & Roblin, R. O. Cell density-dependent secretion of plasminogen activator by 3T3 cells. J. Cell. Physiol. 91: 31-37.

Christman, J. K., Silverstein, S. C. & Acs, G. Plasminogen activators. In: Proteinases in Mammalian Cells and Tissues (Barrett, A. J. ed.), pp. 91-149, North-Holland Publishing Co., Amsterdam.

Granelli-Piperno, A., Vassalli, J.-D. & Reich, E. Secretion of plasminogen activator by human polymorphonuclear leukocytes. Modulation by glucocorticoids and other effectors. J. Exp. Med. 146: 1693-1706.

Huseby, R. M., Clavin, S. A., Smith, R. E., Hull, R. N. & Smithwick, E. L. Studies on tissue culture plasminogen activator. II. The detection and assay of urokinase and plasminogen activator from LLC-PK$_1$ cultures (porcine) by the synthetic substrate Nα-benzyloxycarbonyl-glycyl-glycyl-arginyl-4-methoxy-2-naphthylamide. Thromb. Res. 10: 679-687.

Loskutoff, D. J. & Edgington, T. S. Synthesis of a fibrinolytic activator and inhibitor by endothelial cells. Proc. Natl. Acad. Sci. U.S.A. 74: 3903-3907.

Nieuwenhuizen, W., Wijngaards, G. & Groeneveld, E. Fluorogenic peptide amide substrates for the estimation of plasminogen activators and plasmin. Anal. Biochem. 83: 143-148.

Nolan, C., Hall, L. S., Barlow, G. H. & Tribby, I. I. E. Plasminogen activator from human embryonic kidney cell cultures. Evidence for a proactivator. Biochim. Biophys. Acta 496: 384-400.

Roblin, R. O., Hammond, M. E., Bensky, N. D., Dvorak, A. M., Dvorak, H. F. & Black, P. H. Generation of macrophage migration inhibitory activity by plasminogen activators. Proc. Natl. Acad. Sci. U.S.A. 74: 1570-1574.

Wu, M.-C., Arimura, G. K. & Yunis, A. A. Purification and characterization of a plasminogen activator secreted by cultured human pancreatic carcinoma cells. Biochemistry 16: 1908-1913.

1978

Åstedt, B., Lundgren, E., Roos, G. & Abu Sinna, G. Release of various molecular forms of plasminogen activators during culture of human ovarian tumours. Thromb. Res. 13: 1031-1037.

Bigbee, W. L. & Jensen, R. H. Characterization of plasminogen activator in human cervical cells. Biochim. Biophys. Acta 540: 285-29.

Cammer, W., Bloom, B. R., Norton, W. T. & Gordon, S. Degradation of basic protein in myelin by neutral proteases secreted by stimulated macrophages - possible mechanism of inflammatory demyelination. Proc. Natl. Acad. Sci. U.S.A. 75: 1554-1558.

Holmberg, L., Lecander, I., Persson, B. & Åstedt , B. An inhibitor from placenta specifically binds urokinase and inhibits plasminogen activator released from ovarian carcinoma in tissue culture. Biochim. Biophys. Acta 544: 128-137.

Loskutoff, D. J. & Paul, D. Intracellular plasminogen activator activity in growing and quiescent cells. J. Cell Physiol. 97: 9-16.

Roblin, R. O., Young, P. L. & Bell, T. E. Concomitant secretion by transformed SVW138-VA13-2RA cells of plasminogen activator(s) and substance(s) which prevent their detection. Biochem. Biophys. Res. Commun. 82: 165-172.

Schroder, E. W., Chou, I.-N., Jaken, S. & Black, P. H. Interferon inhibits the release of plasminogen activator from SV3T3 cells. Nature 276: 828-829.

Werb, Z. Biochemical actions of glucocorticoids on macrophages in culture. Specific inhibition of elastase, collagenase, and plasminogen activator secretion and effects on other metabolic functions. J. Exp. Med. 147: 1695-1712.

Werb, Z. & Aggeler, J. Proteases induce secretion of collagenase and plasminogen activator by fibroblasts. Proc. Natl. Acad. Sci. U.S.A. 75: 1839-1843.

Wunschmannhenderson, B. & Astrup, T. Inhibition of plasminogen activator production in organ cultures by cycloheximide. Experientia 34: 1288-1289.

Zimmerman, M., Quigley, J. P., Ashe, B., Dorn, C., Goldfarb, R. & Troll, W. Direct fluorescent assay of urokinase and plasminogen activators of normal and malignant cells: kinetics and inhibitor profiles. Proc. Natl. Acad. Sci. U.S.A. 75: 750-753.

1979

Åstedt, B. No crossreaction between circulating plasminogen activator and urokinase. Thromb. Res. 14: 535-539.

Coleman, P., Kettner, C. & Shaw, E. Inactivation of the plasminogen activator from HeLa cells by peptides of arginine chloromethyl ketone. Biochim. Biophys. Acta 569: 41-51.

Danø, K. & Reich, E. Plasminogen activator from cells transformed by an oncogenic virus. Inhibitors of the activation reaction. Biochim. Biophys. Acta 566: 138-151.

Drapier, J.-C., Tenu, J.-P., Lemaire, G. & Petit, J.-F. Regulation of plasminogen activator sectretion in mouse peritoneal macrophages. I. Role of serum studied by a new spectrophotometric assay for plasminogen activators. Biochemie 61: 463-471.

Failly-Crepin, C. & Uriel, J. An electrophoretic assay for plasminogen activator. Biochimie 61: 567-571.

Jaken, S. & Black, P. H. Differences in intracellular distribution of plasminogen activator in growing, confluent, and transformed 3T3-cells. Proc. Natl. Acad. Sci. U.S.A. 76: 246-250.

Lewis, L. J. Plasminogen activator (urokinase) from cultured cells. Thromb. Haemost. 42: 895-900.

Paul, D. C., Bobbitt, J. L., Williams, D. C. & Hull, R. N. Immunocytochemical localization of plasminogen activator on porcine kidney cell strain: LLC-PK$_1$ (LP$_{100}$). J. Histochem. Cytochem. 27: 1035-1040.

Entry 1.24

UROKINASE

Summary

EC Number: 3.4.21.31

Earlier names: Cytokinase (from kidney cells); fibrinokinase.

Distribution: Urine; kidney cells.

Source: Urine; culture medium of kidney cells.

Action: Activates plasminogen by cleaving an -Arg-Val- bond, and also degrades protamine and casein. Hydrolyzes some lysyl and arginyl substrates.

Requirements: pH optimum about 7.8.

Substrate, usual: Ac-Lys-OMe; plasminogen, protamine.

Substrate, special: Glt-Gly-Arg-NMec, Z-Gly-Gly-Arg-NMec [1], Ac-Gly-Lys-OMe, Ac-Gly-Lys-O[^{3}H]Me.

Inhibitors: Dip-F, Gbz-OPhNO$_2$, Ac-Gly-Lys-CH$_2$Cl, Pro-Gly-Arg-CH$_2$Cl [2]. Benzamidine. (Not aprotinin, soybean trypsin inhibitor, α_2-macroglobulin or antithrombin.)

Molecular properties: Species of high (47,000) and low (33,400) mol.wt. occur. In each, a 33,400 mol.wt. chain carries the catalytic Ser and His residues, but the larger form has a second, 18,600 mol.wt. chain. The higher mol.wt. form can be converted into the lower by an autolytic process, but the molar activities of the two forms are indistinguishable. pI probably about 7.0.

Comment: This enzyme could be described as "renal plasminogen activator". Kidney cells in culture seem to produce the same enzyme, but it is not clear whether other types of cell synthesize this or other types of plasminogen activator.

References
[1] Zimmerman et al. Proc. Natl. Acad. Sci. 75: 750-753, 1978.
[2] Kettner & Shaw Biochim. Biophys. Acta 569: 31-40, 1979.

Bibliography

1947

Astrup, T. & Permin, P. M. Fibrinolysis in the animal organism. Nature 159: 681-682.

1948

Astrup, T. & Permin, P. M. Fibrinokinase and fibrinolytic enzymes. Nature 161: 689–690.

1950

Lewis, J. H. & Ferguson, J. H. Studies on a proteolytic enzyme system of the blood. II. Fibrinolysokinase activators for profibrinolysin. J. Clin. Invest. 29: 1059–1068.

1951

Astrup, T. Fibrinokinase. Acta Physiol. Scand. 24: 267–277.

1957

Sherry, S. & Alkjaersig, N. Biochemical, experimental, and clinical studies of proteolytic enzymes: with particular reference to the fibrinolytic enzyme of human plasma. Ann. N. Y. Acad. Sci. 68: 52–66.

1961

Hansen, P. F., Jorgensen, M., Kjeldgaard, N. O. & Plough, J. Urokinase – an activator of plasminogen from human urine. Angiology 12: 367–371.

1964

Lack, C. H. & Ali, S. Y. Tissue activator of plasminogen. Nature 201: 1030–1031.

Lorand, L. & Mozen, M. M. Ester–hydrolysing activity of urokinase preparations. Nature 201: 392–393.

Sherry, S., Alkjaersig, N. & Fletcher, A. P. Assay of urokinase preparations with the synthetic substrate acetyl–L–lysine methyl ester. J. Lab. Clin. Med. 64: 145–153.

1965

Ali, S. Y. & Lack, C. H. Studies on the tissue activator of plasminogen. Distribution of activator and proteolytic activity in the subcellular fractions of rabbit kidney. Biochem. J. 96: 63–74.

Burges, R. A., Brammer, K. W. & Coombes, J. D. Molecular weight of urokinase. Nature 208: 894 only.

Lesuk, A., Terminiello, L. & Traver, J. H. Crystalline human urokinase: some properties. Science 147: 880–882.

Lorand, L. & Condit, E. V. Ester hydrolysis by urokinase. Biochemistry 4: 265–270.

1966

Boomgaard, J., Deggeiler, K., Bleyenberg, A. & Vreeken, J. Studies on urokinase. The measurement of urokinase. Clin. Chim. Acta 13: 498–505.

Hamberg, U. & Savolainen, M.-L. Studies on plasminogen activators. II. A method for the separation of purified urokinase from small samples of urine. Acta Chem. Scand. B 20: 2551–2558.

Lesuk, A., Terminiello, L. & Traver, J. H. Sephadex gel-filtration

of human urokinase preparations. Fed. Proc. Fed. Am. Soc. Exp.
Biol. 25: 194 only.
White, W. F., Barlow, G. H. & Mozen, M. M. The isolation and
characterization of plasminogen activators (urokinase) from
human urine. Biochemistry 5: 2160-2169.

1967

Lesuk, A., Terminiello, L., Traver, J. H. & Groff, J. L.
Biochemical and biophysical studies of human urokinase. Thromb.
Diath. Haemorrh. 18: 293-294.
Robbins, K. C., Summaria, L., Hsieh, B. & Shah, K. J. The peptide
chains of human plasmin. Mechanism of activation of human
plasminogen to plasmin. J. Biol. Chem. 242: 2333-2342.
Walton, P. L. The hydrolysis of α-N-acetylglycyl-L-lysine methyl
ester by urokinase. Biochim. Biophys. Acta. 132: 104-114.

1968

Ali, S. Y. & Evans, L. Purification of rabbit kidney cytokinase
and a comparison of its properties with human urokinase.
Biochem. J. 107: 293-303.
Hamberg, U. & Savolainen, M.-L. Studies on plasminogen activators.
III. Assay of fibrinolytic activator enzymes from plasma and
urine. Acta Chem. Scand. B 22: 1452-1460.
Kucinski, C. S., Fletcher, A. P. & Sherry, S. Effect of urokinase
antiserum on plasminogen activators: demonstration of
immunological dissimilarity between plasma plasminogen activator
and urokinase. J. Clin. Invest. 47: 1238-1253.

1969

Bernik, M. B. & Kwaan, H. C. Plasminogen activator activity in
cultures from human tissues. An immunological and histochemical
study. J. Clin. Invest. 48: 1740-1753.
Hamberg, U. The fibrinolytic activation mechanism in human
plasma. Proc. Roy. Soc. B 173: 293-309.
Kok, P. & Astrup, T. Isolation and purification of a tissue
plasminogen activator and its comparison with urokinase.
Biochemistry 8: 79-86.

1970

Ali, S. Y. Tissue activator of plasminogen (cytokinase). Methods
Enzymol. 19: 834-838.
Ball, A. P. & Day, E. D. Immunological identification of two
urokinases. Thromb. Diath. Haemorrh. 24: 475-486.
Ball, A. P. & Day, E. D. Urokinase-induced fibrinolysis of [125]I-
fibrin. Conditions for first-order activation kinetics, density-
gradient separation of split products, and a microassay.
Thromb. Diath. Haemorrh. 24: 463-474.
Day, E. D. & Ball, A. P. Immunological identification of three
urokinases. Thromb. Diath. Haemorrh. 24: 487-494.
Landmann, H. & Markwardt, F. Irreversible synthetische Inhibitoren
der Urokinase. Experientia 26: 145-147.

White, W. F. & Barlow, G. H.   Urinary plasminogen activator.
Methods Enzymol. 19: 665-672.

1971

Aoki, N. & von Kaulla, K. N.   Dissimilarity of human vascular
plasminogen activator and human urokinase.   J. Lab. Clin. Med.
78: 354-362.

1972

Aoki, N. & von Kaulla, K. N.   Potentiation by synthetic
fibrinolytic agents of vascular activator- and urokinase-induced
lysis of human plasma clots as demonstrated by
thrombelastography.   Thromb. Diath. Haemorrh. 27: 246-251.
Barlow, G. H. & Lazar. L.   Characterization of the plasminogen
activator isolated from human embryo kidney cells: comparison
with urokinase.   Thromb. Res. 1: 201-208.
Heimburger, N.   Basic mechanism of action of streptokinase and
urokinase.   Thromb. Diath. Haemorrh. 27: 21-30.
Kok, P. & Astrup, T.   Differentiation between plasminogen
activators by means of epsilon-aminocaproic acid.   Thromb.
Diath. Haemorrh. 27: 77-87.
Marsh, N. A. & Arocha-Pinango, C. L.   Evaluation of the fibrin
plate method for estimating plasminogen activators.   Thromb.
Diath. Haemorrh. 28: 75-88.
Thorsen, S., Glas-Greenwalt, P. & Astrup, T.   Differences in the
binding to fibrin of urokinase and tissue plasminogen activator.
Thromb. Diath. Haemorrh. 28: 65-74.

1973

Bernik, M. B.   Increased plasminogen activator (urokinase) in
tissue culture after fibrin deposition.   J. Clin. Invest. 52:
823-834.
Petkov, D., Christova, E. & Karadjova, M.   Amidase activity of
urokinase. I. Hydrolysis of α-N-acetyl-L-lysine p-nitroanilide.
Thromb. Diath. Haemorrh. 29: 276-285.
Rickli, E. E. & Otavsky, W. I.   Release of an N-terminal peptide
from human plasminogen during activation with urokinase.
Biochim. Biophys. Acta 295: 381-384.
Wiman, B.   Primary structure of peptides released during
activation of human plasminogen by urokinase.   Eur. J. Biochem.
39: 1-9.

1974

Aoki, N.   Preparation of plasminogen activator from vascular trees
of human cadavers. Its comparison with urokinase.   J. Biochem.
75: 731-741.
Bernik, M. B., White, W. F., Oller, E. P. & Kwaan, H. C.
Immunological identity of plasminogen activator in human urine,
heart, blood vessels, and tissue culture.   J. Lab. Clin. Med.
84: 546-558.
Claeys, H. & Vermylen, J.   Physico-chemical and proenzyme

properties of NH_2-terminal glutamic acid and NH_2-terminal lysine human plasminogen. Biochim. Biophys. Acta 342: 351-359.

Thorsen, S. & Astrup, T. Substrate composition and the effect of ε-aminocaproic acid on tissue plasminogen activator and urokinase-induced fibrinolysis. Thromb. Diath. Haemorrh. 32: 306-324.

1975

Barlow, G. H., Rueter, A. & Tribby, I. Production of plasminogen activator by tissue culture techniques. In: Proteases and Biological Control (Reich, E., Rifkin, D. B. & Shaw, E. eds), pp. 325-341, Cold Spring Harbor Laboratory.

Danø, K. & Reich, E. Inhibitors of plasminogen activation. In: Proteases and Biological Control (Reich, E., Rifkin, D. B. & Shaw, E. eds), pp. 357-366, Cold Spring Harbor Laboratory, New York.

Petkov, D., Christova, E., Pojarlieff, I. & Stambolieva, N. Structure-activity relationship in the urokinase hydrolysis of N-acetyl-L-lysine anilides. Eur. J. Biochem. 51: 25-32.

Soberano, M. E., Ong, E. B., Johnson, A. J., Schoellmann, G. & Levy, M. Inhibition studies of urokinase with diisopropylphosphorofluoridate (DFP) and p-nitrophenyl-p'-guanidinobenzoate (NPGB). Fed. Proc. Fed. Am. Soc. Exp. Biol. 34: 860 only.

Sodetz, J. M. & Castellino, F. J. A comparison of mechanisms of activation of rabbit plasminogen isozymes by urokinase. In: Proteases and Biological Control (Reich, E., Rifkin, D. B. & Shaw, E. eds), pp. 311-324, Cold Spring Harbor Laboratory, New York.

Sodetz, J. M. & Castellino, F. J. The mechanism of activation of rabbit plasminogen by urokinase. J. Biol. Chem. 250: 3041-3049.

Summaria, L., Arzadon, L., Bernabe, P. & Robbins, K. C. The activation of plasminogen to plasmin by urokinase in the presence of the plasmin inhibitor trasylol. The preparation of plasmin with the same NH_2-terminal heavy (A) chain sequence as the parent zymogen. J. Biol. Chem. 250: 3988-3995.

1976

Åstedt, B. & Holmberg, L. Immunological identity of urokinase and ovarian carcinoma plasminogen activator released in tissue culture. Nature 261: 595-597.

Barlow, G. H. Urinary and kidney cell plasminogen activator (urokinase). Methods Enzymol. 45: 239-244.

Christman, J. K. & Acs, G. Uniqueness of plasminogen activators. Nature 264: 681 only.

Holmberg, L., Bladh, B. & Åstedt, B. Purification of urokinase by affinity chromatography. Biochim. Biophys. Acta 445: 215-222.

Imanari, T., Kaizu, T., Yoshida, H., Yates, K., Pierce, J. V. & Pisano, J. J. Radiochemical assays for human urinary, salivary, and plasma kallikreins. In: Chemistry and Biology of the Kallikrein-Kinin System in Health and Disease, DHEW Publication

No. (NIH) 76-791 (Pisano, J. J. & Austen, K. F. eds), pp. 205-213, U.S. Government Printing Office, Washington.

Imanari, T., Wilcox, G. M. & Pisano, J. J. Sensitive radiochemical esterolytic assays for urokinase. Clin. Chim. Acta 71: 267-276.

Ong, E. B., Johnson, A. J. & Schoellmann, G. Identification of an active site histidine in urokinase. Biochim. Biophys. Acta 429: 252-257.

Soberano, M. E., Ong, E. B. & Johnson, A. J. The effects of inhibitors on the catalytic conversion of urokinase. Thromb. Res. 9: 675-681.

Soberano, M. E., Ong, E. B., Johnson, A. J., Levy, M. & Schoellman, G. Purification and characterization of two forms of urokinase. Biochim. Biophys. Acta 445: 763-773.

Tamura, Y. & Fujii, S. Purification of human plasma kallikrein and urokinase by affinity chromatography. J. Biochem. 80: 507-511.

Violand, B. N. & Castellino, F. J. Mechanism of the urokinase-catalyzed activation of human plasminogen. J. Biol. Chem. 251: 3906-3912.

1977

Åstedt, B., Bladh, B. & Holmberg, L. Some characteristics of urokinase released in organ culture of human kidney. Experientia 33: 589-593.

Christensen, U. Kinetic studies of the urokinase-catalysed conversion of NH_2-terminal glutamic acid plasminogen to plasmin. Biochim. Biophys. Acta 481: 638-647.

Christman, J. K., Silverstein, S. C. & Acs, G. Plasminogen activators. In: Proteinases in Mammalian Cells and Tissues (Barrett, A. J. ed.), pp. 91-149, North-Holland Publishing Co., Amsterdam.

Claeson, G., Aurell, L., Karlsson, G. & Friberger, P. Substrate structure and activity relationship. In: New Methods for the Analysis of Coagulation using Chromogenic Substrates (Witt, I., ed.), pp.37-54, Walter de Gruyter, Berlin.

Huseby, R. M., Clavin, S. A., Smith, R. E., Hull, R. N. & Smithwick, E. L. Studies on tissue culture plasminogen activator. II. The detection and assay of urokinase and plasminogen activator from LLC-PK_1 cultures (porcine) by the synthetic substrate Nα-benzyloxycarbonyl-glycyl-glycyl-arginyl-4-methoxy-2-naphthylamide. Thromb. Res. 10: 679-687.

Morita, T., Kato, H., Iwanaga, S., Takada, K., Kimura, T. & Sakakibara, S. New fluorogenic substrates for α-thrombin, factor X, kallikreins and urokinase. J. Biochem. 82: 1495-1498.

Nieuwenhuizen, W., Wijngaards, G. & Groeneveld, E. Fluorogenic peptide amide substrates for the estimation of plasminogen activators and plasmin. Anal. Biochem. 83: 143-148.

Nieuwenhuizen, W., Wijngaards, G. & Groeneveld, E. Synthetic substrates and the discrimination between urokinase and tissue plasminogen activator activity. Thromb. Res. 11: 87-89.

Nolan, C., Hall, L. S., Barlow, G. H. & Tribby, I. I. E.

Plasminogen activator from human embryonic kidney cell cultures. Evidence for a proactivator. Biochim. Biophys. Acta 496: 384-400.

Ong, E. B., Soberano, M. E., Johnson, A. J. & Schoellmann, G. Studies on the biochemistry of urokinase. Thromb. Haemostas. 38: 801-808.

Svendsen, L. Estimation of urokinase activity by means of a highly susceptible synthetic chromogenic peptide substrate. In: New Methods for the Analysis of Coagulation using Chromogenic Substrates (Witt I., ed.), pp. 251-259, Walter de Gruyter, Berlin.

1978

Bigbee, W. L., Weintraub, H. B. & Jensen, R. H. Sensitive fluorescence assays for urokinase using synthetic peptide 4-methoxy-β-naphthylamide substrates. Anal. Biochem. 88: 114-122.

Friedman, R. B., Kwaan, H. C. & Szczecinski, M. Improved sensitivity of chromogenic substrate assays for urokinase and plasmin. Thromb. Res. 12: 37-46.

Holmberg, L., Lecander, I., Persson, B. & Åstedt, B. An inhibitor from placenta specifically binds urokinase and inhibits plasminogen activator released from ovarian carcinoma in tissue culture. Biochim. Biophys. Acta 544: 128-137.

Nieuwenhuizen, W., Wijngaards, G. & Groeneveld, E. Fluorogenic substrates for sensitive and differential estimation of urokinase and tissue plasminogen activator. Haemostasis 7: 146-149.

Zimmerman, M., Quigley, J. P., Ashe, B., Dorn, C., Goldfarb, R. & Troll, W. Direct fluorescent assay of urokinase and plasminogen activators of normal and malignant cells: kinetics and inhibitor profiles. Proc. Natl. Acad. Sci. U.S.A. 75: 750-753.

1979

Åstedt, B. No crossreaction between circulating plasminogen activator and urokinase. Thromb. Res. 14: 535-539.

Åstedt, B., Holmberg, L., Wagner, G., Richter, P. & Ploug, J. Purification of urokinase by a beta-naphthamidine affinity column. Thromb. Haemost. 42: 924-928.

Coleman, P., Kettner, C. & Shaw, E. Inactivation of the plasminogen activator from HeLa cells by peptides of arginine chloromethyl ketone. Biochim. Biophys. Acta 569: 41-51.

Kettner, C. & Shaw, E. The susceptibility of urokinase to affinity labeling by peptides of arginine chloromethyl ketone. Biochim. Biophys. Acta 569: 31-40.

Lewis, L. J. Plasminogen activator (urokinase) from cultured cells. Thromb. Haemost. 42: 895-900.

Paul, D. C., Bobbitt, J. L., Williams, D. C. & Hull, R. N. Immunocytochemical localization of plasminogen activator on porcine kidney cell strain: LLC-PK$_1$ (LP$_{100}$). J. Histochem. Cytochem. 27: 1035-1040.

Shakespeare, M. & Wolf, P. The demonstration of urokinase antigen in whole blood. Thromb. Res. 14: 825-835.

Entry 1.25

COMPLEMENT COMPONENT C̄1r

Summary

EC Number: (3.4.21.-)

Earlier names: -

Distribution: The inactive precursor, C1r, occurs in plasma of man
(about 100 μg/ml) and other mammals, largely as the C1 complex
(C1r,C1s,C1q). C̄1r appears when the complement system is
activated by the "classical" route, i.e. following the
interaction of C1q with immune complexes.

Source: Activated plasma.

Action: Activates C1s, probably by a single peptide bond cleavage.
Does not activate C1r except in the presence of C1s (or even Dip-
C̄1s) and Ca²⁺. There is controversy as to whether pure C̄1r has
esterolytic activity [1], but hydrolysis of Ac-Arg-OMe, Ac-Gly-
Lys-OMe and Z-Lys-OPhNO₂ has been reported [2].

Requirements: Neutral pH.

Substrate, usual: C1s, often in a hemolytic, complement-activation
assay.

Inhibitors: Dip-F, Pms-F, Tos-Lys-CH₂Cl, 6-aminohexanoic acid,
polyanetholesulfonate. C̄1-inhibitor. (Inhibition by C1-inhibitor
has been questioned, and leupeptin, hirudin and aprotinin found
not inhibitory [2].)

Molecular properties: The proenzyme, C1r, is labile at 56°C; it
has a mol.wt. of about 170,000, being composed of two identical,
non-covalently linked chains (mol.wt. 83,000), each cleaved to
fragments of 27,000 (a chain) and 56,000 mol.wt. (b chain) in
the (autocatalytic?) activation. No peptide is released, but the
pI rises. The light (a) chains are labelled by radioactive Dip-
F. At concentrations above 0.5 μM, C̄1r exists in an aggregated
form. C̄1r has A280,1% 11.7.

Comment: Formation of C̄1r, which acts early in the classical
pathway of complement activation, is inhibited by Ca²⁺,
polyanetholesulfonate, C̄1 inhibitor, Dip-F, Pms-F and Gbz-
OPhNO₂. C1r is activated by C1q after the latter has undergone
a conformational change that follows adsorption of the C1
complex to antigen-antibody complexes.

References
[1] Sim et al. Biochem. J. 163: 219-227, 1977.
[2] Andrews & Baillie J. Immunol. 123: 1403-1408, 1979.

Bibliography

1964

Haines, A. L. & Lepow, I. H. Studies on human C'1-esterase. I.
Purification and enzymatic properties. J. Immunol. 92: 456-467.
Haines, A. L. & Lepow, I. H. Studies on human C'1-esterase. II.
Function of purified C'1-esterase in the human complement
system. J. Immunol. 92: 468-478.
Naff, G. B., Pensky, J. & Lepow, I. N. The macromolecular nature
of the first component of human complement. J. Exp. Med. 119:
593-613.

1968

Naff, G. B. & Ratnof, O. D. The enzymatic nature of C'1r.
Conversion of C'1s to C'1 esterase and digestion of amino acid
esters by C'1r. J. Exp. Med. 128: 571-593.

1969

Ratnoff, O. D., Pensky, J., Ogston, D. & Naff, G. B. The
inhibition of plasmin, plasma kallikrein, plasma permeability
factor, and the C'1r subcomponent of the first component of
complement by serum C'1 esterase inhibitor. J. Exp. Med. 129:
315-331.

1974

Sakai, K. & Stroud, R. M. The activation of C1s with purified
C1r. Immunochemistry 11: 191-196.
Valet, G. & Cooper, N. R. Isolation and characterization of the
proenzyme form of the C1r subunit of the first complement
component. J. Immunol. 112: 1667-1673.

1975

Takahashi, K., Nagasawa, S. & Koyama, J. A gross structure of an
activated form of a subunit of the first component of human
complement, C1r. FEBS Lett. 55: 156-160.

1976

Cooper, N. R. & Ziccardi, R. J. The nature and reactions of
complement enzymes. In: Proteolysis and Physiological
Regulation (Ribbons, D. W. & Brew, K. eds), pp. 167-187,
Academic Press, New York.
Gigli, I., Porter, R. R. & Sim, R. B. The unactivated form of the
first component of human complement, C1. Biochem. J. 157: 541-
548.
Ziccardi, R. J. & Cooper, N. R. Activation of C1r by proteolytic
cleavage. J. Immunol. 116: 504-509.
Ziccardi, R. J. & Cooper, N. R. Physicochemical and functional

characterization of the Clr subunit of the first complement
component. J. Immunol. 116: 496-503.

1977

Porter, R. R. Structure and activation of the early components of
complement. Fed. Proc. Fed. Am. Soc. Exp. Biol. 36: 2191-2196.
Porter, R. R. The biochemistry of complement. Biochem. Soc.
Trans. 5: 1659-1674.
Reboul, A., Arlaud, G. J., Sim, R. B. & Colomb, M. G. A
simplified procedure for the purification of Cl-inactivator
from human plasma. Interaction with complement subcomponents Clr
and Cls. FEBS Lett. 79: 45-50.
Sim, R. B., Porter, R. R., Reid, K. B. M. & Gigli, I. The
structure and enzymic activities of the Clr and Cls subcomponents
of Cl, the first component of human serum complement. Biochem.
J. 163: 219-227.
Tamura, Y., Hirado, M., Okamura, K., Minato, Y. & Fujii, S.
Synthetic inhibitors of trypsin, plasmin, thrombin, C_1r and C_1
esterase. Biochim. Biophys. Acta 484: 417-422.

1978

Dodds, A. W., Sim, R. B., Porter, R. R. & Kerr, M. A. Activation
of the first component of human complement (Cl) by antibody-
antigen aggregates. Biochem. J. 175: 383-390.
Okamura, K. & Fujii, S. Isolation and characterization of
different forms of Clr, a subcomponent of the first component
of human complement. Biochim. Biophys. Acta 534: 258-266.
Ziccardi, R. J. & Cooper, N. R. Modulation of the antigenicity of
Clr and Cls by Cl inactivator. J. Immunol. 121: 2148-2152.

1979

Andrews, J. M. & Baillie, R. D. The enzymatic nature of human
Clr: a subcomponent of the first component of complement. J.
Immunol. 123: 1403-1408.
Arland, G. J., Reboul, A., Sim, R. B. & Colomb, M. G. Interaction
of Cl-inhibitor with the Clr and Cls subcomponents in human
Cl. Biochim. Biophys. Acta 576: 151-162.
Esser, A. F., Batholomew, R. M., Parce, J. W. & McConnell, M. The
physical state of membrane lipids modulates the activation of
the first component of complement. J. Biol. Chem. 254: 1768-
1770.
Lachmann, P. J. Complement. In: The Antigens (Sela, M. ed.),
vol. 5, pp. 284-353, Academic Press, New York.
Mori, Y., Koketsu, M., Abe, N. & Koyama, J. Purification and some
properties of rabbit Clr. J. Biochem. 85: 1023-1028.
Sim, R. B., Arlaud, G. J. & Colomb, M. G. Cl Inhibitor-dependent
dissociation of human complement component Cl bound to immune
complexes. Biochem. J. 179: 449-457.

Entry 1.26

COMPLEMENT COMPONENT C̄1s

Summary

EC Number: (3.4.21.-)

Earlier names: C1 esterase.

Distribution: The inactive precursor, C1s, occurs in plasma of man (about 100 µg/ml) and other mammals, largely as the C1 complex (C1r,C1s,C1q).

Source: Activation of precursor from plasma.

Action: Activates C4 by cleavage of the α chain. C4b then binds C2 which is also cleaved by C̄1s, leaving C2a bound in the C4b,2a complex, the "classical" C3 convertase (#1.27). Also hydrolyzes N-blocked esters of e.g. tyrosine and lysine, so has catalytic properties of both trypsin and chymotrypsin [1].

Requirements: Neutral pH.

Substrate, usual: Complement component C4 (destruction of hemolytic activity being followed).

Substrate, special: Boc-Lys-OPhNO$_2$ more sensitive than Boc-Tyr-OPhNO$_2$.

Inhibitors: Dip-F, benzamidine; moderately sensitive to leupeptin, but not chymostatin. C̄1-inhibitor, many other inhibitors of trypsin or chymotrypsin (but not soybean trypsin inhibitor).

Molecular properties: The precursor, C1s, is stable at 56°C; it has a mol.wt. of 83,000, and may form a dimer in the presence of Ca^{2+}. Activation by C̄1r is by cleavage of the single chain to disulfide-linked subunits of mol.wt. 27,000 (a chain) and 56,000 (b chain). The active site serine is in the light chain. Trypsin and plasmin also activate, in vitro. C̄1r has A280,1% 9.4.

Comment: C1s closely resembles the monomer of C1r in mol.wt., amino acid composition, mol.wts of cleavage products, etc. so the two proteins may well be homologous.

References
[1] Sim et al. Biochem. J. 163: 219-227, 1977.

Bibliography

1956

Becker, E. L. Concerning the mechanism of complement action. 1. Inhibition of complement activity by diisopropyl fluorophosphate. J. Immunol. 77: 462-468.

Lepow, I. H., Ratnoff, O. D., Rosen, F. S. & Pillmer, L. Observations on a pro-esterase associated with partially purified first component of human complement (C'l). Proc. Soc. Exp. Biol. Med. 92: 32-37.

1964

Haines, A. L. & Lepow, I. H. Studies on human C'l-esterase. I. Purification and enzymatic properties. J. Immunol. 92: 456-467.

Haines, A. L. & Lepow, I. H. Studies on human C'l-esterase. II. Function of purified C'l-esterase in the human complement system. J. Immunol. 92: 468-478.

1968

Naff, G. B. & Ratnof, O. D. The enzymatic nature of C'lr. Conversion of C'ls to C'l esterase and digestion of amino acid esters by C'lr. J. Exp. Med. 128: 571-593.

1973

Barkas, T., Scott, G. K. & Fothergill, J. E. Purification, characterization and active-site studies on human serum complement subcomponent Cls. Biochem. Soc. Trans. 1: 1219-1220.

Sakai, K. & Stroud, R. M. Purification, molecular properties, and activation of Cl proesterase, Cls. J. Immunol. 110: 1010-1020.

1974

Bing, D. H., Cory, M. & Doll, M. The inactivation of human $\overline{Cl}$ by benzamidine and pyridinium sulfonylfluorides. J. Immunol. 113: 584-590.

Sakai, K. & Stroud, R. M. The activation of Cls with purified Clr. Immunochemistry 11: 191-196.

Valet, G. & Cooper, N. R. Isolation and characterization of the proenzyme form of the Cls subunit of the first complement component. J. Immunol. 112: 339-350.

1975

Fearon, D. T. & Austen, K. F. Inhibition of complement-derived enzymes. Ann. N. Y. Acad. Sci. 256: 441-450.

Müller-Eberhard, H. J. Complement. Annu. Rev. Biochem. 44: 697-724.

Scott, G. K. & Fothergill, J. E. Proteolytic specificity of human complement subcomponent Cls. Biochem. Soc. Trans. 3: 934-935.

Sumi, H., Izumiya, N. & Muramatu, N. A simple rapid separation of Cl-esterase using an immunoadsorbent column. J. Biochem. 78: 481-484.

Takahashi, K., Nagasawa, S. & Koyama, J. The NH_2-terminal sequences of a subunit of the first component of human complement, C1s, and its activated form, C1s. FEBS Lett. 50: 330-333.

1976

Cooper, N. R. & Ziccardi, R. J. The nature and reactions of complement enzymes. In: Proteolysis and Physiological Regulation (Ribbons, D. W. & Brew, K. eds), pp. 167-187, Academic Press, New York.

Gigli, I., Porter, R. R. & Sim, R. B. The unactivated form of the first component of human complement, C1. Biochem. J. 157: 541-548.

Ishizagi, E., Yoshioka, Y., Mori, Y. & Koyama, J. Isolation of two forms of activated C1s, a subcomponent of the first component of rabbit complement. J. Biochem. 80: 1423-1427.

Müller-Eberhard, H. The serum complement system. In: Textbook of Immunopathology (Miescher, P. A. & Müller-Eberhard, H. J. eds), pp. 45-73, Grune & Stratton, New York.

Ogston, D., Murray, J. & Crawford, G. P. M. Inhibition of the activated C1s subunit of the first component of complement by antithrombin III in the presence of heparin. Thromb. Res. 9: 217-222.

Takashashi, K., Tamoto, K. & Koyama, J. Effects of protease inhibitors of actinomycetes on the first component of human complement. J. Antibiot. 24: 983-985.

1977

Andrews, J. M., Rosen, F. S., Silverberg, S. J., Cory, M., Schneeberger, E. E. & Bing, D. H. Inhibition of C1s-induced vascular leakage in guinea pigs by substituted benzamidine and pyridinium compounds. J. Immunol. 118: 466-471.

Arlaud, G. J., Reboul, A. & Colomb, M. G. Proenzymic C1s associated with catalytic amounts of C1r. Study of the activation process. Biochim. Biophys. Acta 485: 227-235.

Arlaud, G. J., Reboul, A., Meyer, C. M. & Colomb, M. G. Purification of proenzymic and activated human C1s free of C1r. Effect of calcium and ionic strength on activated C1s. Biochim. Biophys. Acta 485: 215-226.

Porter, R. R. Structure and activation of the early components of complement. Fed. Proc. Fed. Am. Soc. Exp. Biol. 36: 2191-2196.

Porter, R. R. The biochemistry of complement. Biochem. Soc. Trans. 5: 1659-1674.

Reboul, A., Arlaud, G. J., Sim, R. B. & Colomb, M. G. A simplified procedure for the purification of C1-inactivator from human plasma. Interaction with complement subcomponents C1r and C1s. FEBS Lett. 79: 45-50.

Sim, R. B., Porter, R. R., Reid, K. B. M. & Gigli, I. The structure and enzymic activities of the C1r and C1s subcomponents of C1, the first component of human serum complement. Biochem. J. 163: 219-227.

Tamura, Y., Hirado, M., Okamura, K., Minato, Y. & Fujii, S.

Synthetic inhibitors of trypsin, plasmin, thrombin, $C_1\bar{r}$ and C_1 esterase. Biochim. Biophys. Acta 484: 417-422.

1978

Andrews, J. M., Roman, D. P. & Bing, D. N. Inhibition of four human serine proteases by substituted benzamidines. J. Med. Chem. 21: 1202-1207.

Dodds, A. W., Sim, R. B., Porter, R. R. & Kerr, M. A. Activation of the first component of human complement (C1) by antibody-antigen aggregates. Biochem. J. 175: 383-390.

Kerr, M. A. & Porter, R. R. The purification and properties of the second component of human complement. Biochem. J. 171: 99-107.

Porter, R. R. & Reid, K. B. M. The biochemistry of complement. Nature 275: 699-704.

Ziccardi, R. J. & Cooper, N. R. Modulation of the antigenicity of C1r and C1s by C1 inactivator. J. Immunol. 121: 2148-2152.

1979

Arland, G. J., Reboul, A., Sim, R. B. & Colomb, M. G. Interaction of C1-inhibitor with the C1r and C1s subcomponents in human C1. Biochim. Biophys. Acta 576: 151-162.

Campbell, R. D., Booth, N. A. & Fothergill, J. E. The purification and characterization of subcomponent C1s of the first component of bovine complement. Biochem. J. 183: 579-588.

Lachmann, P. J. Complement. In: The Antigens (Sela, M. ed.), vol. 5, pp. 284-353, Academic Press, New York.

Laurell, A.-B., Martensson, U. & Sjöholm, A. G. Quantitation of C1r C1s-C1 inactivator complexes by electroimmunoassay. Acta Pathol. Microbiol. Scand. C 87: 79-81.

Reboul, A., Thielens, N., Villiers, M. B. & Colomb, M. G. Purification of human complement subcomponent C4. C4 cleavage by C1s. FEBS Lett. 103: 156-161.

Sim, R. B., Arlaud, G. J. & Colomb, M. G. $C\bar{1}$ Inhibitor-dependent dissociation of human complement component $C\bar{1}$ bound to immune complexes. Biochem. J. 179: 449-457.

1980

Takahashi, K., Nagasawa, S. & Koyama, J. Effect of $C\bar{1}$-inhibitor, benzamidine, and heat treatment on the hydrolytic activity of a subcomponent of the first component of human complement, $C\bar{1}s$. Biochim. Biophys. Acta 611: 196-204.

COMPLEMENT COMPONENT C3 CONVERTASE (CLASSICAL)

Summary

EC Number: (3.4.21.-)

Earlier names: Abbreviation: $\overline{C42}$.

Distribution: The precursor components occur in plasma of man and other mammals.

Source: Generated from C4 and C2 by the action of $\overline{C1s}$ alone or in $\overline{C1}$.

Action: Cleaves C3 at the carboxyl side of $-Arg_{77}-$ in the α chain to yield C3a (mol.wt. 8900) and C3b (mol.wt. 171,000).

Requirements: Neutral pH. Mg^{2+}.

Substrate, usual: C3 (the only known protein substrate for $\overline{C42}$) in a hemolytic complement activation assay.

Substrate, special: Ac-Gly-Lys-OMe.

Inhibitors: Ac-Gly-Lys-OMe competitively inhibits the cleavage of C3.

Molecular properties: A bi-molecular complex of the larger fragments of C4 (C4b) and C2 (C2b), mol.wt. about 280,000, which is formed in the presence of Mg^{2+}. Catalytic site probably on the C2b, with C4b modulating specificity.

Comment: The C2b component exhibits proteinase (C3-cleaving) activity when in the C4b,2b complex, whereas isolated C2b has only esterase activity. C3 convertase has a half-life of only 5 min at 37°C. Stable preparations of the C3 and C5 convertases can be prepared by use of C2 oxidized with iodine.

It is through the C4b component that C3 convertase binds to membranes and becomes cytolytic, but in cell-free solution the enzyme can still activate C3. The C3b fragment is able to bind to membranes and immune complexes, or associate with its activating enzyme (C3 convertase) to generate C5 convertase (#1.28).

Human complement components C2 and factor B are genetically linked to the HLA-B locus, and are almost certainly homologous with each other.

Bibliography

1966

Stroud, R. M., Mayer, M. M., Miller, J. A. & McKenzie, A. T.
C'2a^d, an inactive derivative of C'2 released during decay of
EAC'4,2a. Immunochemistry 3: 163-176.

1967

Müller-Eberhard, H. J., Polley, M. J. & Calcott, M. A. Formation
and functional significance of a molecular complex derived from
the second and the fourth component of human complement. J.
Exp. Med. 125: 359-380.
Polley, M. J. & Müller-Eberhard, H. J. Enhancement of the
hemolytic activity of the second component of human complement
by oxidation. J. Exp. Med. 126: 1013-1025.

1975

Cooper, N. R. Enzymatic activity of the second component of
complement. Biochemistry 14: 4245-4251.
Fearon, D. T. & Austen, K. F. Inhibition of complement-derived
enzymes. Ann. N. Y. Acad. Sci. 256: 441-450.
Hugli, T. E. Human anaphylatoxin (C3a) from the third component
of complement. Primary structure. J. Biol. Chem. 250: 8293-
8301.
Müller-Eberhard, H. J. Initiation of membrane attack by
complement: assembly and control of C3 and C5 convertase. In:
Proteases and Biological Control (Reich, E., Rifkin, D. B. &
Shaw, E. eds), pp. 229-241, Cold Spring Harbor Laboratory, New
York.

1976

Medicus, R. G., Götze, O. & Müller-Eberhard, H. J. The serine
protease nature of the C3 and C5 convertases of the classical
and alternative complement pathways. Scand. J. Immunol. 5:
1049-1055.

1977

Bolotin, C., Morris, S., Tack, B. & Prahl, J. Purification and
structural analysis of the fourth component of human complement.
Biochemistry 16: 2008-2015.
Nagasawa, S. & Stroud, R. M. Purification and limited proteolysis
of C2. Fed. Proc. Fed. Am. Soc. Exp. Biol. 36: 1208 only.
Porter, R. R. Structure and activation of the early components of
complement. Fed. Proc. Fed. Am. Soc. Exp. Biol. 36: 2191-2196.

1978

Kerr, M. A. & Porter, R. R. The purification and properties of
the second component of human complement. Biochem. J. 171: 99-
107.
Lachmann, P. J. & Hobart. M. J. Complement technology. In:
Handbook of Experimental Immunology (Weir, D. M. ed.), 3rd edn.,

pp. 5A.1-5A.23, Blackwell, London.
Porter, R. R. & Reid, K. B. M.   The biochemistry of complement.
   Nature 275: 699-704.

1979

Campbell, R. D., Booth, N. A. & Fothergill, J. E.   The purification
   and characterization of subcomponent C1s of the first component
   of bovine complement.   Biochem. J. 183: 579-588.
Lachmann, P. J. Complement.   In: The Antigens (Sela, M. ed.),
   vol. 5, pp. 284-353, Academic Press, New York.
Strunk, R. C. & Giclas, P. C.   Control of human classical
   complement pathway C3 convertase: inhibition of C3 cleavage by
   C5 or cell-bound C3b.   Fed. Proc. Fed. Am. Soc. Exp. Biol. 38:
   1011 only.

Entry 1.28

COMPLEMENT COMPONENT C5 CONVERTASE

Summary

EC Number: (3.4.21.-)

Earlier names: Abbreviation: C423.

Distribution: Precursors occur in plasma of man and other mammals.

Source: Activated precursors, which can assemble on cell membranes or free in solution.

Action: Cleaves C5 at an arginyl bond in the larger (α) chain, to liberate C5a (15,000 mol.wt.) and C5b (160,000 mol.wt.), in complement activation by the classical pathway.

Requirements: Neutral pH. Mg²⁺.

Substrate, usual: C5 (the only known protein substrate for C423) in a hemolytic, complement-activation assay.

Inhibitors: -

Molecular properties: The C5 convertase comprises C4b,2b acting in conjunction with C3b, not necessarily in a stable complex.

Comment: As with C3 convertase, the catalytic site seems to be located on the C2b component. It seems that C3b modifies the specificity of C42 by binding C5. C5 convertase is a labile enzyme, having a half-life of only 10 min at 37°C. As for C3 convertase, the instability is attributed to the dissociation of C2b from the complex. Stable preparations of the C3 and C5 convertases can be prepared by use of C2 oxidized with iodine.
The generation of the C5 convertase represents the completion of the "triggered enzyme cascade" of complement activation. The remaining steps, which comprise the "terminal pathway" and involve components with no known catalytic activity (C5, C6, C7, C8 and C9), constitute the membrane attack mechanism.

Bibliography

1967

Müller-Eberhard, H. J., Polley, M. J. & Calcott, M. A. Formation and functional significance of a molecular complex derived from the second and the fourth component of human complement. J. Exp. Med. 125: 359-380.

1975

Müller-Eberhard, H. J.   Initiation of membrane attack by complement: assembly and control of C3 and C5 convertase.   In: Proteases and Biological Control (Reich, E., Rifkin, D. B. & Shaw, E. eds), pp. 229-241, Cold Spring Harbor Laboratory, New York.

1976

Medicus, R. G., Götze, O. & Müller-Eberhard, H. J.   The serine protease nature of the C3 and C5 convertases of the classical and alternative complement pathways.   Scand. J. Immunol. 5: 1049-1055.

1978

Lachmann, P. J. & Hobart. M. J.   Complement technology.   In: Handbook of Experimental Immunology (Weir, D. M. ed.), 3rd edn., pp. 5A.1-5A.23, Blackwell, London.
Porter, R. R. & Reid, K. B. M.   The biochemistry of complement. Nature 275: 699-704.

1979

Gerard, C. & Hugli, T. E.   Anaphylatoxin from the fifth component of porcine complement. Purification and partial chemical characterization.   J. Biol. Chem. 254: 6346-6351.
Goldberger, G., Brenner, M. J. & Colten, H. R.   Kinetics of synthesis and secretion of a precursor of the fourth component of complement (Pro-C4).   Fed. Proc. Fed. Am. Soc. Exp. Biol. 38: 1011 only.
Lachmann, P. J.   Complement.   In: The Antigens (Sela, M. ed.), vol. 5, pp. 284-353, Academic Press, New York.
Matthews, W. J., Jr., Marino, J. T., Jr., Goldberger, G., Gash, D. & Colten, H. R.   Feedback inhibition of the biosynthesis of the fourth component of complement (C4).   Fed. Proc. Fed. Am. Soc. Exp. Biol. 38: 1011 only.
Reboul, A., Thielens, N., Villiers, M. B. & Colomb, M. G.   Purification of human complement subcomponent C4. C4 cleavage by C1s.   FEBS Lett. 103: 156-161.
Tack, B. F., Morris, S. C. & Prahl, J. W.   Third component of human complement: structural analysis of the polypeptide chains of C3 and C3b.   Biochemistry 18: 1497-1503.
Villa, M. L., Porta, C., Masserini, C., Lodola, M. A., Clerici, E. & Vaglini, M.   Complement proteins C3 and C4, immunoadherence and hemolysis in malignant melanoma.   Fed. Proc. Fed. Am. Soc. Exp. Biol. 38: 1292 only.

Entry 1.29

COMPLEMENT FACTOR D

Summary

EC Number: (3.4.21.-)

Earlier names: C3 proactivator convertase (C3PAse); factor $\overline{D}$.

Distribution: Occurs in human plasma (about 1 μg/ml), apparently already in the active form.

Source: Plasma.

Action: Acts on complement factor B in C3b,B,Mg^{2+}-complex to generate Ba (30,000 mol.wt.) and Bb (62,000 mol.wt.) or $\overline{C3b,Bb}$, probably by cleavage of an -Arg-Lys- bond [1]. Also acts on the cobra venom factor (CVF)-factor B complex, but has no action on uncomplexed factor B.

The substrate specificity of factor D on peptide esters is distinct from those of $\overline{C1r}$, $\overline{C1s}$, plasmin, urokinase, trypsin and chymotrypsin. Characteristically, arginine esters are prefered to those of lysine and tyrosine.

Requirements: Neutral pH; presence of C3b and Mg^{2+} for formation of substrate complex.

Substrate, usual: Suitably complexed factor B, in a hemolytic, complement-activation assay.

Substrate, special: Bz-Ile-Glu-Gly-Arg-NPhNO$_2$ (S-2222) (weak activity) [2].

Inhibitors: Dip-F (which inhibits both esterolytic and hemolytic activities).

Molecular properties: Mol.wt. about 25,000, a single chain with N-terminal isoleucine.

Comment: A component of the alternative pathway of complement activation, but possibly homologous with the catalytically active b chains of $\overline{C1r}$ and $\overline{C1s}$. Cleavage of factor B in the C3b,B complex is clearly analogous to the cleavage by $\overline{C1s}$ of C2 in C4b,2, in the classical pathway, and yields the alternative C3/C5 convertase.

References
[1] Lesavre et al. J. Immunol. 123: 529-534, 1979.
[2] Davis et al. Biochemistry 18: 5082-5087, 1979.

Bibliography

1971

Götze, O. & Muller-Eberhard, H. J. The C3-activator system: an alternate pathway of complement activation. J. Exp. Med. 134: 90s-108s.

1972

Müller-Eberhard, H. J. & Götze, O. C3 proactivator convertase and its mode of action. J. Exp. Med. 135: 1003-1008.

1973

Cooper, N. R. Formation and function of a complex of the C3 proactivator with a protein from cobra venom. J. Exp. Med. 137: 451-460.

Fearon, D. T., Austen, K. F. & Ruddy, S. Formation of a hemolytically active cellular intermediate by the interaction between properdin factor B and D and the activated third component of complement. J. Exp. Med. 138: 1305-1313.

Hunsicker, L. G., Ruddy, S. & Austen, K. F. Alternate complement pathway: factors involved in cobra venom factor (CoVF) activation of the third component of complement (C3). J. Immunol. 110: 128-138.

1974

Brade, V., Nicholson, A., Bitter-Suermann, D. & Hadding, U. Formation of the C3-cleaving properdin enzyme on zymosan. Demonstration that factor D is replaceable by proteolytic enzymes. J. Immunol. 113: 1735-1743.

Fearon, D. T., Austen, K. F. & Ruddy, S. Properdin factor D. II. Activation to D̄ by properdin. J. Exp. Med. 140: 426-436.

Fearon, D. T., Austen, K. F. & Ruddy, S. Properdin factor D: characterization of its active site and isolation of the precursor form. J. Exp. Med. 139: 355-366.

1975

Fearon, D. T. & Austen, K. F. Properdin: initiation of alternative complement pathway. Proc. Natl. Acad. Sci. U.S.A. 72: 3220-3224.

Götze, O. Proteases of the properdin system. In: Proteases and Biological Control (Reich, E., Rifkin, D. B. & Shaw, E. eds), pp. 255-272, Cold Spring Harbor Laboratory, New York.

Schreiber, R. D., Medicus, R. G., Götze, O. & Muller-Eberhard, H. J. Properdin- and nephritic factor-dependent C3 convertases: requirement of native C3 for enzyme formation and the function of bound C3b as properdin receptor. J. Exp. Med. 142: 760-772.

Vogt, W., Schmidt, G., Dieminger, L. & Lynen, R. Formation and composition of the C3 activating enzyme complex of the properdin system. Sequential assembly of its components on solid-phase trypsin-agarose. Z. Immun.-Forsch. 149: 440-455.

1976

Cooper, N. R. & Ziccardi, R. J. The nature and reactions of complement enzymes. In: Proteolysis and Physiological Regulation (Ribbons, D. W. & Brew, K., eds), pp. 167-187, Academic Press, New York.

Martin, A., Lachmann, P. J., Halbwachs, L. & Hobart, M. J. Haemolytic diffusion plate assays for factors B and D of the alternative pathway of complement activation. Immunochemistry 13: 317-324.

Schreiber, R. D., Götze, O. & Müller-Eberhard, H. J. Alternative pathway of complement: demonstration and characterization of initiating factor and its properdin-independent function. J. Exp. Med. 144: 1062-1093.

Volanikis, J. E., Schultz, D. R. & Stroud, R. M. Evidence that C1s participates in the alternative complement pathway. Int. Arch. Allergy Appl. Immunol. 50: 68-80.

1978

Lachmann, P. J. & Hobart. M. J. Complement technology. In: Handbook of Experimental Immunology (Weir, D. M. ed.), 3rd edn., pp. 5A.1-5A.23, Blackwell, London.

Lesavre, P. H. & Muller-Eberhard, H. J. Mechanism of action of factor D of the alternative complement pathway. J. Exp. Med. 148: 1498-1509.

Pangburn, M. K. & Müller-Eberhard, H. J. Complement C3 convertase: cell surface restriction of β1H control and generation of restriction on neuraminidase-treated cells. Proc. Natl. Acad. Sci. U.S.A. 75: 2416-2420.

Schreiber, R. D., Pangburn, M. K., Lesavre, P. H. & Müller-Eberhard, H. J. Initiation of the alternative pathway of complement: recognition of activators by bound C3b and assembly of the entire pathway from six isolated proteins. Proc. Natl. Acad. Sci. U.S.A. 75: 3948-3952.

1979

Davis, A. E., III., Zalut, C., Rosen, F. S. & Alper, C. A. Human factor D of the alternative complement pathway. Physicochemical characteristics and N-terminal amino acid sequence. Biochemistry 18: 5082-5087.

Lachmann, P. J. Complement. In: The Antigens (Sela, M. ed.), vol. 5, pp. 284-353, Academic Press, New York.

Lesavre, P. H., Hugli, T. E., Esser, A. F. & Müller-Eberhard, H. J. The alternative pathway of C3/C5 convertase: chemical basis of factor B activation. J. Immunol. 123: 529-534.

COMPLEMENT COMPONENT C3/C5 CONVERTASE (ALTERNATIVE)

Summary

EC Number: (3.4.21.-)

Earlier names: C3 proactivator, glycine-rich β-glycoprotein (equivalent to properdin factor B), glycine-rich γ-glycoprotein (equivalent to factor Bb), heat-labile factor. Abbreviation: C3b,Bb.

Distribution: Plasma (200 µg factor B/ml in man). C3 and factor B are synthesized by macrophages and perhaps other cells.

Source: Plasma.

Action: Cleaves the α chain of C3, a 185,000 mol.wt. β-globulin once called factor A, to yield C3a and C3b. The split occurs on the carboxyl side of $-Arg_{77}-$ (which becomes the essential C-terminal residue of C3a). Uncomplexed factor B has esterase activity against Ac-Gly-Lys-OMe, Z-Lys-OMe and Ac-Lys-OMe, but no detected proteolytic (C3-cleaving) activity.

Complement C5 also is cleaved at the same point as by the classical C5 convertase (see Comments).

Requirements: Acts at neutral pH. Mg^{2+} is required for complexing of C3b with Bb.

Substrate, usual: C3, in a complement activation assay.

Substrate, special: Ac-Gly-Lys-OMe.

Inhibitors: Dip-F [1].

Molecular properties: This enzyme is C3b,Bb. Factor B, mol.wt. 93,000, is a single chain β-globulin. It is destroyed in serum by heating at 50°C for 30 min, and was recognized early as a heat-labile factor in the complement system. Factor B is greatly stabilized when complexed with cobra venom factor (a snake analogue of C3b) to yield CVF,B, a C3/C5 convertase unaffected by C3b-inactivator (#1.31).

Factor D, and some trypsin-like enzymes, cleave factor B to Bb (63,000) plus Ba (30,000). Component Bb has the catalytic site in the active C3b,Bb complex, which is stabilized by properdin, a 223,000 mol.wt. γ-globulin. Instability in the absence of properdin is due to dissociation of factor Bb, in inactive form, from C3b, which is able to bind more factor B.

Comment: The role of factor B (as Bb) in the C3/C5 convertase of the alternative pathway is strongly analogous with that of C2 (as C2a) in the classical pathway. Both enzymes cleave esters, but peptide bonds only in association with other components (normally C3b).

The alternative C5 convertase may be a complex containing extra C3b (i.e. $C3b_n,Bb$), with the additional C3b conferring C5 convertase activity as it does in the classical $C4b,2a,3b$.

Human C2 and factor B are both genetically linked to the HLA-B locus; they are thought to be derived from a single ancestral protein, and to be structurally homologous.

References
[1] Medicus et al. Scand J. Immunol. 5: 1049–1055, 1976.

Bibliography

1968

Pensky, J., Hinz, C. F., Todd, E. W., Wedgewood, R. J., Boyer, J. T. & Lepow, I. H. Properties of highly purified human properdin. J. Immunol. 100: 142–158.

1970

Boenisch, T. & Alper, C. A. Isolation and properties of a glycine-rich γ-glycoprotein of human serum. Biochim. Biophys. Acta 214: 135–140.
Boenisch, T. & Alper, C. A. Isolation and properties of a glycine-rich β-glycoprotein of human serum. Biochim. Biophys. Acta 221: 529–535.

1971

Götze, O. & Muller-Eberhard, H. J. The C3-activator system: an alternate pathway of complement activation. J. Exp. Med. 134: 90s–108s.
Goodkofsky, I. & Lepow, I. H. Functional relationship of factor B in the properdin system to C3 proactivator of human serum. J. Immunol. 107: 1200–1204.
Müller-Eberhard, H. J. & Fjellström, K.-E. Isolation of the anticomplementary protein from cobra venom and its mode of action. J. Immunol. 107: 1666–1672.

1972

Muller-Eberhard, H. J. & Götze, O. C3 proactivator convertase and its mode of action. J. Exp. Med. 135: 1003–1008.

1973

Alper, C. A., Goodkofsky, I. & Lepow, I. H. The relationship of glycine-rich β-glycoprotein to factor B in the properdin system and to the cobra factor-binding protein of human serum. J. Exp. Med. 137: 424–437.
Brade, V., Lee, G. D., Nicholson, A., Shin, H. S. & Mayer, M. M. The reaction of zymosan with the properdin system in normal and

C4-deficient guinea pig serum: demonstration of C3- and C5-cleaving multi-unit enzymes, both containing factor B, and accelleration of their formation by the classical complement pathway. J. Immunol. 111: 1389-1400.

Cooper, N. R. Formation and function of a complex of the C3 proactivator with a protein from cobra venom. J. Exp. Med. 137: 451-460.

Fearon, D. T., Austen, K. F. & Ruddy, S. Formation of a hemolytically active cellular intermediate by the interaction between properdin factor B and D and the activated third component of complement. J. Exp. Med. 138: 1305-1313.

1974

Brade, V., Nicholson, A., Bitter-Suermann, D. & Hadding, U. Formation of the C3-cleaving properdin enzyme on zymosan. Demonstration that factor D is replaceable by proteolytic enzymes. J. Immunol. 113: 1735-1743.

Minta, J. O. & Lepow, I. H. Studies on the sub-unit structure of human properdin. Immunochemistry 11: 361-368.

Vogt, W., Dieminger, L., Lynen, R. & Schmidt, G. Formation and composition of the complex with cobra venom factor that cleaves the third component of complement. Hoppe-Seyler's Z. Physiol. Chem. 355: 171-183.

1975

Fearon, D. T. & Austen, K. F. Properdin: binding to C3b and stabilization of the C3b-dependent C3 convertase. J. Exp. Med. 142: 856-863.

Fearon, D. T. & Austen, K. F. Properdin: initiation of alternative complement pathway. Proc. Natl. Acad. Sci. U.S.A. 72: 3220-3224.

Götze, O. Proteases of the properdin system. In: Proteases and Biological Control (Reich, E., Rifkin, D. B. & Shaw, E. eds), pp. 255-272, Cold Spring Harbor Laboratory, New York.

Müller-Eberhard, H. J. Complement. Annu. Rev. Biochem. 44: 697-724.

Schreiber, R. D., Medicus, R. G., Götze, O. & Müller-Eberhard, H. J. Properdin and nephritic factor-dependent C3 convertases: requirement of native C3 for enzyme formation and the function of bound C3b as properdin receptor. J. Exp. Med. 142: 760-772.

Vogt, W., Schmidt, G., Dieminger, L. & Lynen, R. Formation and composition of the C3 activating enzyme complex of the properdin system. Sequential assembly of its components on solid-phase trypsin-agarose. Z. Immun.-Forsch. 149: 440-455.

1976

Bentley, C., Bitter-Suermann, D., Hadding, U. & Brade, V. In vitro synthesis of factor B of the alternative pathway of complement activation by mouse peritoneal macrophages. Eur. J. Immunol. 6: 393-398.

Chapitis, J. & Lepow, I. H. Multiple sedimenting species of

properdin in human serum and interaction of purified properdin with the third component of complement. J. Exp. Med. 143: 241-257.

Cooper, N. R. & Ziccardi, R. J. The nature and reactions of complement enzymes. In: Proteolysis and Physiological Regulation (Ribbons, D. W. & Brew, K. eds), pp. 167-187, Academic Press, New York.

Corbin, N. C. & Hugli, T. E. The primary structure of porcine C3a anaphylatoxin. J. Immunol. 117: 990-995.

Daha, M. R., Fearon, D. T. & Austen, K. F. C3 nephritic factor (C3NeF): stabilization of fluid phase and cell bound alternative pathway convertase. J. Immunol. 116: 1-7.

Daha, M. R., Fearon, D. T. & Austen, K. F. C3 requirements for formation of alternative pathway C5 convertase. J. Immunol. 117: 630-634.

Martin, A., Lachmann, P. J., Halbwachs, L. & Hobart, M. J. Haemolytic diffusion plate assays for factors B and D of the alternative pathway of complement activation. Immunochemistry 13: 317-324.

Medicus, R. G., Götze, O. & Müller-Eberhard, H. J. Alternative pathway of complement: recruitment of precursor properdin by the labile C3/C5 convertase and the potentiation of the pathway. J. Exp. Med. 144: 1076-1093.

Medicus, R. G., Götze, O. & Müller-Eberhard, H. J. The serine protease nature of the C3 and C5 convertases of the classical and alternative complement pathways. Scand. J. Immunol. 5: 1049-1055.

Medicus, R. G., Schreiber, R. D., Götze, O. & Müller-Eberhard, H. J. A molecular concept of the properdin pathway. Proc. Natl. Acad. Sci. U.S.A. 73: 612-616.

Vogt, W., Lynen, R., Schmidt, G. & Dieminger, L. Inefficient utilization of the fifth component of guinea pig complement in cobra venom factor-induced complement lysis; efficient utilization of human C5. J. Immunol. 116: 1753 only.

1977

Fearon, D. T. & Austen, K. F. Activation of the alternative complement pathway with rabbit erythrocytes by circumvention of the regulatory action of endogenous control proteins. J. Exp. Med. 146: 22-33.

Vogt, W., Dames, W., Schmidt, G. & Dieminger, L. Complement activation by the properdin system: formation of a stoichiometric, C3 cleaving complex of properdin factor B with C3b. Immunochemistry 14: 201-205.

1978

Fearon, D. T. Regulation by membrane sialic acid of β1H-dependent decay-dissociation of amplification C3 convertase of the alternative complement pathway. Proc. Natl. Acad. Sci. U.S.A. 75: 1971-1975.

Kerr, M. A. & Porter, R. R. The purification and properties of the second component of human complement. Biochem. J. 171: 99–107.

Pangburn, M. K. & Müller-Eberhard, H. J. Complement C3 convertase: cell surface restriction of β1H control and generation of restriction on neuraminidase-treated cells. Proc. Natl. Acad. Sci. U.S.A. 75: 2416–2420.

Schreiber, R. D., Pangburn, M. K., Lesavre, P. H. & Müller-Eberhard, H. J. Initiation of the alternative pathway of complement: recognition of activators by bound C3b and assembly of the entire pathway from six isolated proteins. Proc. Natl. Acad. Sci. U.S.A. 75: 3948–3952.

1979

Bianco, C., Götze, O. & Cohn, Z. A. Regulation of macrophage migration by products of the complement system. Proc. Natl. Acad. Sci. U.S.A. 76: 888–891.

Lachmann, P. J. Complement. In: The Antigens (Sela, M. ed.), vol. 5, pp. 284–353, Academic Press, New York.

Entry 1.31

COMPLEMENT COMPONENT C3b INACTIVATOR

Summary

EC Number: (3.4.21.-)

Earlier names: Conglutinogen-activating factor, factor C.
Abbreviations: KAF, C3bINA (but note that KAF has also been used
for the "kinase-activating factor", #2.08).

Distribution: Occurs in human plasma at 30-50 µg/ml, together with
β1H globulin (560 µg/ml).

Source: Plasma.

Action: The C3b α chain is cleaved at two points, giving fragments
of 3000, 43,000 and 68,000 mol.wt., the larger ones still
disulfide-linked. The modified C3b is no longer active in
hemolysis or adherence reactions, and cannot form C4,2,3 or
C3b,Bb. Inactive on C3. C4b is also inactivated by cleavage of
the α chain, and cell-bound C4b in the C4,2 complex may also be
susceptible to attack.

Requirements: Neutral pH. Action on C3b requires β1H globulin,
which displaces factor Bb competitively from the C3b,Bb
complex, making C3b susceptible to cleavage. C4 binding protein
serves a comparable function in the inactivation of C4b.

Substrate, usual: Cell-bound C3b, in a hemolytic assay.

Inhibitors: Dip-F. C3 nephritic factor can stabilize C3b,B and
C3b,Bb against dissociation by β1H, and thus against C3bINA [1].

Molecular properties: A β-globulin, mol.wt. 93,000, consisting of
two disulfide-linked chains of 55,000 and 42,000.

Comment: C3b inactivator (C3bINA) and β1H (a non-proteolytic
glycoprotein of 150,000 mol.wt.) act as regulators of C3-
convertase formation in the fluid phase, and thereby help to
confine enzyme formation to the surface of the particulate
activators. The binding of β1H to C3b (with displacement of Bb)
generates an efficient binding site for C3bINA. Sialic acid
residues on non-activating cells, e.g. sheep erythrocytes,
appear to facilitate the β1H-mediated action of C3bINA, but
sheep erythrocytes treated with neuraminidase or NaIO₄ acquire
the capacity to activate the human alternative pathway,
apparently because bound C3b is no longer susceptible to attack
by C3bINA.

Hereditary deficiency of C3bINA has been described: the result is chronic depletion of C3 due to the continuous activation of complement by the alternative pathway.

References
[1] Lachmann In: The Antigens (Sela ed.), vol. 5, pp. 284-353, Academic Press, New York, 1979.

Bibliography

1953

Pillemer, L., Lepow, I. H., & Blum, L. The requirement for a hydrazine-sensitive serum factor and heat-labile serum factors in the inactivation of human C'3 by zymosan. J. Immunol. 71: 339-345.

1967

Tamura, N. & Nelson, R. A. Three naturally-occurring inhibitors of components of complement in guinea pig and rabbit serum. J. Immunol. 99: 582-589.

1968

Gigli, I. & Nelson, R. A. Complement dependent immune phagocytosis. I. Requirements for C'1, C'4, C'2, C'3. Exp. Cell Res. 51: 45-67.
Lachmann, P. J. & Müller-Eberhard, H. J. The demonstration in human serum of "conglutinogen-activating factor" and its effect on the third component of complement. J. Immunol. 100: 691-698.

1969

Ruddy, S. & Austen, K. F. C3 inactivator of man. 1. Hemolytic measurement by the inactivation of cell-bound C3. J. Immunol. 102: 533-543.

1971

Abramson, N., Alper, C. A., Lachmann, P. J., Rosen, F. S. & Jandl, J. H. Deficiency of C3 inactivator in man. J. Immunol. 107: 19-27.
Ruddy, S. & Austen. K. F. C3b inactivator of man. II. Fragments produced by C3b inactivator cleavage of cell-bound or fluid phase C3b. J. Immunol. 107: 742-750.

1972

Alper, C. A., Rosen, F. S. & Lachmann, P. J. Inactivator of the third component of complement as an inhibitor in the properdin pathway. Proc. Natl. Acad. Sci. U.S.A. 69: 2910-2913.

1973

Nicol, P. A. E. & Lachmann, P. J. The alternate pathway of complement activation. The role of C3 and its inactivator (KAF). Immunology 24: 259-275.
Ruddy, S. & Austen, K. F. C3b inactivator of man. II. Fragments

produced by C3b inactivator cleavage of cell-bound or fluid phase C3b. J. Immunol. 107: 742-750.

Ruddy, S., Fearon, D. T. & Austen, K. F. Participation of factors B, D, E and C3b inactivator in cobra venom (CVF)-induced activation of the alternate C pathway. J. Immunol. 111: 289-290.

1975

Brecy, H. & Hartmann, L. Distinction between hereditary and acquired angioneurotic oedema according to the complement system. Biomedicine 23: 328-334.

Cooper, N. R. Isolation and analysis of the mechanism of action of an inactivator of C4b in normal human serum. J. Exp. Med. 141: 890-903.

Gitlin, J. D., Rosen, F. S. & Lachmann, P. J. The mechanism of action of the C3b inactivator (conglutinogen-activating factor) on its naturally occurring substrate, the major fragment of the third component of complement (C3b). J. Exp. Med. 141: 1221-1226.

Lachmann, P. J. & Halbwachs, L. The influence of C3b inactivator (KAF) concentration on the ability of serum to support complement activation. Clin. Exp. Immunol. 21: 109-114.

Shiraishi, S. & Stroud, R. M. Cleavage products of C4b produced by enzymes in human serum. Immunochemistry 12: 935-939.

1976

Cooper, N. R. & Ziccardi, R. J. The nature and reactions of complement enzymes. In: Proteolysis and Physiological Regulation (Ribbons, D. W. & Brew, K. eds), pp. 167-187, Academic Press, New York.

Daha, M. R., Fearon, D. T. & Austen, K. F. C3 requirements for formation of alternative pathway C5 convertase. J. Immunol. 117: 630-634.

Medicus, R. G., Schreiber, R. D., Götze, O. & Müller-Eberhard, H. J. A molecular concept of the properdin pathway. Proc. Natl. Acad. Sci. U.S.A. 73: 612-616.

Whaley, K. & Ruddy, S. Modulation of C3b hemolytic activity by a plasma protein distinct from C3b inactivator. Science 193: 1011-1013.

Whaley, K. & Ruddy, S. Modulation of the alternative complement pathway by β1H globulin. J. Exp. Med. 144: 1147-1163.

1977

Fearon, D. T. Purification of C3b inactivator and demonstration of its two polypeptide chain structure. J. Immunol. 119: 1248-1252.

Fearon, D. T. & Austen, K. F. Activation of the alternative complement pathway due to resistance of zymosan-bound amplification convertase to endogenous regulatory mechanisms. Proc. Natl. Acad. Sci. U.S.A. 74: 1683-1687.

Fearon, D. T. & Austen, K. F. Activation of the alternative complement pathway with rabbit erythrocytes by circumvention of the regulatory action of endogenous control proteins. J. Exp. Med. 146: 22-33.

Nagasawa, S. & Stroud, R. M. Mechanism of action of the C3b inactivator: requirement for a high molecular weight cofactor (C3b-C4bINA cofactor) and production of a new C3b derivative (C3b'). Immunochemistry 14: 749-756.

Pangburn, M. K., Schreiber, R. D. & Müller-Eberhard, H. J. Human complement C3b inactivator: isolation, characterization and demonstration of an absolute requirement for the serum protein β1H for cleavage of C3b and C4b in solution. J. Exp. Med. 146: 257-270.

1978

Fearon, D. T. Regulation by membrane sialic acid of β1H-dependent decay-dissociation of amplification C3 convertase of the alternative complement pathway. Proc. Natl. Acad. Sci. U.S.A. 75: 1971-1975.

Fujita, T., Gigli, I. & Nussenzweig, V. Human C4-binding protein. II. Role in proteolysis of C4b by C3b-inactivator. J. Exp. Med. 148: 1044-1051.

Lachmann, P. J. & Hobart. M. J. Complement technology. In: Handbook of Experimental Immunology (Weir, D. M. ed.), 3rd edn., pp. 5A.1-5A.23, Blackwell, London.

Nydegger, U., Fearon, D. T. & Austen, K. F. Modulation of the alternative pathway in dilute whole serum by introduction of purified components and control proteins. J. Immunol. 120: 1788-1789.

Pangburn, M. K. & Müller-Eberhard, H. J. Complement C3 convertase: cell surface restriction of β1H control and generation of restriction on neuraminidase-treated cells. Proc. Natl. Acad. Sci. U.S.A. 75: 2416-2420.

1979

Fujita, T. & Nussenzweig, V. The role of C4-binding protein and β1H in proteolysis of C4b and C3b. J. Exp. Med. 150: 267-276.

Gaither, T. A., Hammer, C. H. & Frank, M. M. Studies on the molecular mechanisms of C3b inactivation and a simplified assay of β1H and the C3b inactivator (C3bINA). J. Immunol. 123: 1195-1204.

Lachmann, P. J. Complement. In: The Antigens (Sela, M. ed.), vol. 5, pp. 284-353, Academic Press, New York.

Law, S. K., Fearon, D. T. & Levine, R. P. Action of the C3b-inactivator on cell-bound C3b. J. Immunol. 122: 759-765.

Entry 1.32

RED CELL NEUTRAL ENDOPEPTIDASE

Summary

EC Number: 3.4.21.24

Earlier names: -

Distribution: Sheep red cell stroma.

Source: Sheep red cell stroma.

Action: Cleaves hydrophobic regions of low molecular weight peptides; does not degrade large proteins.

Requirements: pH near 7.

Substrate, usual: HCO-Met-Val.

Inhibitors: Dip-F (10^{-5}M). (Not EDTA.)

Molecular properties: Molecular weight 340,000, probably six identical subunits.

Comment: The original report seems not to have been confirmed, although most investigators have worked with other species (e.g. ref. 1).

References
[1] Tökes & Chambers Biochim. Biophys. Acta 389: 325-338, 1975.

Bibliography

1972

Witheiler, J. & Wilson, D. B. The purification and characterization of a novel peptidase from sheep red cells. J. Biol. Chem. 247: 2217-2221.

Entry 1.33
LEUKOCYTE ELASTASE

Summary

EC Number: 3.4.21.11 (shared by all mammalian elastases).

Earlier names: Neutrophil elastase, lysosomal elastase.

Distribution: Azurophil granules of human neutrophil polymorphonuclear leukocytes, and analogous organelles of some other mammals.

Source: Azurophil granules of human (and some animal) neutrophil leukocytes. Sometimes human leukemic cells are used.

Action: Digests elastin and proteins generally. Cleaves synthetic ester and amide substrates, especially those containing Val or Ala.

Requirements: pH optimum about 8.5.

Substrate, usual: Elastin. Boc-Ala-OPhNO$_2$, Z-Ala-2-ONap, Ac-(Ala)$_3$-NPhNO$_2$.

Substrate, special: MeO-Suc-Ala-Ala-Pro-Val-NMec or -SBzl for maximal sensitivity of assays [1]. Ac-ONapAs-D with a diazonium salt to stain for activity (although this substrate is also sensitive to cathepsin G and chymase). Ac-Ala-Ala-Aval-NPhNO$_2$ for active site titration.

Inhibitors: Dip-F, Pms-F, MeOSuc-Ala-Ala-Pro-Val-CH$_2$Cl, elastatinal, elasnin [2], oleic acid. Soybean trypsin inhibitor, aprotinin, bronchial proteinase inhibitor, human seminal plasma inhibitor I, chick pea inhibitor.

Molecular properties: Glycoprotein, mol.wt. about 30,000. pI about 10. A280,1% 9.85. At least three forms can be distinguished by electrophoresis. N-terminal sequence is strongly homologous with pancreatic elastase (#1.05).

Comment: Has been grouped with other elastases under EC 3.4.21.11, although there are clear differences between the enzymes in specificity and inhibitor sensitivity.

References
[1] Castillo et al. Anal. Biochem. 99: 53-64, 1979.
[2] Ōmura et al. Biochem. Biophys. Res. Commun. 83: 704-709, 1978.

Bibliography

1922

Nye, R. N. Studies on the pneumonic exudate. V. The relation of
pneumonic lung protease activity to hydrogen ion concentration,
and a consideration of the origin of the enzyme. J. Exp. Med.
35: 153-160.

1931

Husfeldt, E. Proteolytic Enzyme in den Leucocyten des Menschen.
Hoppe-Seyler's Z. Physiol. Chem. 194: 137-165.

1966

Kueppers, F. & Bearn, A. G. A possible experimental approach to
the association of hereditary αl-antitrypsin deficiency and
pulmonary emphysema. Proc. Soc. Exp. Biol. Med. 121: 1207-1209.

1968

Janoff, A. & Scherer, J. Mediators of inflammation in leukocyte
lysosomes. IX. Elastinolytic activity in granules of human
polymorphonuclear leukocytes. J. Exp. Med. 128: 1137-1151.
Janoff, A. & Zeligs, J. D. Vascular injury and lysis of basement
membrane in vitro by neutral protease of human leukocytes.
Science 161: 702-704.

1969

Janoff, A. Alanine p-nitrophenyl esterase activity of human
leucocyte granules. Biochem. J. 114: 157-159.

1970

Janoff, A. Inhibition of human granulocyte elastase by gold
sodium thiomalate. Biochem. Pharmacol. 19: 626-628.
Janoff, A. Mediators of tissue damage in leukocyte lysosomes. X.
Further studies on human granulocyte elastase. Lab. Invest. 22:
228-236.

1971

Janoff, A. & Basch, R. S. Further studies on elastase-like
esterases in human leukocyte granules. Proc. Soc. Exp. Biol.
Med. 136: 1045-1049.
Ohlsson, K. Properties of leucocyte protease. Clin. Chim. Acta
32: 399-405.

1972

Hochstrasse, K., Schuster, R., Reichert, R. & Heimberger, N.
Detection and quantitative determination of complexes between
leucocyte proteases and α-1-antitrypsin in body secretions and
body fluids. Hoppe-Seyler's Z. Physiol. Chem. 353: 1120-1124.
Janoff, A. Human granulocyte elastase. Further delineation of its
role in connective tissue damage. Am. J. Pathol. 68: 579-591.
Janoff, A. Neutrophil proteases in inflammation. Annu. Rev. Med.
23: 177-190.

Lieberman, J. & Kaneshiro, W.   Inhibition of leucocyte elastase
from purulent sputum by alpha$_1$-antitrypsin.   J. Lab. Clin.
Invest. 80: 88-101.

1973

Janoff, A.   Purification of human granulocyte elastase by affinity
chromatography.   Lab. Invest. 29: 458-464.
Janoff, A. & Blondin, J.   The effect of human granulocyte elastase
on bacterial suspensions.   Lab. Invest. 29: 454-457.

1974

Janoff, A. & Blondin, J.   The role of elastase in the digestion of
E. coli proteins by human polymorphonuclear leukocytes. 1.
Experiments in vitro.   Proc. Soc. Exp. Biol. Med. 145: 1427-
1430.
Ohlsson, K. & Olsson, I.   The neutral proteases of human
granulocytes. Isolation and partial characterization of
granulocyte elastases.   Eur. J. Biochem. 42: 519-527.
Rindler-Ludwig, R., Schmalzl, F. & Braunsteiner, H.   Esterases in
human neutrophil granulocytes: evidence for their protease
nature.   Br. J. Haematol. 27: 57-64.
Schmidt, W. & Havemann, K.   Isolation of elastase-like and
chymotrypsin-like neutral proteases from human granulocytes.
Hoppe-Seyler's Z. Physiol. Chem. 355: 1077-1082.
Sweetman, F. & Ornstein, L.   Electrophoresis of elastase-like
esterases from human neutrophils.   J. Histochem. Cytochem. 22:
327-339.

1975

Bielefeld, D. R., Senior, R. M. & Yu, S. Y.   A new method for
determination of elastolytic activity using [^{14}C]-labelled
elastin and its application to leukocyte elastase.   Biochem.
Biophys. Res. Commun. 67: 1553-1559.
Dewald, B., Rindler-Ludwig, R., Bretz, U. & Baggiolini, M.
Subcellular localization and heterogeneity of neutral proteases
in neutrophilic polymorphonuclear leukocytes.   J. Exp. Med. 141:
709-723.
Dubin, A., Koj, A. & Chudzik, J.   Isolation and some molecular
parameters of elastase-like neutral proteinases from horse blood
leucocytes.   Biochem. J. 153: 389-396.
Feinstein, G. & Janoff, A.   A rapid method of purification of
human granulocyte cationic neutral proteases: purification and
further chracterization of human granulocyte elastase.   Biochim.
Biophys. Acta 403: 493-505.
Janoff, A.   At least three human neutrophil lysosomal proteases
are capable of degrading joint connective tissues.   Ann. N. Y.
Acad. Sci. 256: 402-408.
Koj, A., Chudzik, J. & Dubin, A.   Substrate specificity and
modification of the active centre of elastase-like neutral
proteinases from horse blood leucocytes.   Biochem. J. 153: 397-
402.

Malemud, C. J. & Janoff, A. Human polymorphonuclear leukocyte elastase and cathepsin G mediate the degradation of lapine articular cartilage proteoglycan. Ann. N. Y. Acad. Sci. 256: 254-262.

Malemud, C. J. & Janoff, A. Identification of neutral proteases in human neutrophil granules that degrade articular cartilage proteoglycan. Arthritis Rheum. 18: 361-368.

Ohlsson, K. αl-Antitrypsin and α2-macroglobulin. Interactions with human neutrophil collagenase and elastase. Ann. N.Y. Acad. Sci. 256: 409-419.

Ohlsson, K. Granulocyte collagenase and elastase and their interactions with α_1-antitrypsin and α_2-macroglobulin. In: Proteases in Biological Control (Reich, E., Rifkin, D. B. & Shaw, E. eds), pp. 591-602, Cold Spring Harbor Laboratory, New York.

Ohlsson, K. & Delshammar, M. Interactions between granulocyte elastase and collagenase and the plasma proteinase inhibitors in vitro and in vivo. In: Dynamics of Connective Tissue Macromolecules (Burleigh, P. M. C. & Poole, A. R. eds), pp. 259-275, North-Holland Publishing Co., Amsterdam.

Ohlsson, K. & Tegner, H. Granulocyte collagenase, elastase and plasma protease inhibitors in purulent sputum. Eur. J. Clin. Invest. 5: 221-227.

Plow, E. F. & Edgington, T. S. An alternative pathway for fibrinolysis. I. The cleavage of fibringen by leukocyte proteases at physiologic pH. J. Clin. Invest. 56: 30-38.

Powers, J. C., Carroll, D. L. & Tuhy, P. M. Synthetic active site-directed inhibitors of elastolytic proteases. Ann. N. Y. Acad. Sci. 256: 420-425.

Schmidt, W., Egbring, R. & Havemann, K. Effect of elastase-like and chymotrypsin-like neutral proteases from human granulocytes on isolated clotting factors. Thromb. Res. 6: 315-326.

Taylor, J. C. & Crawford, I. P. Purification and preliminary characterization of human leucocyte elastase. Arch. Biochem. Biophys. 169: 91-101.

Tuhy, P. M. & Powers, J. C. Inhibition of human leukocyte elastase by peptide chloromethyl ketones. FEBS Lett. 50: 359-361.

1976

Ardelt, W., Koj, A., Chudzik, J. & Dubin, A. Inactivation of some pancreatic and leucocyte elastases by peptide chloromethyl ketones and alkyl isocyanates. FEBS Lett. 67: 156-160.

Ardelt, W., Tomczak, Z., Księżny, S. & Dudek-Wojciechowska, G. Neutral elastolytic proteinase from canine leucocytes. Purification and characterisation. Biochim. Biophys. Acta 445: 683-693.

Baugh, R. J. & Travis, J. Human leukocyte granule elastase: rapid isolation and characterization. Biochemistry 15: 836-841.

Blondin, J. & Janoff, A. Role of lysosomal elastase in digestion of Escherichia coli proteins by human polymorphonuclear leukocytes - experiments with living leukocytes. J. Clin. Invest. 58: 971-979.

Delshammar, M. & Ohlsson, K. Isolation and partial characterization of elastase from dog granulocytes. Eur. J. Biochem. 69: 125-131.

Dubin, A., Koj, A. & Chudzik, J. Isolation and some molecular parameters of elastase-like neutral proteinases from horse blood leucocytes. Biochem. J. 153: 389-396.

Feinstein, G. & Janoff, A. The use of polyacrylamide gel electrophoresis to quantitate activity of leucocyte elastase and to detect its activity in complexes of biological interest. Anal. Biochem. 71: 358-366.

Feinstein, G. & Janoff, A. Variations in elastase like esterase activities in human leucocytes during cell maturation. Proc. Soc. Exp. Biol. Med. 152: 36-41.

Feinstein, G., Malemud, C. J. & Janoff, A. The inhibition of human leucocyte elastase and chymotrypsin-like protease by elastatinal and chymostatin. Biochim. Biophys. Acta 429: 925-932.

Hugli, T. E., Taylor, J. C. & Crawford, I. P. Selective cleavage of human C3 by human leukocyte elastase (HLE). J. Immunol. 116: 1737 only.

Janoff, A., Feinstein, G., Malemud, C. J. & Elias, J. M. Degradation of cartilage proteoglycan by human leukocyte granule neutral proteases - a model of joint injury. I. Penetration of enzyme into rabbit articular cartilage and release of $^{35}SO_4$-labelled material from the tissue. J. Clin. Invest. 57: 615-624.

Johnson, U., Ohlsson, K. & Olsson, I. Effects of granulocyte neutral proteases on complement components. Scand. J. Immunol. 5: 421-426.

Keiser, H., Greenwald, R. A., Feinstein, G. & Janoff, A. Degradation of cartilage proteoglycan by human leukocyte granule neutral proteases - a model of joint injury. II. Degradation of isolated bovine nasal cartilage proteoglycan. J. Clin. Invest. 57: 625-632.

Koj, A., Chudzik, J. & Dubin, A. Substrate specificity and modificaton of the active centre of elastase-like neutral proteinases from horse blood leucocytes. Biochem. J. 153: 397-402.

Kruze, D., Menninger, H., Fehr, K. & Böni, A. Purification and some properties of a neutral protease from human leukocyte granules and its comparison with pancreatic elastase. Biochim. Biophys. Acta 438: 503-513.

Odeberg, H. & Olsson, I. Microbicidal mechanisms of human granulocytes: synergistic effects of granulocyte elastase and myeloperoxidase or chymotrypsin-like cationic protein. Infect. Immun. 14: 1276-1283.

Ohlsson, K. Collagenase and elastase released during peritonitis are complexed by plasma protease inhibitors. Surgery 79: 652-

657.

Ohlsson, K. & Laurell, C.-B. The disappearance of enzyme-inhibitor complexes from the circulation of man. Clin. Sci. Mol. Med. 51: 87-92.

Ohlsson, K. & Skude, G. Demonstration and semiquantitative determination of complexes of various proteases and human $\alpha 2$-macroglobulin. Clin. Chim. Acta 66: 1-7.

Ohlsson, K., Olsson, I., Delshammar, M. & Schiessler, H. Elastases from human canine granulocytes. I. Some proteolytic and esterolytic properties. Hoppe-Seyler's Z. Physiol. Chem. 357: 1245-1250.

Ohlsson, K. & Tegner, H. Inhibition of elastase from granulocytes by a low molecular weight bronchial protease inhibitor. Scand. J. Clin. Lab. Invest. 36: 437-445.

Powers, J. C. Inhibitors of elastase and pulmonary emphysema. Trends Biochem. Sci. 1: 211-214.

Senior, R. M., Bielefeld, D. R. & Starcher, B. C. Comparison of the elastolytic effects of human leukocyte elastase and porcine pancreatic elastase. Biochem. Biophys. Res. Commun. 72: 1327-1334.

Solomon, A., Schmidt, W. & Havemann, K. Bence Jones proteins and light chains of immunoglobulins. XIII. Effect of elastase-like and chymotrypsin-like neutral proteases derived from human granulocytes on Bence Jones proteins. J. Immunol. 117: 1010-1014.

Starkey, P. M. & Barrett, A. J. Human lysosomal elastase. Catalytic and immunological properties. Biochem. J. 155: 265-271.

Starkey, P. M. & Barrett, A. J. Neutral proteinases of human spleen. Purification and criteria for homogeneity of elastase and cathepsin G. Biochem. J. 155: 255-263.

Thorne, K. J. I., Oliver, R. C. & Barrett, A. J. Lysis and killing of bacteria by lysosomal proteinases. Infect. Immun. 14: 555-563.

Vischer, T. L., Bretz, U. & Baggiolini, M. In vitro stimulation of lymphocytes by neutral proteinases from human polymorphonuclear leukocyte granules. J. Exp. Med. 144: 863-872.

1977

Ashe, B. M. & Zimmerman, M. Specific inhibition of human granulocyte elastase by cis-unsaturated fatty acids and activation by the corresponding alcohols. Biochem. Biophys. Res. Commun. 75: 194-199.

Blow, A. M. J. The action of human lysosomal elastase on the oxidized insulin B chain. Biochem. J. 161: 13-16.

Brozna, J. P., Senior, R. M., Kreutzer, D. L. & Ward, P. A. Chemotactic factor inactivators of human granulocytes. J. Clin. Invest. 60: 1280-1288.

Dorn, C. P., Zimmerman, M., Yang, S. S., Yurewicz, E. C., Ashe, B. M., Frankshun, R. & Jones, H. Proteinase inhibitors. 1.

Inhibitors of elastase. J. Med. Chem. 20: 1464-1468.
Egbring, R., Schmidt, W., Fuchs, G. & Havemann, K. Demonstration
   of granulocytic proteases in plasma of patients with acute
   leukemia and septicemia with coagulation defects. Blood 49:
   219-231.
Lestienne, P., Dimicoli, J.-L. & Bieth, J. Trifluoroacetyl
   tripeptides as potent inhibitors of human leukocyte elastase.
   J. Biol. Chem. 252: 5931-5933.
Ohlsson, K. & Olsson, I. The extracellular release of granulocyte
   collagenase and elastase during phagocytosis and inflammatory
   processes. Scand. J. Haematol. 19: 145-152.
Ohlsson, K., Tegner, H. & Åkesson, U. Isolation and partial
   characterization of a low molecular weight acid stable protease
   inhibitor from human bronchial secretion. Hoppe-Seyler's Z.
   Physiol. Chem. 358: 583-589.
Powers, J. C., Gupton, B. F., Harley, A. D., Nishino, N. &
   Whitley, R. J. Specificity of porcine pancreatic elastase,
   human leucocyte elastase and cathepsin G. Inhibition with
   peptide chloromethyl ketones. Biochim. Biophys. Acta 485: 156-
   166.
Roughley, P. J. The degradation of cartilage proteolycans by
   tissue proteinases. Proteoglycan heterogeneity and the pathway
   of proteolytic degradation. Biochem. J. 167: 639-646.
Roughley, P. J. The degradation of proteoglycan by leucocyte
   elastase. Biochem. Soc. Trans. 5: 443-445.
Roughley, P. J. & Barrett, A. J. The degradation of cartilage
   proteoglycans by tissue proteinases. Proteoglycan structure and
   its susceptibility to proteolysis. Biochem. J. 167: 629-637.
Schiessler, H., Ohlsson, K., Olsson, I., Arnhold, M., Birk, Y. &
   Fritz, H. Elastases from human and canine granulocytes. II.
   Interaction with protease inhibitors of animal, plant, and
   microbial origin. Hoppe-Seyler's Z. Physiol. Chem. 358: 53-58.
Seemüller, U., Meier, M., Ohlsson, K., Müller, H.-P. & Fritz, H.
   Isolation and characterization of a low molecular weight
   inhibitor (of chymotrypsin and human granulocyte elastase and
   cathepsin G) from leeches. Hoppe-Seyler's Z. Physiol. Chem.
   358: 1105-1117.
Senior, R. M., Tegner, H., Kuhn, C., Ohlsson, K., Starcher, B. C.
   & Pierce, J. A. The induction of pulmonary emphysema with human
   leukocyte elastase. Am. Rev. Respir. Dis. 116: 469-475.
Starkey, P. M. Elastase and cathepsin G: the serine proteinases
   of human neutrophil leucocytes and spleen. In: Proteinases in
   Mammalian Cells and Tissues (Barrett, A. J. ed.), pp. 57-89,
   North-Holland Publishing Co., Amsterdam.
Starkey, P. M. The effect of human neutrophil elastase and
   cathepsin G on the collagen of cartilage, tendon, and cornea.
   Acta Biol. Med. Ger. 36: 1549-1554.
Starkey, P. M., Barrett, A. J. & Burleigh, P. M. The degradation
   of articular collagen by neutrophil proteinases. Biochim.
   Biophys. Acta 483: 386-397.

Taylor, J. C. & Kueppers, F. Electrophoretic mobility of leukocyte elastase of normal subjects and patients with chronic obstructive pulmonary disease. Am. Rev. Respir. Dis. 116: 531-536.

Taylor, J. C., Crawford, I. P. & Hugli, T. E. Limited degradation of the third component (C3) of human complement by human leukocyte elastase (HLE): partial characterization of C3 fragments. Biochemistry 16: 3390-3396.

Twumasi, D. Y. & Liener, I. E. Proteases from purulent sputum. Purification and properties of the elastase and chymotrypsin-like enzymes. J. Biol. Chem. 252: 1917-1926.

Zimmerman, M. & Ashe, B. M. Substrate specificity of the elastase and the chymotrypsin-like enzyme of the human granulocyte. Biochim. Biophys. Acta 480: 241-245.

1978

Baggiolini, M., Bretz, U. & Dewald, B. Subcellular localization of granulocyte enzymes. In: Neutral Proteases of Human Polymorphonuclear Leukocytes (Havemann, K. & Janoff, A. eds), pp. 3-17, Urban & Schwartzenberg, Baltimore.

Barrett, A. J. The possible role of neutrophilic proteinases in damage to articular cartilage. Agents Actions 8: 11-18.

Bieth, J. Elastases: structure, function and pathological role. Front. Matrix Biol. 6: 1-82.

Blondin, J., Janoff, A. & Powers, J. C. Digestion of E. coli proteins by human neutrophil elastase and chymotrypsin-like enzyme (cathepsin G): experiments with a cell-free system and living leukocytes. In: Neutral Proteases of Human Polymorphonuclear Leukocytes (Havemann, K. & Janoff, A. eds), pp. 39-55, Urban & Schwartzenberg, Baltimore.

Blue, M.-L. & Janoff, A. Possible mechanisms of emphysema in cigarette smokers: release of elastase from human polymorphonuclear leukocytes by cigarette smoke condensate in vitro. Am. Rev. Respir. Dis. 117: 317-325.

Bretz, U. Stimulation of lymphocytes by polymorphonuclear leukocyte elastase and cathepsin G, in vitro. In: Neutral Proteases of Human Polymorphonuclear Leukocytes (Havemann, K. & Janoff, A. eds), pp. 323-327, Urban & Schwartzenberg, Baltimore.

Christner, P., Weinbaum, G., Sloan, B. & Rosenbloom, J. Degradation of tropoelastin by proteases. Anal. Biochem. 88: 682-688.

Davies, M., Barrett, A. J., Travis, J., Sanders, J. & Coles, G. A. The degradation of human glomerular basement membrane with purified lysosomal proteinases: evidence for the pathogenic role of the polymorphonuclear leucocyte in glomerulonephritis. Clin. Sci. Mol. Med. 54: 233-240.

Egbring, R. & Havemann, K. Possible role of polymorphonuclear granulocyte proteases in blood coagulation. In: Neutral Proteases of Human Polymorphonuclear Leukocytes (Havemann, K. & Janoff, A. eds), pp. 442-455, Urban & Schwartzenberg, Baltimore.

Fletcher, D. S., Williams, H. R. & Lin, T.-Y. Effects of human

polymorphonuclear leukocyte collagenase on sub-component Clq of
the first component of human complement. Biochim. Biophys. Acta
540: 270-277.

Folds, J. D., Prince, H. & Spitznagel, J. K. Limited cleavage of
human immunoglobulins by elastase of human neutrophil
polymorphonuclear granulocytes. Possible modulator of immune
complex disease. Lab. Invest. 39: 313-321.

Fritz, H. Necessity of a critical consideration of the homogeneity
of PMN proteases applied to biological assay systems: failure to
detect intrinsic kininogenase activity in PMN elastase. In:
Neutral Proteases of Human Polymorphonuclear Leukocytes
(Havemann, K. & Janoff, A. eds), pp. 261-263, Urban &
Schwartzenberg, Baltimore.

Galdston, M. Enhanced activity of human neutrophil elastase
complexed to alpha$_2$-macroglobulin in serum of alpha$_1$-antitrypsin
deficient individuals. In: Neutral Proteases of Human
Polymorphonuclear Leukocytes (Havemann, K. & Janoff, A. eds),
pp. 191-194, Urban & Schwartzenberg, Baltimore.

Gramse, M., Bingenheimer, C. & Havemann, K. Granulocyte elastase
and its interference with fibrinogen. In: Neutral Proteases of
Human Polymorphonuclear Leukocytes (Havemann, K. & Janoff, A.
eds), pp. 347-355, Urban & Schwartzenberg, Baltimore.

Gramse, M., Bingenheimer, C., Schmidt, W., Egbring, R. & Havemann,
K. Degradation products of fibrinogen by elastase-like neutral
protease from human granulocytes. Characterization and effects
on blood coagulation in vitro. J. Clin. Invest. 61: 1027-1033.

Hadding, U., Löffler, C., Gramse, M. & Havemann, K. Influence of
elastase-like protease on guinea pig C3, factor B of the
properdin system and the tumor cell line EL4. In: Neutral
Proteases of Human Polymorphonuclear Leukocytes (Havemann, K. &
Janoff, A. eds), pp. 287-291, Urban & Schwartzenberg, Baltimore.

Havemann, K., Schmidt, W., Bogdahn, U. & Gramse, M. Effect of
polymorphonuclear granulocyte proteases on immunocompetent
cells. In: Neutral Proteases of Human Polymorphonuclear
Leukocytes (Havemann, K. & Janoff, A. eds), pp. 306-322, Urban &
Schwartzenberg, Baltimore.

Janoff, A. Granulocyte elastase: role in arthritis and in
pulmonary emphysema. In: Neutral Proteases of Human
Polymorphonuclear Leukocytes (Havemann, K. & Janoff, A. eds),
pp. 390-417, Urban & Schwartzenberg, Baltimore.

Janoff, A. & Feinstein, G. Granulocyte elastase: isolation and
characterization. In: Neutral Proteases of Human
Polymorphonuclear Leukocytes (Havemann, K. & Janoff, A. eds),
pp. 102-113, Urban & Schwartzenberg, Baltimore.

Lestienne, P. & Bieth, J. G. The inhibition of human leukocyte
elastase by basic pancreatic trypsin inhibitor. Arch. Biochem.
Biophys. 190: 358-360.

Lestienne, P., Dimicoli, J.-L. & Bieth, J. On the inhibition of
human leukocyte elastase by trifluoroacetyl tripeptides. J.
Biol. Chem. 253: 3459-3460.

Lively, M. O. & Powers, J. C.　Specificity and reactivity of human
granulocyte elastase and cathepsin G, porcine pancreatic
elastase, bovine chymotrypsin and trypsin toward inhibition with
sulfonyl fluorides.　Biochim. Biophys. Acta 525: 171-179.
Lonky, S. A., Marsh, J. & Wohl, H.　Stimulation of human
granulocyte elastase by platelet factor 4 and heparin.　Biochem.
Biophys. Res. Commun. 85: 1113-1118.
Ohlsson, K.　Purification and properties of granulocyte collagenase
and elastase.　In: Neutral Proteases of Human Polymorphonuclear
Leukocytes (Havemann, K. & Janoff, A. eds), pp. 89-101, Urban &
Schwartzenberg, Baltimore.
Ohlsson, K. & Olsson, A.-S.　Immunoreactive granulocyte elastase
in human serum.　Hoppe-Seyler's Z. Physiol. Chem. 359: 1531-
1539.
Ohno, H., Saheki, T., Awaya, J., Nakagawa, A. & Omura, S.
Isolation and characterization of elasnin, a new human
granulocyte elastase inhibitor produced by a strain of
Streptomyces.　J. Antibiot. 31: 1116-1123.
Olsson, I., Odeberg, H., Weiss, J. & Elsbach, P.　Bactericidal
cationic proteins of human granulocytes.　In: Neutral Proteases
of Human Polymorphonuclear Leukocytes (Havemann, K. & Janoff, A.
eds), pp. 18-32, Urban & Schwartzenberg, Baltimore.
Omura, S., Ohno, H., Saheki, T., Yoshida, M. & Nakagawa, A.
Elasnin, a new human granulocyte elastase inhibitor produced by
a strain of Streptomyces.　Biochem. Biophys. Res. Commun. 83:
704-709.
Powers, J. C., Gupton, B. F., Lively, M. O., Nishono, N. &
Whitley, R. J.　Synthetic inhibitors of granulocyte elastase and
cathepsin G.　In: Neutral Proteases of Human Polymorphonuclear
Leukocytes (Havemann, K. & Janoff, A. eds), pp. 221-233, Urban &
Schwartzenberg, Baltimore.
Schiessler, H., Hochstrasse, K. & Ohlsson, K.　Acid stable
inhibitors of granulocyte neutral proteases in human mucous
secretions: biochemistry and possible biological function.　In:
Neutral Proteases of Human Polymorphonuclear Leukocytes
(Havemann, K. & Janoff, A. eds), pp. 195-207, Urban &
Schwartzenberg, Baltimore.
Schmidt, M. E., Douglas, S. D., Quie, P., Nelson, R. D., Schmidt,
W. & Havemann, K.　Effect of neutral granulocyte proteases on
human immunocompetent cells: action of elastase-like protease
and chymotrypsin-like protease on mononuclear phagocytes.　In:
Neutral Proteases of Human Polymorphonuclear Leukocytes
(Havemann, K. & Janoff, A. eds), pp. 298-305, Urban &
Schwartzenberg, Baltimore.
Schmidt, W.　Differential release of elastase and chymotrypsin
from polymorphonuclear leukocytes.　In: Neutral Proteases of
Human Polymorphonuclear Leukocytes (Havemann, K. & Janoff, A.
eds), pp. 77-79, Urban & Schwartzenberg, Baltimore.
Solomon, A.　Possible role of PMN proteases in immunoglobulin
degradation and amyloid formation.　In: Neutral Proteases of

Human Polymorphonuclear Leukocytes (Havemann, K. & Janoff, A. eds), pp. 423-438, Urban & Schwartzenberg, Baltimore.

Taylor, J. C. Properties of human granulocyte elastase from normal subjects and from patients with chronic obstructive pulmonary disease. In: Neutral Proteases of Human Polymorphonuclear Leukocytes (Havemann, K. & Janoff, A. eds), pp. 129-135, Urban & Schwartzenberg, Baltimore.

Taylor, J. C. & Tlougan, J. Isoelectric focusing of human granulocyte lysosomal elastase in thin-layer polyacrylamide gels. Anal. Biochem. 90: 481-487.

Travis, J., Baugh, R., Giles, P. J., Johnson, D., Bowen, J. & Reilly, C. F. Human leukocyte elastase and cathepsin G: isolation, characterization and interaction with plasma proteinase inhibitors. In: Neutral Proteases of Human Polymorphonuclear Leukocytes (Havemann, K. & Janoff, A. eds), pp. 118-128, Urban & Schwartzenberg, Baltimore.

Vassalli, J.-D., Granelli-Piperno, A., Griscelli, C. & Reich, E. Specific protease deficiency in polymorphonuclear leukocytes of Chédiak-Higashi syndrome and beige mice. J. Exp. Med. 147: 1285-1290.

Ward, P. A., Kreutzer, D. L. & Senior, R. M. The modulation of leukotaxis by neutral proteases and other factors from neutrophils. In: Neutral Proteases of Human Polymorphonuclear Leukocytes (Havemann, K. & Janoff, A. eds), pp. 276-286, Urban & Schwartzenberg, Baltimore.

Wasi, S., Movat, H. Z., Pass, E. & Chan, J. Y. C. Production, conversion and destruction of kinins by human neutrophil leukocyte proteases. In: Neutral Proteases of Human Polymorphonuclear Leukocytes (Havemann, K. & Janoff, A. eds), pp. 245-260, Urban & Schwartzenberg, Baltimore.

Winninger, C., Lestienne, P., Dimicoli, J.-L. & Bieth, J. NMR and enzymatic investigation of the interaction between elastase and sodium trifluoroacetate. Biochim. Biophys. Acta 526: 227-234.

Zimmerman, M., Ashe, B. M., Dorn, C. P., Yang, S.-S. & Jones, H. Synthetic and natural inhibitors of human granulocyte elastase. In: Neutral Proteases of Human Polymorphonuclear Leukocytes (Havemann, K. & Janoff, A. eds), pp. 234-237, Urban & Schwartzenberg, Baltimore.

1979

Campbell, E. J., White, R. R., Senior, R. M., Rodriguez, R. J. & Kuhn, C. Receptor-mediated binding and internalization of leukocyte elastase by alveolar macrophages in vitro. J. Clin. Invest. 64: 824-833.

Carp, H. & Janoff, A. In vitro suppression of serum elastase-inhibitory capacity by reactive oxygen species generated by phagocytosing polymorphonuclear leukocytes. J. Clin. Invest. 63: 793-797.

Castillo, M. J., Nakajima, K., Zimmerman, M. & Powers, J. C. Sensitive substrates for human leukocyte and porcine pancreatic elastase: a study of the merits of various chromophoric and

fluorogenic leaving groups in assays for serine proteases.
Anal. Biochem. **99**: 53-64.

Del Mar, E. G., Brodrick, J. W., Geokas, M. C. & Largmann, C.
Effect of oxidation of methionine in a peptide substrate for
human elastases: a model for inactivation of α_1-protease
inhibitor. Biochem. Biophys. Res. Commun. 88: 346-350.

Galdston, M., Levytska, V., Liener, I. E. & Twumasi, D. Y.
Degradation of tropoelastin and elastin substrates by human
neutrophil elastase, free and bound to alpha$_2$-macroglobulin in
serum of the M and Z (Pi) phenotypes for alpha$_1$-antitrypsin.
Am. Rev. Respir. Dis. 119: 435-441.

Levy, H. & Feinstein, G. The digestion of the oxidized B chain of
insulin by human neutrophile proteases: elastase and
chymotrypsin-like protease. Biochim. Biophys. Acta 567: 35-42.

Martodam, R. R., Baugh, R. J., Twumasi, D. Y. & Liener, I. E. A
rapid procedure for the large scale purification of elastase and
cathepsin G from human sputum. Prep. Biochem. 9: 15-31.

Martodam, R. R., Twumasi, D. Y., Liener, I. E., Powers, J. C.,
Nishino, N. & Krejcarek, G. Albumin microspheres as carrier of
an inhibitor of leukocyte elastase: potential therapeutic agent
for emphysema. Proc. Natl. Acad. Sci. U.S.A. 76: 2128-2132.

Nakajima, K., Powers, J. C., Ashe, B. M. & Zimmerman, M. Mapping
the extended substrate binding site of cathepsin G and human
leukocyte elastase. Studies with peptide substrates related to
the α_1-protease inhibitor reactive site. J. Biol. Chem. 254:
4027-4032.

Rodriguez, J. R., Seals, J. E., Radin, A., Lin, J. S., Mandl, I. &
Turino, G. M. Neutrophil lysosomal elastase activity in normal
subjects and in patients with chronic obstructive pulmonary
disease. Am. Rev. Respir. Dis. 119: 409-417.

Entry 1.34

PLATELET ELASTASE-LIKE ENZYME

Summary

EC Number: (3.4.21.-)

Earlier names: Platelet elastase.

Distribution: Human platelets, in α-granules and a membrane fraction.

Source: Human platelets.

Action: Hydrolysis of Ac-(Ala)$_3$-OMe; degradation of solubilized kappa elastin. Slight activity against ^{125}I-labelled insoluble elastin.

Requirements: Neutral pH.

Substrate, usual: ^{125}I-Labelled elastin.

Substrate, special: Ac-(Ala)$_3$-OMe.

Inhibitors: (Inhibition characteristics not having been reported, the present classification as a serine proteinase is clearly presumptive.)

Molecular properties: Mol.wt. 26,000; also latent form of 28,000. Reported to be antigenically distinct from lysosomal elastase, and to have a different (Val-) N-terminus, and no carbohydrate prosthetic group.

Comment: This enzyme is difficult to detect, because platelets contain much less elastase activivty than the neutrophil leucocytes which normally contaminate platelet preparations.

Bibliography

1969

Robert, B. & Robert, L. Determination of elastolytic activity with ^{125}I and ^{131}I labelled elastin. Eur. J. Biochem. 11: 62-67.

1970

Legrand, Y., Robert, B., Szigeti, M. Pignaud, G., Caen, J. & Robert, L. Etudes sur une protease élastinolytique des plaquettes sanguines humaine. Atherosclerosis 12: 451-465.

Robert, B., Szigeti, M., Robert, L., Legrand, Y., Pignaud, G. &

Caen, J. Release of elastolytic activity from blood platelets. Nature 227: 1248-1249.

1972

Loeven, W. A. Elastolytic activity in blood. In: Pulmonary Emphysema and Proteolysis (Mittman, C. ed.), pp. 355-360, Academic Press, New York.

Robert, B., Legrand, Y., Soria, C., Caen, J. & Robert, L. Etude sur les protéases des plaquettes sanguines humaines: purification d'une élastase. C. R. Acad. Sci. D (Paris) 274: 1749-1752.

1973

Legrand, Y., Caen, J., Booyse, F. M., Rafelson, M. E., Robert, B. & Robert, L. Studies on a human blood platelet protease with elastolytic activity. Biochim. Biophys. Acta 309: 406-413.

Legrand, Y., Pignaud, G., Fauvel, F. & Caen, J. Variation of the proelastase/elastase ratio in human blood platelet populations. Haemostasis 1: 169-174.

1975

Legrand, Y. & Robert, B. Purification and characterization of human blood platelet elastase. Protides Biol. Fluids Proc. Colloqu. 22: 419-424.

Legrand, Y., Pignaud, G., Caen, J. P., Robert, B. & Robert, L. Separation of human blood platelet elastase and proelastase by affinity chromatography. Biochem. Biophys. Res. Commun. 63: 224-231.

1976

Ehrlich, H. P. & Gordon, J. L. Proteinases in platelets. In: Platelets in Biology and Pathology (Gordon, J. ed.), pp. 353-372, North-Holland Publishing Co., Amsterdam.

1977

Legrand, Y., Caen, J. P., Robert, L. & Wautier, J. L. Platelet elastase and leukocyte elastase are two different entities. Thromb. Haemost. 37: 580-582.

Legrand, Y., Pignaud, G. & Caen, J. P. Human blood platelet elastase and proelastase. Activation of proelastase and release of elastase after adhesion of platelets to collagen. Haemostasis 6: 180-189.

Legrand, Y., Pignaud, G. & Caen, J. P. Purification of platelet proteases: activation of proelastase by a trypsin-like enzyme. FEBS Lett. 76: 294-298.

1978

Legrand, Y., Caen, J. & Robert, L. Substrate specificity of leukocyte and platelet elastases. Thromb. Haemost. 39: 785-786.

Starkey, P. M., Gordon, J. L., Ehrlich, H. P. & Barrett, A. J. Do platelets contain elastase? Thromb. Haemost. 39: 542-543.

1980

Hornebeck, W., Starkey, P. M., Gordon, J. L., Legrand, Y.,
Pignaud, G., Robert, L., Caen, J. P., Ehrlich, H. P. & Barrett,
A. J. The elastase-like enzyme of platelets. Thrombos.
Haemostas. 42: 1681-1683.

Entry 1.35

TISSUE ELASTASE

Summary

EC Number: (3.4.21.-)

Earlier names: -

Distribution: Aorta of man, dog, pig, mouse. Culture medium of rabbit aortic smooth muscle cells. Human breast tumor tissue.

Source: Pig aorta.

Action: Digests elastin and synthetic substrates with lower specific activity than pig pancreatic elastase.

Requirements: Neutral pH.

Substrate, usual: Elastin. $Suc-(Ala)_3-NPhNO_2$.

Inhibitors: α_1-Proteinase inhibitor, α_2-macroglobulin.

Molecular properties: Mol.wt. 23,000.

Comment: The pig aortic enzyme is reported to show partial immunological identity with pancreatic elastase.

Bibliography

1963

Ebel, A. & Fontaine, R. Mise en évidence d'élastase dans la paroi artérielle normale et pathologique. Path. Biol. (Paris) 11: 885-889.

1965

Citterio, C. & Cunego, A. Determinazione del principio elastolitico nelle arterie cerebrali umane. G. Geront 13: 353-360.

1969

Loeven, W. A. Elastolytic enzyme in the vessel wall. J. Atheroscler. Res. 9: 35-43.

1974

Robert, B., Derouette, J.-C. & Robert, L. Mise en évidence de protéases à activité élastolytique dans les extraits d'aortes humaines et animales. C. R. Acad. Sci. D (Paris) 278: 3251-

3254.

1975

Hornebeck, W., Derouette, J.-C. & Robert, L. Isolation,
purification and properties of aortic elastase. FEBS Lett. 58:
66-70.

Robert, B. & Robert, L. Aortic elastase. Its role in the
degradation of arterial elastin. Protides Biol. Fluids Proc.
Colloqu. 22: 413-418.

1976

Hornebeck, W. & Robert, L. Isolation and properties of an aortic
elastase. Protides Biol. Fluids Proc. Colloqu. 23: 205-209.

Hornebeck, W., Derouette, J.-C., Roland, J., Chatelet, F.,
Bouissou, H. & Robert, L. Corrélation entre l'age,
l'arteriosclérose et l'activité élastinolytique de la paroi
aortique humaine. C. R. Acad. Sci. D (Paris) 282: 2003-2006.

1977

Hornebeck, W. & Robert, L. Elastase-like enzymes in aortas and
human breast carcinomas: quantitative variations with age and
pathology. In: Elastin and Elastic Tissue (Sandberg, L. B.,
Gray, W. R. & Franzblau, C. eds), pp. 145-163, Plenum Press, New
York.

Hornebeck, W., Derouette, J. C., Brechemier, D., Adnet, J. J. &
Robert, L. Elastogenesis and elastinolytic activity in human
breast cancer. Biomedicine 26: 48-52.

1978

Hornebeck, W., Adnet, J. J. & Robert, L. Age dependent variation
of elastin and elastase in aorta and human breast cancers. Exp.
Geront. 13: 293-298.

Entry 1.36

CATHEPSIN G

Summary

EC Number: 3.4.21.20

Earlier names: Chymotrypsin-like enzyme of neutrophil leucocytes.

Distribution: Azurophil granules of human neutrophil leukocytes.

Source: Azurophil granules of human neutrophil leukocytes (about 3 $\mu g/10^6$ cells).

Action: Digests many proteins. Reported to generate chemotactic fragments from complement .components C3 and C5. Acts on some synthetic substrates of chymotrypsin, specificity being closer to chymotrypsin C than A.

Requirements: pH optimum about 7.5.

Substrate, usual: Bz-Phe-2-ONap, Bz-Tyr-OEt.

Inhibitors: Dip-F, Z-Phe-CH$_2$Br, Z-Gly-Leu-Phe-CH$_2$Cl, chymostatin. Soybean and lima bean trypsin inhibitors, α_1-antichymotrypsin, α_2-macroglobulin, α_1-proteinase inhibitor.

Molecular properties: Molecular weight about 30,000, single polypeptide chain containing three disulfide bonds. pI about 10 associated with high arginine content [1]. A280,1% 6.64.

Comment: Only cellular distribution distinguishes this enzyme from mast cell chymase I (#1.42), as far as present knowledge of properties goes.

References
[1] Travis et al. In: Neutral Proteases of Human Polymorphonuclear Leukocytes. Biochemistry, Physiology and Clinical Significance (Havermann, K. & Janoff, A. eds), pp. 118-128, Urban & Schwartzenberg, Baltimore, 1978.

Bibliography

1973

Odeberg, H., Olsson, I. & Venge, P. Cationic proteins of human granulocytes. IV. Esterase activity. Lab. Invest. 32: 86-90.
Rindler, R. & Braunsteiner, H. Soluble proteins from human leukocyte granules. Esterase activity of cationic proteins. Blut 27: 26-32.

1974

Gerber, A. C., Carson, J. H. & Hadorn, B. Partial purification and characterization of a chymotrypsin-like enzyme from human neutrophil leucocytes. Biochim. Biophys. Acta 364: 103–112.

Rindler, R., Schmalzl, F. & Braunsteiner, H. Isolierung und Charakterisierung der chymotrypsinähnlichen Protease aus neutrophilen Granulozyten des Menschen. Schweiz Med. Wochenschr. 104: 132–133.

Rindler-Ludwig, R., Schmalzl, F. & Braunsteiner, H. Esterases in human neutrophil granulocytes: evidence for their protease nature. Br. J. Haematol. 27: 57–64.

Schmidt, W. & Havemann, K. Isolation of elastase-like and chymotrypsin-like neutral proteases from human granulocytes. Hoppe-Seyler's Z. Physiol. Chem. 355: 1077–1082.

1975

Dewald, B., Rindler-Ludwig, R., Bretz, U. & Baggiolini, M. Subcellular localisation and heterogeneity of neutral proteases in neutrophilic polymorphonuclear leucocytes. J. Exp. Med. 141: 709–723.

Feinstein, G. & Janoff, A. A rapid method for purification of human granulocyte cationic neutral proteases: purification and characterization of human granulocyte chymotrypsin-like enzyme. Biochim. Biophys. Acta 403: 477–492.

Janoff, A. At least three human neutrophil lysosomal proteases are capable of degrading joint connective tissues. Ann. N. Y. Acad. Sci. 256: 402–408.

Malemud, C. J. & Janoff, A. Identification of neutral proteases in human neutrophil granules that degrade articular cartilage proteoglycan. Arthritis Rheum. 18: 361–368.

Ohlsson, K. Granulocyte collagenase and elastase and their interactions with α_1-antitrypsin and α_2-macroglobulin. In: Proteases in Biological Control (Reich, E., Rifkin, D. B. & Shaw, E. eds), pp. 591–602, Cold Spring Harbor Laboratory, New York.

Rindler-Ludwig, R. & Braunsteiner, H. Cationic proteins from human neutrophil granulocytes. Evidence for their chymotrypsin-like properties. Biochim. Biochem. Acta 379: 606–617.

Schmidt, W., Egbring, R. & Havemann, K. Effect of elastase-like and chymotrypsin-like neutral proteases from human granulocytes on isolated clotting factors. Thromb. Res. 6: 315–326.

Venge, P. & Olsson, I. Cationic proteins of human granulocytes. VI. Effects on the complement system and mediation of chemotactic activity. J. Immunol. 115: 1505–1508.

Venge, P., Olsson, I. & Odeberg, H. Cationic proteins of human granulocytes. V. Interaction with plasma protease inhibitors. Scand. J. Clin. Lab. Invest. 35: 737–744.

1976

Feinstein, G., Malemud, C. J. & Janoff, A. The inhibition of human leucocyte elastase and chymotrypsin-like protease by elastatinal and chymostatin. Biochim. Biophys. Acta 429: 925-932.

Odeberg, H. & Olsson, I. Microbicidal mechanisms of human granulocytes: synergistic effects of granulocyte elastase and myeloperoxidase or chymotrypsin-like cationic protein. Infect. Immun. 14: 1276-1283.

Ohlsson, K. & Åkesson, U. α_1-Antichymotrypsin interaction with cationic proteins from granulocytes. Clin. Chim. Acta 73: 285-291.

Solomon, A., Schmidt, W. & Havemann, K. Bence Jones proteins and light chains of immunoglobulins. XIII. Effect of elastase-like and chymotrypsin-like neutral proteases derived from human granulocytes on Bence Jones proteins. J. Immunol. 117: 1010-1014.

Starkey, P. M. & Barrett, A. J. Human cathepsin G. Catalytic and immunological properties. Biochem. J. 155: 273-278.

Starkey, P. M. & Barrett, A. J. Neutral proteinases of human spleen. Purification and criteria for homogeneity of elastase and cathepsin G. Biochem. J. 155: 255-263.

Vischer, T. L., Bretz, U. & Baggiolini, M. In vitro stimulation of lymphocytes by neutral proteinases from human polymorphonuclear leukocyte granules. J. Exp. Med. 144: 863-872.

                              1977

Blow, A. M. J. & Barrett, A. J. The action of human cathepsin G on the oxidized insulin B chain. Biochem. J. 161: 17-19.

Brozna, J. P., Senior, R. M., Kreutzer, D. L. & Ward, P. A. Chemotactic factor inactivators of human granulocytes. J. Clin. Invest. 60: 1280-1288.

Egbring, R., Schmidt, W., Fuchs, G. & Havemann, K. Demonstration of granulocytic proteases in plasma of patients with acute leukemia and septicemia with coagulation defects. Blood 49: 219-231.

Powers, J. C., Gupton, B. F., Harley, A. D., Nishino, N. & Whitley, R. J. Specificity of porcine pancreatic elastase, human leucocyte elastase and cathepsin G. Inhibition with peptide chloromethyl ketones. Biochim. Biophys. Acta 485: 156-166.

Roughley, P. J. The degradation of cartilage proteolycans by tissue proteinases. Proteoglycan heterogeneity and the pathway of proteolytic degradation. Biochem. J. 167: 639-646.

Roughley, P. J. & Barrett, A. J. The degradation of cartilage proteoglycans by tissue proteinases. Proteoglycan structure and its susceptibility to proteolysis. Biochem. J. 167: 629-637.

Schmidt, W. & Havemann, K. Chymotrypsin-like neutral protease from lysosomes of human polymorphonuclear leucocytes. Hoppe-Seyler's Z. Physiol. Chem. 358: 555-564.

Seemüller, U., Meier, M., Ohlsson, K., Müller, H.-P. & Fritz, H. Isolation and characterization of a low molecular weight inhibitor (of chymotrypsin and human granulocyte elastase and

cathepsin G) from leeches. Hoppe-Seyler's Z. Physiol. Chem.
358: 1105-1117.
Starkey, P. M. Elastase and cathepsin G: the serine proteinases
of human neutrophil leucocytes and spleen. In: Proteinases in
Mammalian Cells and Tissues (Barrett, A. J. ed.), pp. 57-89,
North-Holland Publishing Co., Amsterdam.
Starkey, P. M. The effect of human neutrophil elastase and
cathepsin G on the collagen of cartilage, tendon, and cornea.
Acta Biol. Med. Ger. 36: 1549-1554.
Starkey, P. M., Barrett, A. J. & Burleigh, P. M. The degradation
of articular collagen by neutrophil proteinases. Biochim.
Biophys. Acta 483: 386-397.
Tegner, H., Ohlsson, K. & Olsson, I. The interaction between a
low molecular weight protease inhibitor of bronchial mucus and
chymotrypsin-like cationic proteins of granulocytes. Hoppe-
Seyler's Z. Physiol. Chem. 358: 431-433.
Tsung, P.-K., Kegeles, S. W. & Becker, E. L. Some differences in
intracellular distribution and properties of the chymotrypsin-
like esterase activity of human and rabbit neutrophils.
Biochim. Biophys. Acta 499: 212-227.
Twumasi, D. Y. & Liener, I. E. Proteases from purulent sputum.
Purification and properties of the elastase and chymotrypsin-like
enzymes. J. Biol. Chem. 252: 1917-1926.
Zimmerman, M. & Ashe, B. M. Substrate specificity of the elastase
and the chymotrypsin-like enzyme of the human granulocyte.
Biochim. Biophys. Acta 480: 241-245.

1978

Blondin, J., Janoff, A. & Powers, J. C. Digestion of E. coli
proteins by human neutrophil elastase and chymotrypsin-like
enzyme (cathepsin G): experiments with a cell-free system and
living leukocytes. In: Neutral Proteases of Human
Polymorphonuclear Leukocytes (Havemann, K. & Janoff, A. eds),
pp. 39-55, Urban & Schwartzenberg, Baltimore.
Baggiolini, M., Bretz, U. & Dewald, B. Subcellular localization
of granulocyte enzymes. In: Neutral Proteases of Human
Polymorphonuclear Leukocytes (Havemann, K. & Janoff, A. eds),
pp.3-17, Urban & Schwartzenberg, Baltimore.
Bretz, U. Stimulation of lymphocytes by polymorphonuclear
leukocyte elastase and cathepsin G, in vitro. In: Neutral
Proteases of Human Polymorphonuclear Leukocytes (Havemann, K. &
Janoff, A. eds), pp. 323-327, Urban & Schwartzenberg, Baltimore.
Davies, M., Barrett, A. J., Travis, J., Sanders, J. & Coles, G. A.
The degradation of human glomerular basement membrane with
purified lysosomal proteinases: evidence for the pathogenic role
of the polymorphonuclear leucocyte in glomerulonephritis. Clin.
Sci. Mol. Med. 54: 233-240.
Egbring, R. & Havemann, K. Possible role of polymorphonuclear
granulocyte proteases in blood coagulation. In: Neutral
Proteases of Human Polymorphonuclear Leukocytes (Havemann, K. &
Janoff, A. eds), pp. 442-455, Urban & Schwartzenberg, Baltimore.

Havemann, K., Schmidt, W., Bogdahn, U. & Gramse, M. Effect of polymorphonuclear granulocyte proteases on immunocompetent cells. In: Neutral Proteases of Human Polymorphonuclear Leukocytes (Havemann, K. & Janoff, A. eds), pp. 306-322, Urban & Schwartzenberg, Baltimore.

Lively, M. O. & Powers, J. C. Specificity and reactivity of human granulocyte elastase and cathepsin G, porcine pancreatic elastase, bovine chymotrypsin and trypsin toward inhibition with sulfonyl fluorides. Biochim. Biophys. Acta 525: 171-179.

Mannheim, W. Microbial aggregation induced by neutrophil chymotrypsin-like cationic proteins. In: Neutral Proteases of Human Polymorphonuclear Leukocytes (Havemann, K. & Janoff, A. eds), pp. 33-34, Urban & Schwartzenberg, Baltimore.

Olsson, I., Odeberg, H., Weiss, J. & Elsbach, P. Bactericidal cationic proteins of human granulocytes. In: Neutral Proteases of Human Polymorphonuclear Leukocytes (Havemann, K. & Janoff, A. eds), pp. 18-32, Urban & Schwartzenberg, Baltimore.

Powers, J. C., Gupton, B. F., Lively, M. O., Nishono, N. & Whitley, R. J. Synthetic inhibitors of granulocyte elastase and cathepsin G. In: Neutral Proteases of Human Polymorphonculear Leukocytes (Havemann, K. & Janoff, A. eds), pp. 221-233, Urban & Schwartzenberg, Baltimore.

Rindler-Ludwig, R., Bretz, U. & Baggiolini, M. Cathepsin G: the chymotrypsin-like enzyme of human polymorphonuclear leukocytes. In: Neutral Proteases of Human Polymorphonuclear Leukocytes (Havemann, K. & Janoff, A. eds), pp. 138-149, Urban & Schwartzenberg, Baltimore.

Schiessler, H., Hochstrasse, K. & Ohlsson, K. Acid stable inhibitors of granulocyte neutral proteases in human mucous secretions: biochemistry and possible biological function. In: Neutral Proteases of Human Polymorphonuclear Leukocytes (Havemann, K. & Janoff, A. eds), pp.195-207, Urban & Schwartzenberg, Baltimore.

Schmidt, M. E., Douglas, S. D., Quie, P., Nelson, R. D., Schmidt, W. & Havemann, K. Effect of neutral granulocyte proteases on human immunocompetent cells: action of elastase-like protease and chymotrypsin-like protease on mononuclear phagocytes. In: Neutral Proteases of Human Polymorphonuclear Leukocytes (Havemann, K. & Janoff, A. eds), pp. 298-305, Urban & Schwartzenberg, Baltimore.

Schmidt, W. Differential release of elastase and chymotrypsin from polymorphonuclear leukocytes. In: Neutral Proteases of Human Polymorphonuclear Leukocytes (Havemann, K. & Janoff, A. eds), pp. 77-79, Urban & Schwartzenberg, Baltimore.

Schmidt, W. & Havemann, K. Chymotrypsin-like neutral proteases from lysosomes of human polymorphonuclear leukocytes. In: Neutral Proteases of Human Polymorphonuclear Leukocytes (Havemann, K. & Janoff, A. eds), pp. 150-160, Urban & Schwartzenberg, Baltimore.

Travis, J., Baugh, R., Giles, P. J., Johnson, D., Bowen, J. &

Reilly, C. F. Human leukocyte elastase and cathepsin G: isolation, characterization and interaction with plasma proteinase inhibitors. In: Neutral Proteases of Human Polymorphonuclear Leukocytes (Havemann, K. & Janoff, A. eds), pp. 118-128, Urban & Schwartzenberg, Baltimore.

Travis, J., Bowen, J. & Baugh, R. Human α-1-antichymotrypsin: interaction with chymotrypsin-like proteinases. Biochemistry 17: 5651-5656.

Venge, P. Polymorphonuclear leukocyte proteases and their effects on complement components and neutrophil function. In: Neutral Proteases of Human Polymorphonuclear Leukocytes (Havemann, K. & Janoff, A. eds), pp. 264-275, Urban & Schwartzenberg, Baltimore.

Zimmerman, M., Ashe, B. M., Dorn, C. P., Yang, S.-S. & Jones, H. Synthetic and natural inhibitors of human granulocyte elastase. In: Neutral Proteases of Human Polymorphonuclear Leukocytes (Havemann, K. & Janoff, A. eds), pp. 234-237, Urban & Schwartzenberg, Baltimore.

1979

Levy, H. & Feinstein, G. The digestion of the oxidized B chain of insulin by human neutrophile proteases: elastase and chymotrypsin-like protease. Biochim. Biophys. Acta 567: 35-42.

Nakajima, K., Powers, J. C., Ashe, B. M. & Zimmerman, M. Mapping the extended substrate binding site of cathepsin G and human leukocyte elastase. Studies with peptide substrates related to the α_1-protease inhibitor reactive site. J. Biol. Chem. 254: 4027-4032.

Entry 1.37

NEUTROPHIL PROTEINASE 3

Summary

EC Number: (3.4.21.-)

Earlier names: -

Distribution: Human neutrophil leukocyte azurophil granules.

Source: As above.

Action: Acts on casein. More active against Ac-1-ONap than Ac-2-ONapAS-D (unlike leukocyte elastase, #1.33).

Requirements: Neutral pH.

Substrate, usual: Ac-1-ONap, casein.

Inhibitors: Dip-F, Pms-F.

Molecular properties: Less cathodal in electrophoresis, and probably of higher mol.wt., than leukocyte elastase.

Comment: Not fully purified or characterized.

Bibliography

1976

Bretz, U. Serine proteinases from azurophil granules of human polymorphonuclear leukocytes (PMN). Experientia 32: 788 only.

1978

Baggiolini, M., Bretz, U. & Dewald, B. Subcellular localization of granulocyte enzymes. In: Neutral Proteases of Human Polymorphonuclear Leukocytes (Havemann, K. & Janoff, A. eds), pp.3-17, Urban & Schwartzenberg, Baltimore.

Entry 1.38

NEUTROPHIL PROTEINASE 4

Summary

EC Number: (3.4.21.-)

Earlier names: Neutrophil collagenase (but see Comments).

Distribution: Azurophil granules of human neutrophil leucocytes.

Source: Granule fraction of human leukemic granulocytes.

Action: General proteolytic activity against fibrin and proteoglycan; action on collagen is probably on extra-helical peptides, eliminating cross-links.

Requirements: Neutral pH.

Substrate, usual: (Usually quantified by immunoassay).

Inhibitors: Dip-F, but also EDTA. α_1-Proteinase inhibitor, α_2-macroglobulin.

Molecular properties: Mol.wt. 76,000, with chains of 42,000 and 33,000. Not highly cationic, but shows electrophoretic heterogeneity.

Comment: This is the enzyme described only by Ohlsson and co-workers, and originally thought to be identical with the neutrophil collagenase (see #4.02). It differs, however, in its greater abundance in the leukocytes (about 3 $\mu g/10^6$ cells), interaction with α_1-proteinase inhibitor and other inhibitors of serine proteinases, and general proteolytic activity. It is not clear why the enzyme has not been encountered by other groups working on the proteinases of human neutrophils.

Bibliography

1973

Ohlsson, K. & Olsson, I. The neutral proteases of human granulocytes. Isolation and partial characterization of two granulocyte collagenases. Eur. J. Biochem. 36: 473-481.

1975

Ohlsson, K. αl-Antitrypsin and α2-macroglobulin. Interactions with human neutrophil collagenase and elastase. Ann. N. Y. Acad. Sci. 256: 409-419.

Ohlsson, K. Granulocyte collagenase and elastase and their interactions with α_1-antitrypsin and α_2-macroglobulin. In: Proteases in Biological Control (Reich, E., Rifkin, D. B. & Shaw, E. eds), pp. 591–602, Cold Spring Harbor Laboratory, New York.

Ohlsson, K. & Delshammar, M. Interactions between granulocyte elastase and collagenase and the plasma proteinase inhibitors in vitro and in vivo. In: Dynamics of Connective Tissue Macromolecules (Burleigh, P. M. C. & Poole, A. R. eds), pp. 259–275, North-Holland Publishing Co., Amsterdam.

Ohlsson, K. & Tegner, H. Granulocyte collagenase, elastase and plasma protease inhibitors in purulent sputum. Eur. J. Clin. Invest. 5: 221–227.

1976

Ohlsson, K. Collagenase and elastase released during peritonitis are complexed by plasma protease inhibitors. Surgery 79: 652–657.

Ohlsson, K. & Skude, G. Demonstration and semiquantitative determination of complexes of various proteases and human $\alpha 2$-macroglobulin. Clin. Chim. Acta 66: 1–7.

1977

Ohlsson, K. & Olsson, I. Collagenase and elastase during phagocytosis and inflammatory processes. Scand. J. Haematol. 19: 145–152.

Ohlsson, K. & Olsson, I. Neutral proteases of human granulocytes. IV. Interaction between human granulocyte collagenase and plasma protease inhibitors. J. Lab. Clin. Med. 89: 269–277.

Ohlsson, K. & Olsson, I. The extracellular release of granulocyte collagenase and elastase during phagocytosis and inflammatory processes. Scand. J. Haematol. 19: 145–152.

Ohlsson, K., Olsson, I. & Spitznagel, J. K. Localization of chymotrypsin-like cationic protein, collagenase and elastase in azurophil granules of human neutrophilic polymorphonuclear leukocytes. Hoppe-Seyler's Z. Physiol. Chem. 358: 361–366.

1978

Baggiolini, M., Bretz, U. & Dewald, B. Subcellular localization of granulocyte enzymes. In: Neutral Proteases of Human Polymorphonuclear Leukocytes (Havemann, K. & Janoff, A. eds), pp.3–17, Urban & Schwartzenberg, Baltimore.

Ohlsson, K. Purification and properties of granulocyte collagenase and elastase. In: Neutral Proteases of Human Polymorphonuclear Leukocytes (Havemann, K. & Janoff, A. eds), pp. 89–101, Urban & Schwartzenberg, Baltimore.

Entry 1.39

NEUTROPHIL PROTEINASE 5

Summary

EC Number: (3.4.21.-)

Earlier names: Neutral peptide generating protease.

Distribution: Human neutrophil leukocyte cell membranes: partially external.

Source: Human neutrophil leukocytes.

Action: Liberates a biologically active neutral peptide (1000 mol.wt.) from a precursor (85,000-95,000 mol.wt.) in plasma. Also hydrolyzes Tos-Arg-OMe, Bz-Arg-OMe, Z-Lys-OPhNO$_2$. (Not substrates of elastase, #1.33, or cathepsin G, #1.36.)

Requirements: Neutral pH.

Substrate, usual: (Assay by formation of the active peptide, which contracts smooth muscle of isolated guinea pig ileum.)

Inhibitors: Dip-F. Aprotinin.

Molecular properties: Protein of molecular weight about 30,000 in a single chain; pI 7.8-8.3 (thus less cationic than elastase or cathepsin G). Soluble only at high ionic strength.

Comment: The enzyme is distinct from the bradykinin-generating system of neutrophil leukocytes, but the purification procedure is very similar to one which yields elastase and cathepsin G in the hands of others. The enzyme of Nakamura et al. may not be the same as that studied by Wintroub and co-workers.

Bibliography

1974

Wintroub, B. U., Goetzl, E. J. & Austen, K. F. A neutrophil-dependent pathway for the generation of a neutral peptide mediator. Partial characterization of components and conrol by α-1-antitrypsin. J. Exp. Med. 140: 812-824.

1976

Nakamura, S., Yoshinaga, M. & Hayashi, H. Interaction between lymphocytes and inflammatory exudate cells. II. A proteolytic enzyme released by PMN as a possible mediator for enhancement of thymocyte response.   J. Immunol. 117: 1-6.

1977

Wintroub, B. U., Goetzl, E. J. & Austen, K. F.   A neutrophil-
dependent pathway for the generation of a neutral peptide
mediator. II. Subcellular localization of the neutrophil
protease. Immunology 33: 41-49.

1979

Coblyn, J. S., Austen, K. F. & Wintroub, B. U. Purification and
characterization of a human neutrophil neutral protease –
neutral peptide-generating protease. J. Clin. Invest. 63: 998-
1005.

Entry 1.40

RABBIT POLYMORPH SERINE PROTEINASE

Summary

EC Number: (3.4.21.-)

Earlier names: Histonase.

Distribution: Granules of rabbit polymorphonuclear leukocytes.

Source: As above.

Action: Degrades histone and some other proteins.

Requirements: pH optimum 7.0-7.5.

Substrate, usual: Histone.

Inhibitors: Somewhat inhibited by Tos-Phe-CH$_2$Cl, 6-aminohexanoate and soybean trypsin inhibitor.

Molecular properties: pI 4-5 after treatment with heparin.

Comment: The enzyme may be related to the lysosomal elastase of human neutrophil granulocytes (#1.33).
    The properties summarized above are those of the enzyme studied by Davies et al.; the enzymes producing chemotactic peptides (see Bibliography) may not be the same.

Bibliography

1970

Ward, P. A. & Hill, J. H. C5 chemotactic fragments produced by an enzyme in lysosomal granules of neutrophils. J. Immunol. 104: 535-543.

1971

Davies, P., Krakauer, K. & Weissmann, G. Subcellular distribution of neutral proteases in rabbit polymorphonuclear cells. Nature 228: 761-762.
Davies, P., Rita, G. A., Krakauer, K. & Weissmann, G. Characterization of a neutral protease from lysosomes of rabbit polymorphonuclear leucocytes. Biochem. J. 123: 559-569.

1972

Davies, P., Krakauer, K. & Weissmann, G. Calf thymus histone as a substrate for neutral and acid proteases of leucocyte lysosomes

194 SERINE PROTEINASES

and other proteolytic enzymes. Anal. Biochem. 45: 428-440.
1978

Ishida, M., Honda, M. & Hayashi, H. In vitro macrophage chemotactic generation from serum immunoglobulin G by neutrophil seryl protease. Immunology 35: 167-176.

Entry 1.41
HUMAN BONE MARROW SERINE PROTEINASE

Summary

EC Number: (3.4.21.-)

Earlier names: -

Distribution: Human bone marrow, in mitochondria of erythroblasts and granulocytes.

Source: Human bone (rib) marrow.

Action: Endopeptidase activity reported to show specificity for the apo-proteins of pyridoxal enzymes. Cleaves some elastase substrates, including Ac-(Ala)$_3$-OMe, but not elastin.

Requirements: Neutral pH.

Substrate, usual: Apo-ornithine aminotransferase.

Substrate, special: Ac-(Ala)$_3$-OMe, but not Ac-(Ala)$_3$-NPhNO$_2$.

Inhibitors: Dip-F, Pms-F, elastatinal.

Molecular properties: Molecular weight 31,800. Has been crystallized.

Comment: The source tissue is rich in leukocyte elastase, so distinction may be difficult.

Bibliography

1975

Aoki, Y., Takaku, F., Urata, G. & Katunuma, N. New protease inactivating delta-aminolaevulinic-acid synthetase in mitochondria of human bone marrow cells. Biochem. Biophys. Res. Commun. 65: 567-574.

1977

Katunuma, N. & Kominami, E. Group-specific proteinases for apoproteins of pyridoxal enzymes. In: Proteinases in Mammalian Cells and Tissues (Barrett, A.J. ed.), pp. 151-180, North-Holland Publishing Co., Amsterdam.

1978

Aoki, Y. Crystallization and characterization of a new protease in mitochondria of bone marrow cells. J. Biol. Chem. 253: 2026-

2032.

1979

Aoki, Y., Muranaka, S., Nakabayashi, K. & Ueda, Y.
δ-Aminolevulinic acid synthetase in erythroblasts of patients
with pyridoxine-responsive anemia. Hypercatabolism caused by the
increased susceptibility to the controlling protease. J. Clin.
Invest. 64: 1196-1203.

Entry 1.42

CHYMASE I

Summary

EC Number: (3.4.21.-)

Earlier names: Chymotrypsin-like enzyme of mast cells, mast cell
  protease I, skeletal muscle (SK) (group-specific) protease,
  myofibrillar proteinase, rat liver mitochondrial proteinase.
  Note: long ago the name "chymase" was applied to chymosin
  (#3.06), but this usage is now completely obsolete.

Distribution: Granules of dog, rat, cat, bovine and human mast
  cells. Estimated to comprise 15-20% of the protein of granules
  from rat peritoneal mast cells.

Source: Rat and bovine mast cell granules. Mast cell tumors. Skin.
  Muscle. Morris hepatoma.

Action: Degrades histones, casein, muscle proteins, glucagon,
  insulin chains, etc. From rat skeletal muscle, the enzyme was
  originally reported to show specificity for apo-proteins of
  pyridoxal enzymes, but this no longer seems to be true. Acts on
  most synthetic substrates of chymotrypsin.

Requirements: Optimal pH 8-9.

Substrate, usual: Histone, casein. Bz-Tyr-OEt, Ac-Tyr-OEt.

Substrate, special: ClAc-ONapAS-D and PhPr-ONapAS (both chromogenic
  with a diazonium salt).

Inhibitors: Dip-F, Tos-Phe-CH$_2$Cl, Pms-F, chymostatin. Soybean
  trypsin inhibitor, ovoinhibitor.

Molecular properties: A protein of pI 10.6, mol.wt. 26,000,
  consisting of a single polypeptide chain. pI 9.5. Insoluble at
  low salt concentrations, and with a marked tendency to adhere to
  any particulate material. Has been crystallized. The amino acid
  composition has been reported, and the N-terminal sequence shown
  to be homologous to that of bovine chymotrypsin A. The molecule
  seems to contain no carbohydrate, and to have 4 disulfide bonds,
  unlike α-chymotrypsin (5) and chymase II (3) [1].

Comment: Mast cells are widely distributed in tissues, especially
  in the rat, and are so rich in proteolytic activity that they
  have often caused confusion. Thus, activity originating from the
  mast cell granules has been mistakenly attributed to other
  organelles which share their characteristics of high density and

sedimentation rate, and difficult extraction of proteins except
in conditions of, say, high ionic strength. Such organelles
include myofibrils, mitochondria and nuclei.

The chymases closely resemble cathepsin G (#1.36), the
chymotrypsin-like enzyme of (human) neutrophil granulocytes, and
it remains to be established whether (within a single species)
the enzymes are distinct.

References
[1] Everitt & Neurath   Biochimie 61: 653–662, 1979.

Bibliography

1953

Gomori, G. Chloroacetyl esters as histochemical substrates. J.
Histochem. Cytochem. 1: 469–470.

1959

Benditt, E. P. & Arase, M. An enzyme in mast cells with
properties like chymotrypsin. J. Exp. Med. 110: 451–460.

1960

Koszalka, T. R. & Miller, L. L. Proteolytic activity of rat
skeletal muscle. I. Evidence for the existence of an enzyme
active optimally at pH 8.5 to 9.0. J. Biol. Chem. 235: 665–668.
Koszalka, T. R. & Miller, L. L. Proteolytic activity of rat
skeletal muscle. II. Purification and properties of an enzyme
optimally active at pH 8.5 to 9.0. J. Biol. Chem. 235: 669–672.

1963

Lagunoff, D. & Benditt, E. P. Proteolytic enzymes of mast cells.
Ann. N. Y. Acad. Sci. 103: 185–198.

1964

Ende, N., Katayama, Y. & Auditore, J. V. Multiple proteolytic
enzymes in human mast cells. Nature 201: 1197–1198.
Lagunoff, D. & Benditt, E. P. Benzosalicylanilide ester
substrates of proteolytic enzymes. Kinetic and histochemical
studies. I. Chymotrypsin. Biochemistry 3: 1427–1431.

1966

Noguchi, T. & Kandatsu, M. Proteolytic activity in the
myofibrillar fraction of rat skeletal muscle. Agr. Biol. Chem.
(Tokyo) 30: 199–201.
Pastan, I. & Almqvist, S. Purification and properties of a mast
cell protease. J. Biol. Chem. 241: 5090–5094.

1967

Budd, G. C., Darzynkiewicz, Z. & Barnard, E. A. Intracellular
localisation of specific proteases in rat mast cells by electron
microscope autoradiography. Nature 213: 1202–1203.
Darzynkiewicz, Z. & Barnard, E. A. Specific proteases of the rat

mast cell. Nature 213: 1198-1202.

1968

Lagunoff, D. The properties of mast cell proteases. Biochem. Pharmacol. (1968) Suppl. 221-227.

1969

Noguchi, T. & Kandatsu, M. Inhibition of autolytic breakdown of muscle proteins by the sarcoplasm and the serum of rat. Agr. Biol. Chem. (Tokyo) 33: 1226-1228.

1970

Noguchi, T. & Kandatsu, M. Autolytic breakdown of rat skeletal muscle proteins in the alkaline pH range. Agr. Biol. Chem. (Tokyo) 34: 390-394.

1971

Holmes, D., Parsons, M. E., Park, D. C. & Pennington, R. J. An alkaline proteinase in muscle homogenates. Biochem. J. 125: 98P only.

Kawiak, J., Vensell, W. H., Komender, J. & Barnard, E. A. Non-pancreatic proteases of the chymotrypsin family. I. A chymotrypsin-like protease from rat mast cells. Biochim. Biophys. Acta 235: 172-187.

Lloyd, A. G., Morrison, T. & Elmore, D. T. Comparative studies on rat mast-cell-granule protease and a bovine liver-capsule protease. Biochem. J. 123: 43P only.

Lloyd, A. G., Morrison, T. & Elmore, D. T. Observations on rat mast-cell granule protease. Biochem. J. 123: 42P only.

Noguchi, T. & Kandatsu, M. Purification and properties of a new alkaline proteinase of rat skeletal muscle. Agr. Biol. Chem. (Tokyo) 35: 1092-1100.

Vensel, W. H., Komender, J. & Barnard, E. A. Non-pancreatic proteases of the chymotrypsin family. II. Two proteases from a mouse mast cell tumor. Biochim. Biophys. Acta 250: 395-407.

1972

Hagermark, O., Rajka, G. & Bergqvist, U. Experimental itch in human skin elicited by rat mast cell chymase. Acta Derm. Venereol. (Stockh.) 52: 125-128.

Kominami, E., Kobayashi, K., Kominami, S. & Katunuma, N. Properties of a specific protease for pyridoxal enzymes and its biological role. J. Biol. Chem. 247: 6848-6855.

1973

Park, D. C., Parsons, M. E. & Pennington, R. J. Evidence for mast-cell origin of proteinase in skeletal-muscle homogenates. Biochem. Soc. Trans. 1: 730-733.

1974

Mayer, M., Amin, R. & Shafrir, E. Rat myofibrillar protease: enzyme properties and adaptive changes in conditions of muscle protein degradation. Arch. Biochem. Biophys. 161: 20-25.

1975

Banno, Y., Shiotani, T., Towatari, T., Yoshikawa, D., Katsunuma, T., Afting, E.-G. & Katunuma, N. Studies on new intracellular proteases in various organs of rat. 3. Control of group-specific protease under physiological conditions. Eur. J. Biochem. 52: 59-63.

Katunuma, N., Kominami, E., Kobayashi, K., Banno, Y., Suzuki, K., Chichibu, K., Hamaguchi, Y. & Katsunuma, T. Studies on new intracellular proteases in various organs of rat. 1. Purification and comparison of their properties. Eur. J. Biochem. 52: 37-50.

Kominami, E., Banno, Y., Chichibu, K., Shiotani, T., Hagaguchi, Y. & Katunuma, N. Studies on new intracellular proteases in various organs of rat. 2. Mode of limited proteolysis. Eur. J. Biochem. 52: 51-57.

Matsuda, Y. & Fischer, E. H. Pattern of degradation of skeletal muscle glycogen phosphorylase by a rat muscle group specific protease. In: Intracellular Protein Turnover (Schimke, R. T. & Katunuma, N. eds), pp. 213-222, Academic Press, New York.

1976

Birkedal-Hansen, H., Cobb, C. M., Taylor, R. E. & Fullmer, H. M. Activation of fibroblast procollagenase by mast cell proteases. Biochim. Biophys. Acta 438: 273-286.

Haas, R., Weiss, E., Jusic, M. & Heinrich, P. C. Limited proteolysis of the very-lysine-rich histone H1 by a membrane-bound proteinase from rat liver. FEBS Lett. 63: 117-119.

Heinrich, P. C., Raydt, G., Puschendorf, B. & Jusič, M. Subcellular distribution of histone-degrading enzyme activities from rat liver. Eur. J. Biochem. 62: 37-43.

Jusič, M., Seifert, S., Weiss, E., Haas, R. & Heinrich, P. C. Isolation and characterization of membrane-bound proteinase from rat liver. Arch. Biochem. Biophys. 177: 355-363.

Kominami, E. & Katunuma, N. Studies on new intracellular proteases in various organs of rat. Participation of proteases in degradation of ornithine aminotransferase in vitro and in vivo. Eur. J. Biochem. 62: 425-430.

Lagunoff, D. & Pritzl, P. Characterization of rat mast cell granule proteins. Arch. Biochem. Biophys. 173: 554-563.

Mayer, M., Amin, R. & Shafrir, E. Adaptive behaviour of muscle protease in protein wasting conditions. In: Intracellular Protein Catabolism (Hanson, H. & Bohley, P. eds), pp. 510-518, J. A. Barth, Leipzig.

Noguchi, T. & Kandatsu, M. Some properties of alkaline protease in rat muscle compared with that in peritoneal cavity cells. Agr. Biol. Chem. (Tokyo) 40: 927-933.

Orenstein, N. S., Hammond, M. E., Dvorak, H. F. & Feder, J. Esterase and protease activity of purified guinea pig basophil granules. Biochem. Biophys. Res. Commun. 72: 230-235.

1977

Drabikowski, W., Górecka, A. & Jakubiec-Puka, A. Endogenous proteinases in vertebrate skeletal muscle. Int. J. Biochem. 8: 61-71.

Haas, R., Nagasawa, T. & Heinrich, P. C. A proteinase in rat liver mitochondria: its submitochondrial localization and specificity towards endogenous substrates. Hoppe-Seyler's Z. Physiol. Chem. 358: 241-242.

Haas, R., Nagasawa, T. & Heinrich, P. C. The localization of a proteinase within rat liver mitochondria. Biochem. Biophys. Res. Commun. 74: 1060-1065.

Katunuma, N. & Kominami, E. Group-specific proteinases for apoproteins of pyridoxal enzymes. In: Proteinases in Mammalian Cells and Tissues (Barrett, A.J. ed.), pp. 151-180, North-Holland Publishing Co., Amsterdam.

Katunuma, N., Sanada, Y., Kominami, E., Kobayashi, K. & Banno, Y. Intracellular protein catabolism and new serine proteases. Acta Biol. Med. Ger. 36: 1537-1546.

Uchida, K., Murakami, U. & Hiratuka, T. Purification of cardiac myosin from rat heart. Proteolytic cleavage and its inhibition. J. Biochem. 82: 469-476.

Wylie, A. R. G., Lonsdale-Eccles, J. D., Blumson, N. L. & Elmore, D. T. The effect of the proteinase from rat peritoneal mast cells on prothrombin and Factor X. Biochem. Soc. Trans. 5: 1449-1452.

Yurt, R. & Austen, K. F. Preparative purification of the rat mast cell chymase. Characterization and interaction with granule components. J. Exp. Med. 146: 1405-1419.

1978

Banno, Y., Morris, H. P. & Katunuma, N. The serine protease from rat liver and hepatoma 8999. Location and role in mitochondrial protein degradation. J. Biochem. 83: 1545-1554.

Haas, R. & Heinrich, P. C. The localization of an intracellular membrane-bound proteinase from rat liver. Eur. J. Biochem. 91: 171-178.

Haas, R., Heinrich, P. C., Tesch, R. & Witt, I. Cleavage specificity of the serine proteinase from rat liver mitochondria. Biochem. Biophys. Res. Commun. 85: 1039-1046.

Katunuma, N., Yasogawa, N., Kito, K., Sanada, Y., Kawai, H. & Miyoshi, K. Abnormal expression of a serine protease in human dystrophic muscle. J. Biochem. 83: 625-628.

Kobayashi, K., Sanada, Y. & Katunuma, N. Selective cleavage of peptide bonds by a serine protease from rat skeletal muscle. J. Biochem. 84: 477-481.

Murakami, U. & Uchida, K. Purification and characterization of a myosin-cleaving protease from rat heart myofibrils. Biochim. Biophys. Acta 525: 219-229.

Sanada, Y., Yasogawa, N. & Katunuma, N. Crystallization and amino acid composition of a serine protease from rat skeletal muscle.

Biochem. Biophys. Res. Commun. 82: 108-113.

Sanada, Y., Yasogawa, N. & Katunuma, N. Serine protease in mice with hereditary muscular dystrophy. J. Biochem. 83: 27-33.

Seppä, H. E. J. Rat skin main neutral protease: immunohistochemical localization. J. Invest. Dermatol. 71: 311-315.

Seppä, H. E. J. & Järvinen, M. Rat skin main neutral protease: purification and properties. J. Invest. Dermatol. 70: 84-89.

Subramanian, M., Koppikar, S. V. & Fatterpaker, P. Rapid purification of mitochondrial proteases: use of affinity chromatography. Indian J. Biochem. Biophys. 15: 214-218.

Woodbury, R. G., Everitt, M., Sanada, Y., Katunuma, N., Lagunoff, D. & Neurath, H. A major serine protease in rat skeletal muscle: evidence for its mast cell origin. Proc. Natl. Acad. Sci. U.S.A. 75: 5311-5313.

Woodbury, R. G., Gruzenski, G. M. & Lagunoff, D. Immunofluorescent localization of a serine protease in rat small intestine. Proc. Natl. Acad. Sci. U.S.A. 75: 2785-2789.

Yasogawa, N., Sanada, Y. & Katunuma, N. Susceptibilities of various myofibrillar proteins to muscle serine protease. J. Biochem. 83: 1355-1360.

1979

Banno, Y., Morris, H. P. & Katunuma, N. Purification, characterization and localization of serine protease of Morris hepatoma 8999. Eur. J. Biochem. 97: 11-21.

Everitt, M. T. & Neurath, H. Purification and partial characterization of an α-chymotrypsin-like protease of rat peritoneal mast cells. Biochemie 61: 653-662.

Everitt, M. T. & Neurath, H. Rat mast cell chymotrypsin-like enzyme: purification, specificity and interaction with mast cell heparin. Fed. Proc. Fed. Am. Soc. Exp. Biol. 38: 834 only.

Haas, R., Heinrich, P. C. & Sasse, D. Proteolytic enzymes of rat liver mitochondria - evidence for a mast cell origin. FEBS Lett. 103: 168-171.

Sage, H., Woodbury, R. G. & Bornstein, P. Structural studies on human type IV collagen. J. Biol. Chem. 254: 9893-9900.

Sanada, Y., Yasogawa, N. & Katunuma, N. Effect of serine protease on isolated myofibrils. J. Biochem. 85: 481-483.

Seppä, H., Väänänen, K. & Korhonen, K. Effect of mast cell chymase of rat skin on intercellular matrix: a histochemical study. Acta Histochem. 64: 64-70.

Entry 1.43

CHYMASE II

Summary

EC Number: (3.4.21.-)

Earlier names: Rat intestinal muscle group-specific proteinase.
Also known as SI-protease. Mast cell protease II.

Distribution: Rat small intestine, but shown by immunofluorescence
to occur in typical and atypical mast cells of connective
tissues generally.

Source: Rat small intestine (originally muscle layer).

Action: Endopeptidase activity with some selectivity for apo-
proteins of pyridoxal enzymes. Cleaves chymotrypsin substrates
e.g. Ac-Tyr-OEt, and shows slight activity against Ac-(Ala)$_3$-OMe.

Requirements: Optimal pH 8-9.

Substrate, usual: Apo-ornithine aminotransferase.

Inhibitors: Dip-F, Pms-F, chymostatin (not Tos-Phe-CH$_2$Cl, Tos-Lys-
CH$_2$Cl or benzamidine). Potato chymotrypsin inhibitor I, lima
bean and soybean inhibitors, ovoinhibitor (chicken), α_1-
proteinase inhibitor. (Not aprotinin.)

Molecular properties: A basic protein of mol.wt. 24,650, with no
free thiol group, but 3 disulfide bonds. A280,1% 11.0. Has been
crystallized and sequenced; the single polypeptide chain shows
extensive homology with trypsin and chymotrypsin.

Comment: Seems different from the trypsin-like enzyme from the
same source (#1.54). Distinguished from mast cell chymase I
(#1.42) on chemical and immunological grounds, although both are
"chymotrypsin-like". At one time it seemed that intestinal
muscle and mucosa contained distinct enzymes [1], but this now
seems doubtful.

References
[1] Katunuma et al. Eur. J. Biochem. 52: 37-50, 1975.

Bibliography

1972

Kominami, E., Kobayashi, K., Kominami, S. & Katunuma, N.
Properties of a specific protease for pyridoxal enzymes and its

biological role. J. Biol. Chem. 247: 6848-6855.

1975

Banno, Y., Shiotani, T., Towatari, T., Yoshikawa, D., Katsunuma, T., Afting, E.-G. & Katunuma, N. Studies on new intracellular proteases in various organs of rat. 3. Control of group-specific protease under physiological conditions. Eur. J. Biochem. 52: 59-63.

Katunuma, N., Kominami, E., Kobayashi, K., Banno, Y., Suzuki, K., Chichibu, K., Hamaguchi, Y. & Katsunuma, T. Studies on new intracellular proteases in various organs of rat. 1. Purification and comparison of their properties. Eur. J. Biochem. 52: 37-50.

Kominami, E., Banno, Y., Chichibu, K., Shiotani, T., Hamaguchi, Y. & Katunuma, N. Studies on new intracellular proteases in various organs of rat. 2. Mode of limited proteolysis. Eur. J. Biochem. 52: 51-57.

1976

Kominami, E. & Katunuma, N. Studies on new intracellular proteases in various organs of rat. Participation of proteases in degradation of ornithine aminotransferase in vitro and in vivo. Eur. J. Biochem. 62: 425-430.

1977

Katunuma, N. & Kominami, E. Group-specific proteinases for apoproteins of pyridoxal enzymes. In: Proteinases in Mammalian Cells and Tissues (Barrett, A.J. ed.), pp. 151-180, North-Holland Publishing.Co., Amsterdam.

1978

Kobayashi, K. & Katunuma, N. Selective cleavage of peptide bonds by a serine protease from the muscle layer of rat small intestine. J. Biochem. 84: 65-74.

Kobayashi, K., Sanada, Y. & Katunuma, N. Selective cleavage of peptide bonds by a serine protease from rat skeletal muscle. J. Biochem. 84: 477-481.

Woodbury, R. G. & Neurath, H. Purification of an atypical mast cell protease and its levels in developing rats. Biochemistry 17: 4298-4304.

Woodbury, R. G., Gruzenski, G. M. & Lagunoff, D. Immunofluorescent localization of a serine protease in rat small intestine. Proc. Natl. Acad. Sci. U.S.A. 75: 2785-2789.

Woodbury, R. G., Katunuma, N., Kobayashi, K., Titani, K. & Neurath, H. Covalent structure of a group-specific protease from rat small intestine. Biochemistry 17: 811-819.

Entry 1.44

TRYPTASE

Summary

EC Number: (3.4.21.-)

Earlier names: Trypsin-like enzyme of mast cells.

Distribution: Human, dog and mouse mast cell granules. Apparently
very little activity in rat cells.

Source: Dog mastocytoma. Rat mast cell granules. Skin.

Action: Active against casein and some synthetic substrates of
trypsin.

Requirements: Neutral pH.

Substrate, usual: Tos-Arg-OMe.

Substrate, special: Ahx-ONapAS (chromogenic, with a diazonium
salt).

Inhibitors: Dip-F, Tos-Lys-CH$_2$Cl.

Molecular properties: Estimates of mol.wt. range from 35,000 to
160,000.

Comment: Knowledge of this enzyme is very sketchy. Trypsin-like
activity in tissues can be due to many enzymes, so care is
required in identifying tryptase.

Bibliography

1960

Glenner, G. G. & Cohen, L. A. Histochemical demonstration of a
species-specific trypsin-like enzyme in mast cells. Nature 185:
846-847.

1962

Ende, N. & Auditore, J. V. Esterolytic activity in canine and
human mast cells. Am. J. Clin. Pathol. 38: 501-506.
Glenner, G. G., Hopsu, V. K. & Cohen, L. A. Histochemical
demonstration of a trypsin-like esterase activity in mast cells.
J. Histochem. Cytochem. 10: 109-110.

1963

Lagunoff, D. & Benditt, E. P. Proteolytic enzymes of mast cells.

Ann. N. Y. Acad. Sci. 103: 185-198.

1968

Lagunoff, D. The properties of mast cell proteases. Biochem. Pharmacol. (1968) Suppl. 221-227.

1971

Katayama, Y., Auditore, J. V. & Ende, N. Purification and characterization of a mouse mast cell esteroprotease. Exp. Mol. Path. 14: 228-242.

Vensel, W. H., Komender, J. & Barnard, E. A. Non-pancreatic proteases of the chymotrypsin family. II. Two proteases from a mouse mast cell tumor. Biochim. Biophys. Acta 250: 395-407.

1975

Fräki, J. E. & Hopsu-Havu, V. K. Human skin proteases. Separation and characterization of two alkaline proteases, one splitting trypsin and the other chymotrypsin substrates. Arch. Dermatol. Forschung. 253: 261-276.

Gabrielescu, E. Structural integration of neuroprotease activity. Int. Rev. Neurobiol. 17: 189-239.

1976

Birkedal-Hansen, H., Cobb, C. M., Taylor, R. E. & Fullmer, H. M. Activation of fibroblast procollagenase by mast cell proteases. Biochim. Biophys. Acta 438: 273-286.

Orenstein, N. S., Hammond, M. E., Dvorak, H. F. & Feder, J. Esterase and protease activity of purified guinea pig basophil granules. Biochem. Biophys. Res. Commun. 72: 230-235.

1977

Legrand, Y., Pignaud, G. & Caen, J. P. Purification of platelet proteases: activation of proelastase by a trypsin-like enzyme. FEBS Lett. 76: 294-298.

1978

Gabrielescu, E. Neuroproteazele. Editura Academiei Republicii Socialiste România, Bucharest.

Entry 1.45

MACROPHAGE CHYMOTRYPSIN-LIKE PROTEINASE

Summary

EC Number: (3.4.21.-)

Earlier names: -

Distribution: Rat peritoneal macrophages.

Source: Rat peritoneal macrophages (70% pure).

Action: Degrades casein and urea-denatured hemoglobin.

Requirements: Optimum pH 8.5. Activation of the crude enzyme by salt (e.g. 0.25 M NaCl) is attributable to dissociation from a high mol.wt. nucleic acid inhibitor.

Substrate, usual: Casein.

Inhibitors: Dip-F, chymostatin (both rather weak). Endogenous nucleic acid.

Molecular properties: 26,000 mol.wt. Molar NaCl required for extraction.

Comment: The properties of this enzyme closely resemble those of chymase I (#1.42), so contamination of the macrophages by mast cells requires careful exclusion, particularly since macrophages may ingest mast cell granules.

Bibliography

1972

Kato, T., Kojima, K. & Murachi, T. Proteases of macrophages in rat peritoneal exudate, with special reference to the effects of actinomycete protease inhibitors. Biochim. Biophys. Acta 289: 187-193.

1977

Suzuki, Y. & Murachi, T. The occurrence of a neutral protease and its inhibitor in rat peritoneal macrophages. J. Biochem. 82: 215-220.

1978

Nagai, K., Nakamura, T. & Koyama, J. Characterization of macrophage proteases involved in the ingestion of antigen-

antibody complexes by the use of protease inhibitors. FEBS
Lett. **92**: 299-302.
Suzuki, Y. & Marachi, T. A chromatin-bound neutral protease and
its inhibitor in rat peritoneal macrophages. J. Biochem. **84**:
977-984.

Entry 1.46

CHEMOTACTIC SERINE PROTEINASE

Summary

EC Number: (3.4.21.-)

Earlier names: -

Distribution: Rabbit and human skin, epidermis, lymphocytes and various other cell types. Particle-bound within cells.

Source: As above.

Action: Degrades casein, but is extremely active in generating chemotactic factors from complement components. Also cytotoxic.

Requirements: Optimal pH 7.8.

Substrate, usual: Casein.

Substrate, special: [^{3}H]Acetyl-casein.

Inhibitors: Dip-F, Tos-Lys-CH$_2$Cl (not Tos-Phe-CH$_2$Cl). Soybean trypsin inhibitor (not aprotinin). α_2-Macroglobulin, α_1-proteinase inhibitor, Cl inhibitor.

Molecular properties: Mol.wt. 30,000. A high salt concentration is required for extraction.

Comment: In some respects the enzyme resembles mast cell chymase I (#1.42), but it seems to have a wider cellular distribution.

Bibliography

1974

Lazarus, G. S. & Barrett, A. J. Neutral proteinase of rabbit skin: an enzyme capable of degrading skin protein and inducing an inflammatory response. Biochim. Biophys. Acta 350: 1-12.

1976

Levine, N., Hatcher, V. B. & Lazarus, G. S. Proteinases of human epidermis; a possible mechanism for polymorphonuclear leukocyte chemotaxis. Biochim. Biophys. Acta 452: 458-467.

1977

Hatcher, V. B., Lazarus, G. S., Levine, N., Burk, P. G. & Yost, F. J., Jr. Characterization of a chemotactic and cytotoxic proteinase from human skin. Biochim. Biophys. Acta 483: 160-

171.
Lazarus, G. S., Levine, N., Vale, M., Vale, M. & Hatcher, V.
Human chemotactic proteinase. In: Rheumatoid Arthritis.
Cellular Pathology and Pharmacology (Gordon, J. L. & Hazleman,
B. L. eds), pp. 223-231, North-Holland Publishing Co.,
Amsterdam.
Thomas, C. A., Yost, F. J., Jr., Snyderman, R., Hatcher, V. B. &
Lazarus, G. S. Cellular serine proteinase induces chemotaxis by
complement activation. Nature 269: 521-522.

1978
Hatcher, V. G., Oberman, M. S., Lazarus, G. S. & Grayzel, A. I. A
cytotoxic proteinase isolated from human lymphocytes. J.
Immunol. 120: 665-670.

Entry 1.47

ACROSIN

Summary

EC Number: 3.4.21.10

Earlier names: Acrosomal proteinase.

Distribution: Acrosome of spermatozoa of mammals generally, occurring partially as the zymogen; also detected in sheep ovary [1].

Source: Especially boar, ram and human spermatozoa.

Action: Digests proteins and synthetic substrates in a generally trypsin-like manner, but probably with a greater preference for arginyl over lysyl bonds.

Requirements: pH optimum about 8.0.

Substrate, usual: Bz-DL-Arg-NPhNO$_2$, Bz-Arg-OEt. (Bz-DL-Arg-NNap is surprisingly resistant to hydrolysis.)

Inhibitors: Dip-F, Tos-Lys-CH$_2$Cl, Dns-Glu-Gly-Arg-CH$_2$Cl, Gbz-OPhNO$_2$, benzamidine, Zn^{2+}. Natural inhibitors of trypsin.

Molecular properties: Boar: a sialo-glycoprotein (8.3% carbohydrate), mol.wt. 38,000, pI 10.5. N-terminal residue Val. There is an inactive precursor zymogen, proacrosin.

Comment: The enzyme is located on the exterior of the head of the capacitated spermatozoan, and is suspected of helping the sperm penetrate the zona pellucida in fertilization. There are, however, some objections to this idea.

References
[1] Morton In: Immunobiology of Gametes (Edidin & Johnson, eds), pp. 115-155, Alden Press, Oxford, 1976.

Bibliography

1969

Stambaugh, R. & Buckley, J. Identification and subcellular localization of the enzymes effecting penetration of the zona pellucida by rabbit spermatozoa. J. Reprod. Fertil. 19: 423-432.
Stambaugh, R., Brackett, B. G. & Mastroianni, L. Inhibition of in vitro fertilization of rabbit ova by trypsin inhibitors. Biol.

Reprod. 1: 223-227.

1970

Gaddum, P. & Blandau, R. J. Proteolytic reaction of mammalian spermatozoa on gelatin membranes. Science 170: 749-751.

Stambaugh, R. & Buckley, J. Comparative studies of the acrosomal enzymes of rabbit, rhesus monkey, and human spermatozoa. Biol. Reprod. 3: 275-282.

Zaneveld, L. J. D., Robertson, R. T. & Williams, W. L. Synthetic enzyme inhibitors as antifertility agents. FEBS Lett. 11: 345-347.

1971

Garner, D. L., Salisbury, G. W. & Graves, C. N. Electrophoretic fractionation of bovine acrosomal proteins and proteinase. Biol. Reprod. 4: 93-100.

Polakoski, K. L., Zaneveld, L. J. D. & Williams, W. L. An acrosin-acrosin-inhibitor-complex in ejaculated boar sperm. Biochem. Biophys. Res. Commun. 45: 381-386.

Schumacher, G. F. B. Inhibition of rabbit sperm acrosomal protease by human alpha-1-antitrypsin and other protease inhibitors. Contraception 4: 67-78.

Stambaugh, R. & Buckley, J. Acrosomal enzymes of mammalian spermatozozoa effecting fertilization. Fed. Proc. Fed. Am. Soc. Exp. Biol. 30: 1184 only.

1972

Fink, E., Schiessler, H., Arnhold, M. & Fritz, H. [Isolation of a trypsin-like enzyme (acrosin) from boar spermatozoa.] Hoppe-Seyler's Z. Physiol. Chem. 353: 1633-1637.

Fritz, H., Förg-Brey, B., Fink, E., Meier, M., Schiessler, H. & Schirren, C. [Human acrosin: isolation and properties.] Hoppe-Seyler's Z. Physiol. Chem. 353: 1943-1949.

Fritz, H., Förg-Brey, B., Fink, E., Schiessler, H., Jaumann, E. & Arnhold, M. [Characterization of a trypsin-like proteinase (acrosin) from boar spermatozoa by inhibition with different protein proteinase inhibitors. I. Seminal trypsin inhibitors and the trypsin-kallikrein inhibitor from bovine organs.] Hoppe-Seyler's Z. Physiol. Chem. 353: 1007-1009.

Fritz, H., Förg-Brey, B., Meier, M., Arnhold, M. & Tschesche, H. [Human acrosin: inhibition by protein proteinase inhibitors.] Hoppe-Seyler's Z. Physiol. Chem. 353: 1950-1952.

Fritz, H., Förg-Brey, B., Schiessler, H., Arnhold, M. & Fink, E. [Characterization of a trypsin-like proteinase (acrosin) from boar spermatozoa by inhibition with different protein proteinase inbibitors. II. Inhibitors from leeches, soybeans, peanuts, bovine colostrum and sea anemones.] Hoppe-Seyler's Z. Physiol. Chem. 353: 1010-1012.

Fritz, H., Heimburger, N., Meier, M., Arnhold, M., Zaneveld, L. J. D. & Schumacher, G. F. B. [Human acrosin: kinetics of the inhibition by inhibitors from human plasma.] Hoppe-Seyler's Z.

Physiol. Chem. 353: 1953-1956.

Fritz, H., Schiessler, H., Förg-Brey, B., Tschesche, H. & Fink, E.
[Characterization of a trypsin-like proteinase (acrosin) from
boar spermatozoa by inhibition with different protein proteinase
inbibitors. III. Inhibitors from pancreas glands.] Hoppe-
Seyler's Z. Physiol. Chem. 353: 1013-1014.

Garner, D. L. & Graves, C. N. Zymographic analysis of rabbit
acrosomal proteinase. J. Anim. Sci. 35: 241 only.

Meizel, S. Biochemical detection and activation of an inactive
form of trypsin-like enzyme in rabbit testes. J. Reprod.
Fertil. 31: 459-462.

Multamäki, S. & Niemi, M. Trypsin-like proteolytic activity in
acrosomal extract of bull spermatozoa. Int. J. Fertil. 17: 43-
53.

Pederson, H. The acrosome of the human spermatozoon: a new method
for its extraction, and an analysis of its trypsin-like enzyme
activity. J. Reprod. Fertil. 31: 99-107.

Polakoski, K. L., Zaneveld, L. J. D. & Williams, W. L.
Purification of a proteolytic enzyme from rabbit sperm
acrosomes. Biol. Reprod. 6: 23-29.

Schiessler, H., Fritz, H., Arnhold, M., Fink, E. & Tschesche, H.
[Properties of the trypsin-like enzyme (acrosin) from boar
spermatozoa.] Hoppe-Seyler's Z. Physiol. Chem. 353: 1638-1645.

Stambaugh, R. & Buckley, J. Studies on acrosomal proteinase of
rabbit spermatozoa. Biochim. Biophys. Acta 284: 473-477.

Stambaugh, R. & Buckley. J. Histochemical subcellular localisation
of the acrosomal proteinase effecting dissolution of the zona
pellucida using fluorescein-labelled inhibitors. Fertil.
Steril. 23: 348-352.

Williams, W. L. Biochemistry of capacitation of spermatozoa. In:
Biology of Mammalian Fertilization and Implantation (Moghissi,
K. S., Kamran, S. & Hafez, E. S. E. eds), pp. 19-53, Thomas,
Springfield.

Yanagimachi, R. & Teichman, R. J. Cytochemical demonstration of
acrosomal proteinase in mammalian and avian spermatozoa by a
silver proteinate method. Biol. Reprod. 6: 87-97.

Zaneveld, L. J. D., Dragoje, B. M. & Schumacher, G. F. B.
Acrosomal proteinase and proteinase inhibitor of human
spermatozoa. Science 177: 702-703.

Zaneveld, L. J. D., Polakoski, K. L. & Williams, W. L. Properties
of a proteolytic enzyme from rabbit sperm acrosomes. Biol.
Reprod. 6: 30-39.

1973

Connors, E. C., Greenslade, F. C. & Davanzo, J. P. The kinetics
of inhibition of rabbit sperm acrosomal proteinase by l-chloro-
3-tosylamido-7-amino-2-heptanone (TLCK). Biol. Reprod. 9: 57-
60.

Fritz, H., Förg-Brey, B. & Umezawa, H. Leupeptin and antipain.
Strong competitive inhibitors of sperm acrosomal proteinase
(boar acrosin) and kallikreins from porcine organs (pancreas,

submand. glands, urine). Hoppe-Seyler's Z. Physiol. Chem. 354: 1304-1306.

Gilboa, E., Elkana, Y. & Rigbi, M. Purification and properties of human acrosin. Eur. J. Biochem. 39: 85-92.

Meizel, S. & Huang-Yang, Y. H. J. Further studies of an inactive form of a trypsin-like enzyme in rabbit testes. Biochem. Biophys. Res. Commun. 53: 1145-1150.

Polakoski, K. L. & McRorie, R. A. Boar acrosin. II. Classification, inhibition and specificity studies of a proteinase from sperm acrosomes. J. Biol. Chem. 248: 8183-8188.

Polakoski, K. L., McRorie, R. A. & Williams, W. L. Boar acrosin. I. Purification and preliminary characterization of a proteinases from boar sperm acrosomes. J. Biol. Chem. 248: 8178-8182.

Schill, W. B. Acrosin activity in human spermatozoa. Methodological investigations. Arch. Dermol. Forsch. 248: 257-273.

Schirren, C. & Eweis, A. Akrosinaktivität in menschlichen Spermatozoen und Spermatozoendichte im Ejakulat. Andrologia 5: 81-83.

Schleuning, W.-D., Schiessler, H. & Fritz, H. Highly purified acrosomal proteinase (boar acrosin): isolation by affinity chromatography using benzamidine-cellulose and stabilization. Hoppe-Seyler's Z. Physiol. Chem. 354: 550-554.

Suominen, J., Kaufman, M. H. & Setchell, B. P. Prevention of fertilization in vitro by an acrosin inhibitor from rete testis fluid of the ram. J. Reprod. Fertil. 34: 385-388.

Zaneveld, L. J. D., Polakoski, K. L. & Williams, W. L. A proteinase and proteinase inhibitor of mammalian sperm acrosomes. Biol. Reprod. 9: 219-225.

Zaneveld, L. J. D., Schumacher, G. F. B. & Travis, J. Human sperm acrosomal proteinase: antibody inhibition and immunologic dissimilarity to human pancreatic trypsin. Fertil. Steril. 24: 479-484.

Zaneveld, L. J. D., Schumacher, G. F. B., Fritz, H., Fink, E. & Jaumann, E. Interaction of human sperm acrosomal proteinase with human seminal plasma proteinase inhibitors. J. Reprod. Fertil. 32: 525-529.

1974

Fritz, H., Schleuning, W.-D. & Schill, W. B. Biochemistry and clinical significance of the trypsin-like proteinase acrosin from boar and human spermatozoa. In: Proteinase Inhibitors (Fritz, H., Tschesche, H., Greene, L. J. & Truscheit, E. eds), pp. 118-127, Springer-Verlag, Berlin.

Garner, D. L. & Cullison, R. F. Partial purification of bovine acrosin by affinity chromatography. J. Chromatogr. 92: 445-449.

McRorie, R. A. & Williams, W. L. Biochemistry of mammalian fertilization. Annu. Rev. Biochem. 43: 777-803.

Meizel, S., Mukerji, S. K. & Huang-Yang, Y. H. J. Biochemical studies of rabbit testis and epididymal sperm proacrosin. J. Cell Biol. 63: 222a only.

Polakoski, K. L.   Partial purification and characterization of proacrosin from boar sperm. Fed. Proc. Fed. Am. Soc. Exp. Biol. 33: 1380 only.

Polakoski, K. L. & Williams, W. L.   Studies on the purification and characterization of boar acrosin. In: Proteinase Inhibitors (Fritz, H., Tschesche, H., Greene, L. J. & Truscheit, E. eds), pp. 128-135, Springer-Verlag, Berlin.

Schill, W. B.   Quantitative determination of acrosin activity in human spermatozoa. Fertil. Steril. 25: 703-712.

Schill, W. B. & Wolff, H. H.   Ultrastructure of human sperm acrosome and determination of acrosin EC-3.4.21.10 activity under conditions of semen preservation. Int. J. Fertil. 19: 217-223.

Schill, W.-B.   The influence of glycerol on the extractability of acrosin from human spermatozoa. Hoppe-Seyler's Z. Physiol. Chem. 355: 225-228.

Schleuning, W.-D. & Fritz, H.   Some characteristics of highly purified boar sperm acrosin. Hoppe-Seyler's Z. Physiol. Chem. 355: 125-130.

Srivastava, P. N., Munnell, J. F., Yang, C. H. & Foley, C. W. Sequential release of acrosomal membranes and acrosomal enzymes of ram spermatozoa. J. Reprod. Fertil. 36: 363-372.

Stambaugh, R. & Smith, M.   Amino acid content of rabbit acrosomal proteinase and its similarity to human trypsin. Science 186: 745-746.

Zaneveld, L. J. D., Schumacher, G. F. B., Tauber, P. F. & Propping, D.   Proteinase inhibitors and proteinases of human semen. In: Proteinase Inhibitors (Fritz, H., Tschesche, H., Greene, L. J. & Truscheit, E. eds), pp. 136-146, Springer-Verlag, Berlin.

1975

Brown, C. R. & Hartree, E. F.   An acrosin inhibitor in ram spermatozoa that does not originate from the seminal plasma. Hoppe-Seyler's Z. Physiol. Chem. 356: 1909-1913.

Brown, C. R., Andani, Z. & Hartree, E. F.   Studies on ram acrosin. Fluorimetric titration of operational molarity with 4-methylumbelliferyl p-guanidinobenzoate. Biochem. J. 149: 147-154.

Brown, C. R., Andani, Z. & Hartree, E. F.   Studies on ram acrosin. Isolation from spermatozoa, activation by cations and organic solvents and influence of cations on its reaction with inhibitors. Biochem. J. 149: 133-146.

Fritz, H., Schleuning, W.-D., Schiessler, H., Schill, W.-B., Wendt, V. & Winkler, G.   Boar, bull and human sperm acrosin - isolation, properties and biological aspects. In: Proteases and Biological Control (Reich, E., Rifkin, D. B. & Shaw, E. eds), pp. 715-735, Cold Spring Harbor Laboratory, New York.

Garner, D. L.   Improved zymographic detection of bovine acrosin. Anal. Biochem. 67: 688-694.

Garner, D. L., Easton, M. P., Munson, M. E. & Doane, M. A.

Immunofluorescent localisation of bovine acrosin. J. Exp. Zool.
151: 127-131.

Harrison, R. A. P. Aspects of the enzymology of mammalian
spermatozoa. In: The Biology of the Male Gamete (Duckett, J. G.
& Racey, P. A. eds), vol. 7, pp. 301-316, Academic Press, New
York.

Huang-Yang, Y. H. J. & Meizel, S. Purification of rabbit testis
proacrosin and studies of its active form. Biol. Reprod. 12:
232-238.

Meizel, S. & Mukerji, S. K. Proacrosin from rabbit epididymal
spermatozoa: partial purification and initial biochemical
characterization. Biol. Reprod. 13: 83-93.

Morton, D. B. Acrosomal enzymes: immunochemical localization of
acrosin and hyaluronidase in ram spermatozoa. J. Reprod.
Fertil. 45: 375-378.

Mukerji, S. K. & Meizel, S. Conversion of rabbit testis
proacrosin to acrosin. FEBS Lett. 54: 269-273.

Mukerji, S. K. & Meizel, S. The molecular transformation of
rabbit testis proacrosin into acrosin. Arch. Biochem. Biophys.
168: 720-721.

Schiessler, H., Schleuning, W.-D. & Fritz, H. Cleavage specificity
of boar acrosin on polypeptide substrates, ribonuclease and
insulin B-chain. Hoppe-Seyler's Z. Physiol. Chem. 356: 1931-
1936.

Schill, W. B. Acrosin activity of cryo preserved human
spermatozoa. Fertil. Steril. 26: 711-720.

Schill, W. B. & Fritz, H. Nα-Benzoyl-L-arginine ethyl ester
splitting activity (acrosin) in human spermatozoa and seminal
plasma during aging in vitro. Hoppe-Seyler's Z. Physiol. Chem.
356: 83-90.

Schill, W. B., Schleuning, W.-D., Fritz, H., Wendt, V. &
Heimburger, N. Immunofluorescent localization of acrosin in
spermatozoa by boar acrosin antibodies. Naturwissenschaften 62:
540 only.

Schleuning, W. D., Hell, R., Schiessler, H. & Fritz, H. Boar
acrosin. I. Modified isolation procedure, active-site titration
and evidence for the presence of sialic acid. Hoppe-Seyler's Z.
Physiol. Chem. 356: 1915-1921.

Schleuning, W.-D., Kolb, H. J., Hell, R. & Fritz, H. Boar
acrosin. II. Amino acid composition, amino terminal residue and
molecular weight estimations by ultracentrifugation. Hoppe-
Seyler's Z. Physiol. Chem. 356: 1923-1929.

Stambaugh, R. & Smith, M. Studies on the enzymatic and molecular
nature of acrosomal proteinase. In: Proteases in Biological
Control (Reich, E., Rifkin, D. B. & Shaw, E. eds), pp. 707-714,
Cold Spring Harbor Laboratory, New York.

Wendt, V., Leidl, W. & Fritz, H. The influence of various
proteinase inhibitors on the gelatinolytic effect of ejaculated
and uterine boar spermatozoa. Hoppe-Seyler's Z. Physiol. Chem.
356: 1073-1078.

Wendt, V., Leidl, W. & Fritz, H. The lysis effect of bull
spermatozoa on gelatin substrate film. Methodical investigations.
Hoppe-Seyler's Z. Physiol. Chem. 356: 315-323.
Zaneveld, L. J. D., Polakoski, K. L. & Schumacher, G. F. B. The
proteolytic enzyme systems of mammalian genital tract secretions
and spermatozoa. In: Proteases and Biological Control (Reich,
E., Rifkin, D. B. & Shaw, E. eds), pp. 683-706, Cold Spring
Harbor Laboratory, New York.

1976

Allen, G. J. Evidence for the dissimilarity of acrosin and
pancreatic trypsin as revealed by a sensitive immunoassay. J.
Exp. Zool. 200: 185-190.
Bhattacharyya, A. K., Zaneveld, L. J. D., Dragoje, B. M.,
Schumacher, G. F. B. & Travis, J. Inhibition of human sperm
acrosin by synthetic agents. J. Reprod. Fertil. 47: 97-100.
Brantner, J. H., Medicus, R. G. & McRorie, R. A. Fractionation of
proteolytic enzymes by affinity chromatography on Sepharose
aminocaproyl proflavin. J. Chromatog. 129: 97-105.
Brown, C. R. & Hartree, E. F. Comparison of neutral proteinase
activities in cock and ram spermatozoa and observations on a
proacrosin in cock spermatozoa. J. Reprod. Fertil. 46: 155-164.
Brown, C. R. & Hartree, E. F. Effects of acrosin inhibitors on
the soluble and membrane-bound forms of ram acrosin, and a
reappraisal of the role of the enzyme in fertilization. Hoppe-
Seyler's Z. Physiol. Chem. 357: 57-65.
Brown, C. R. & Hartree, E. F. Identification of acrosin in mouse
spermatozoa. J. Reprod. Fertil. 46: 249-251.
Church, K. E. & Graves, C. N. Loss of acrosin from bovine
spermatozoa following cold shock: protective effects of seminal
plasma. Cryobiology 13: 341-346.
Hartree, E. F. Ethics for authors: a case history of acrosin.
Perspect. Biol. Med. 19: 82-91.
Hartree, E. F. & Brown, C. R. Re-examination of the trypsin-like
proteinases of mammalian and avian semen. Ann. Biol. Anim.
Biochim. Biophys. 16: 173 only.
Johnson, L. A., Pursel, V. G., Chaney, N. & Garner, D. L.
Comparison of several extraction procedures for boar spermatozoal
acrosin. Biol. Reprod. 15: 79-83.
McRorie, R. A., Turner, R. B., Bradford, M. M. & Williams, W. L.
Acrolysin, the aminoproteinase catalyzing the initial conversion
of proacrosin to acrosin in mammalian fertilization. Biochem.
Biophys. Res. Commun. 71: 492-498.
Meizel, S. & Lui, C. W. Evidence for the role of a trypsin-like
enzyme in the hamster sperm acrosome reaction. J. Exp. Zool.
195: 137-144.
Meizel, S. & Mukerji, S. K. Biochemical studies of proacrosin and
acrosin from hamster cauda epididymal spermatozoa. Biol.
Reprod. 14: 444-450.
Morton, D. B. Immunoenzymic studies on acrosin and hyaluronidase
in ram spermatozoa. In: Immunobiology of Gametes (Edidin, M. &

Johnson, M. H. eds), pp. 115-155, Alden Press, Oxford.

Morton, D. B. Lysosomal enzymes in mammalian spermatozoa. In: Lysosomes in Biology and Pathology (Dingle, J. T. & Dean, R. T. eds), vol. 5, pp. 203-256, North-Holland Publishing Co., Amsterdam.

Multamäki, S. & Suominen, J. Distribution and removal of the acrosin of bull spermatozoa. Int. J. Fertil. 21: 69-81.

Polakoski, K. L. & Zaneveld, L. J. D. Proacrosin. Methods Enzymol. 45: 325-329.

Schill, W. B. & Fritz, H. Enhancement of sperm acrosin activity by glycerol-pretreatment - quantitative estimations. Arch. Derm. Res. 256: 1-11.

Schleuning, W.-D. & Fritz, H. Sperm acrosin. Methods Enzymol. 45: 330-342.

Schleuning, W.-D., Hell, R. & Fritz, H. Multiple forms of boar acrosin and their relationship to proenzyme activation. Hoppe-Seyler's Z. Physiol. Chem. 357: 207-212.

Schleuning, W.-D., Hell, R. & Fritz, H. Multiple forms of human acrosin: isolation and properties. Hoppe-Seyler's Z. Physiol. Chem. 357: 855-865.

Shams Borhan, G., Schleuning, W.-D., Tschesche, H. & Fritz, H. Occurrence of multiple forms of bull and ram acrosin during proenzyme activation and inhibition of activation by p-nitrophenyl-p'-guanidino-benzoate. Hoppe-Seyler's Z. Physiol. Chem. 357: 667-671.

Shams Borhan, G., Tschesche, H. & Fritz, H. Activation of chymotrypsinogen by boar acrosin and its prevention by antiboar acrosin rabbit γ-globulins. Hoppe-Seyler's Z. Physiol. Chem. 357: 1229-1233.

Stambaugh, R. & Smith, M. Sperm enzymes and their role in fertilization. Prog. Reprod. Biol. 1: 222-232.

Stambaugh, R. & Smith, M. Sperm proteinase release during fertilization of rabbit ova. J. Exp. Zool. 197: 121-125.

Tobias, P. S. & Schumacher, G. F. B. The extraction of acrosin from human spermatozoa. Biol. Reprod. 15: 187-194.

Yang, C. H. & Srivastava, P. N. Inhibition of bull and rabbit sperm enzymes by α-chlorohydrin. J. Reprod. Fertil. 46: 289-293.

1977

Fléchon, J. E., Huneau, D., Brown, C. R. & Harrison, R. A. P. Immunocytochemical localization of acrosin in the anterior segment of the acrosomes of ram, boar and bull spermatozoa. Ann. Biol. Anim. Biochim. Biophys. 17: 749-758.

Garner, D. L. & Easton, M. P. Immunofluorescent localization of acrosin in mammalian spermatozoa. J. Exp. Zool. 200: 157-162.

Garner, D. L., Reamer, S. A., Johnson, L. A. & Lessley, B. A. Failure of immunoperoxidase staining to detect acrosin in the equatorial segment of spermatozoa. J. Exp. Zool. 201: 309-315.

Gwatkin, R. B. L., Wudl, L., Hartree, E. F. & Fink, E. Prevention of fertilization by exposure of hamster eggs to soluble acrosin.

J. Reprod. Fertil. 50: 359-361.

Hartree, E. F. Spermatozoa, eggs and proteinases. Biochem. Soc. Trans. 5: 375-394.

Morton, D. B. The occurrence and function of proteolytic enzymes in the reproductive tract of mammals. In: Proteinases in Mammalian Cells and Tissues (Barrett, A. J. ed.), pp. 445-500, North-Holland Publishing Co., Amsterdam.

Parrish, R. F. & Polakoski, K. L. Effect of polyamines on the activity of acrosin and the activation of proacrosin. Biol. Reprod. 17: 417-422.

Polakoski, K. L. & Parrish, R. F. Boar proacrosin. Purification and preliminary activation studies of proacrosin isolated from ejaculated boar sperm. J. Biol. Chem. 252: 1888-1894.

Polakoski, K. L., Zahler, W. L. & Paulson, J. D. Demonstration of proacrosin and quantitation of acrosin in ejaculated human spermatozoa. Fertil. Steril. 28: 668-670.

Tobias, P. S. & Schumacher, G. F. B. Observation of two proacrosins in extracts of human spermatozoa. Biochem. Biophys. Res. Commun. 74: 434-439.

Zahler, W. L. & Polakoski, K. L. Benzamidine as an inhibitor of proacrosin activation in bull sperm. Biochim. Biophys. Acta 480: 461-468.

                                    1978

Bhattacharyya, A. K. & Zaneveld, L. J. D. Release of acrosin and acrosin inhibitor from human spermotozoa. Fertil. Steril. 30: 70-78.

Brown, C. R. & Harrison, R. A. P. The activation of proacrosin in spermatozoa from ram, bull and boar. Biochim. Biophys. Acta 526: 202-217.

Brown, C. R. & Hartree, E. F. Studies on ram acrosin. Activation of proacrosin accompanying the isolation of acrosin from spermatozoa, and purification of the enzyme by affinity chromatography. Biochem. J. 175: 227-238.

Fink, E. & Wendt, V. Demonstration of acrosin in mouse spermatozoa. J. Reprod. Fertil. 53: 75-76.

Green, D. P. L. The activation of proteolysis in the acrosome reaction of guinea-pig sperm. J. Cell Sci. 32: 153-164.

Green, D. P. L. & Hockaday, A. R. The histochemical localization of acrosin in guinea-pig sperm after the acrosome reaction. J. Cell Sci. 32: 177-184.

Kettner, C., Springhorn, S., Shaw, E., Müller, W. & Fritz, H. Inactivation of boar acrosin by peptidyl-arginyl-chloromethanes. Comparison of the reactivity of acrosin, trypsin and thrombin. Hoppe-Seyler's Z. Physiol. Chem. 359: 1183-1191.

Parrish, R. F. & Polakoski, K. L. An apparent high molecular weight form of boar proacrosin resulting from the presence of a protein that binds to proacrosin. Anal. Biochem. 87: 108-113.

Parrish, R. F. & Polakoski, K. L. Boar m_α-acrosin: purification and characterization of the initial active enzyme resulting from the conversion of boar proacrosin to acrosin. J. Biol. Chem.

253: 8428-8432.
Parrish, R. F., Straus, J. W., Polakoski, K. L. & Dombrose, F. A.
Phospholipid vesicle stimulation of proacrosin activation.
Proc. Natl. Acad. Sci. U.S.A. 75: 149-152.
Zimmerman, R. E. & Burck, R. J. The loss of a low molecular
weight acrosin inhibitor from acrosomes during capacitation.
Proc. Soc. Exp. Biol. Med. 158: 491-495.

1979

Dudkiewicz, A. B., Srivastava, P. N., Yang, C. H. & Williams, W.
L. Extraction of human and rabbit acrosomes: a comparison of
sequential and sonication methods. Andrologia 11: 355-366.
Harrison, R. A. P. & Brown, C. R. The zymogen form of acrosin in
testicular, epididymal and ejaculated spermatozoa from ram.
Gamete Res. 2: 75-87.
Morton, D. B. & McAnulty, P. A. The effect on fertility of
immunizing female sheep with ram sperm acrosin and hyaluronidase.
J. Reprod. Immunol. 1: 61-73.
Mukerji, S. K. & Meizel, S. Rabbit testis proacrosin.
Purification, molecular weight estimation, and amino acid and
carbohydrate composition of the molecule. J. Biol. Chem. 254:
11721-11728.
Parrish, R. F. & Polakoski, K. L. Mammalian sperm proacrosin-
acrosin system. Int. J. Biochem. 10: 391-395.
Parrish, R. F., Goodpasture, J. C., Zaneveld, L. J. D. &
Polakoski, K. L. Polyamine inhibition of the conversion of
human proacrosin to acrosin. J. Reprod. Fertil. 57: 239-243.
Polakoski, K. L., Clegg, C. D. & Parrish, R. F. Identification of
the in vivo sperm proacrosin into acrosin conversion sequence.
Int. J. Biochem. 10: 483-488.
Wincek, T. J., Parrish, R. F. & Polakoski, K. L. Fertilization: a
uterine glycosaminoglycan stimulates the conversion of sperm
proacrosin to acrosin. Science 203: 553-554.

Entry 1.48

SEMININ

Summary

EC Number: (3.4.21.-)

Earlier names: Chymotrypsin-like enzyme of human seminal plasma.

Distribution: Seminal plasma of man and other species, probably originating from the prostate gland.

Source: Human seminal plasma.

Action: Acts on gelatin, casein and other proteins, and has also been reported to hydrolyze Ac-Tyr-OMe and Bz-Tyr-OEt.

Requirements: Optimal pH 7.5-8.0.

Substrate, usual: Casein.

Inhibitors: Hg^{2+}(1mM). (The enzyme is reported to be unaffected by Dip-F, Tos-Lys-CH_2Cl, Tos-Phe-CH_2Cl, EDTA, iodoacetate, cysteine, pepstatin, chymostatin, aprotinin, soybean inhibitors, and α_1-proteinase inhibitor, so classification as a serine proteinase is clearly presumptive).

Molecular properties: Mol.wt. 25,000-35,000.

Bibliography

1955

Lundquist, F., Thorsteinsson, T. & Buus, O. Purification and properties of some enzymes in human seminal plasma. Biochem. J. 59: 69-79.

1972

Fritz, H., Arnhold, M., Förg-Brey, B., Zaneveld, L. J. D. & Schumacher, G. F. B. [Behaviour of the chymotrypsin-like proteinase from human sperm plasma towards protein proteinase inhibitors.] Hoppe-Seyler's Z. Physiol. Chem. 353: 1651-1654.

Syner, F. N. & Moghissi, K. S. Properties of proteolytic enzymes and inhibitors in human semen. In: Biology of Mammalian Fertilization and Implantation (Moghissi, K. S., Kamran, S. & Hafez, E. S. E. eds), pp. 3-18, Thomas, Springfield.

Syner, F. N. & Moghissi, K. S. Purification and properties of a human seminal proteinase. Biochem. J. 126: 1135-1140.

1973

Fritz, H., Förg-Brey, B. & Umezawa, H. Leupeptin and antipain. Strong competitive inhibitors of sperm acrosomal proteinase (boar acrosin) and kallikreins from porcine organs (pancreas, submand. glands, urine). Hoppe-Seyler's Z. Physiol. Chem. 354: 1304-1306.

1974

Suominen, J. The purification and new properties of the neutral proteinase in human semen. Int. J. Fertil. 19: 121-128.

1975

Zaneveld, L. J. D., Polakoski, K. L. & Schumacher, G. F. B. The proteolytic enzyme systems of mammalian genital tract secretions and spermatozoa. In: Proteases and Biological Control (Reich, E., Rifkin, D. B. & Shaw, E. eds), pp. 683-706, Cold Spring Harbor Laboratory, New York.

1977

Morton, D. B. The occurrence and function of proteolytic enzymes in the reproductive tract of mammals. In: Proteinases in Mammalian Cells and Tissues (Barrett, A. J. ed.), pp. 445-500, North-Holland Publishing Co., Amsterdam.

Entry 1.49

SUBMANDIBULAR PROTEINASE A

Summary

EC Number: (3.4.21.-)

Earlier names: Esteroprotease.

Distribution: Submandibular salivary glands (equivalent to submaxillary glands) of rodents (especially males), and perhaps other mammals.

Source: Mouse submandibular gland.

Action: Degrades proteins and some synthetic substrates of trypsin.

Requirements: Optimal pH 7.5-8.0.

Substrate, usual: Tos-Arg-OMe, casein.

Substrate, special: Ac-Met-1-ONap with a diazonium salt, to stain for activity [1].

Inhibitors: Dip-F.

Molecular properties: Mol.wt. 30,000. The amino acid composition has been reported.

Comment: A preparation of this enzyme has been offered commercially as a highly specific agent for the cleavage of arginyl bonds. The submandibular arginine esterases have been resolved electrophoretically into four components, A to D [1,2], of which A and D are the most abundant; form D is the EGF-binding protein (#1.50)

The mouse submandibular gland contains several other trypsin-like proteinases of similar mol.wt., including epidermal growth factor binding protein, submandibular kallikrein and NGF γ-subunit; often these are distinguished by electrophoretic and immunological criteria, rather than enzymological methods [3].

References
[1] Schaller & von Deimling Anal. Biochem. 93: 251-256, 1979.
[2] Schenkein et al. Biochem. Biophys. Res. Commun. 36: 156-165, 1969.
[3] Bothwell et al. J. Biol. Chem. 254: 7287-7294, 1979.

Bibliography

1966

Riekkinen, P. J., Hopsu-Havu, V. K. & Ekfors, T. O. Purification and characteristics of an alkaline protease from rat submandibular gland. Biochim. Biophys. Acta 118: 604-620.

1967

Angeletti, R. A., Angeletti, P. U. & Calissano, P. Testosterone induction of estero-proteolytic activity in the mouse submaxillary gland. Biochim. Biophys. Acta 139: 372-381.

Attardi, D. G., Schlesinger, M. J. & Schlesinger, S. Submaxillary gland of mouse: properties of a purified protein affecting muscle tissue in vitro. Science 156: 1253-1255.

1968

Calissano, P. & Angeletti, P. Testosterone effect on the synthetic rate of two esteropeptidases in the mouse submaxillary gland. Biochim. Biophys. Acta 156: 51-58.

Schenkein, I., Levy, M., Bueker, E. D. & Tokarsky, E. Nerve growth factor of very high yield and specific activity. Science 159: 640-643.

1969

Grossman, A., Lele, K. P., Sheldon, L., Shenkein, I. & Levy, M. The effect of esteroproteases from mouse submaxillary gland on the growth of rat hepatoma cells in tissue culture. Exp. Cell Res. 54: 260-263.

Schenkein, I., Boesman, M., Tokarsky, E., Fishman, L. & Levy, M. Proteases from mouse submaxillary gland. Biochem. Biophys. Res. Commun. 36: 156-165.

1970

Schenkein, I., Levy, M. & Fishman, L. Mouse submaxillary gland proteases. Methods Enzymol. 19: 672-681.

1971

Ekfors, T. O. & Hopsu-Havu, V. K. Immunofluorescent localization of trypsin-like esteropeptidases in the mouse submandibular gland. Histochem. J. 3: 415-420.

1972

Ekfors, T. O., Suominen, J. & Hopsu-Havu, V. K. Increased vascular permeability and activation of plasminogen by the trypsin-like esterases from the mouse submandibular gland. Biochem. Pharmacol. 21: 1370-1373.

1973

Angeletti, M., Sbaraglia, G. & de Campora, E. Biological properties of esteroprotease isolated from the submaxillary gland of mice. Life Sci. 12: 297-305.

1976

Boesman, M., Levy, M. & Schenkein, I. Esteroproteolytic enzymes from the submaxillary gland. Kinetics and other physicochemical properties. Arch. Biochem. Biophys. 175: 463-476.

1977

Schenkein, I., Levy, M., Franklin, E. C. & Frangione, B. Proteolytic enzymes from the mouse submaxillary gland. Specificity restricted to arginine residues. Arch. Biochem. Biophys. 182: 64-70.

1978

Igarashi, K., Torü, K., Nakamura, Y. & Hirose, S. Relationship between polyamine contents and protease activity in the rat submaxillary gland. Biochim. Biophys. Acta 541: 161-169.

Takuma, T., Tanemura, T., Hosoda, S. & Kumegawa, M. Effects of thyroxine and 5α-dihydrotestosterone on the activities of various enzymes in the mouse submandibular gland. Biochim. Biophys. Acta 541: 143-149.

1979

Bothwell, M. A., Wilson, W. H. & Shooter, E. M. The relationship between glandular kallikrein and growth factor-processing proteases of mouse submaxillary gland. J. Biol. Chem. 254: 7287-7294.

Schaller, E. & von Deimling, O. Methionine-α-naphthyl ester, a useful chromogenic substrate for esteroproteases of the mouse submandibular gland. Anal. Biochem. 93: 251-256.

Entry 1.50
EPIDERMAL GROWTH FACTOR BINDING PROTEIN

Summary

EC Number: (3.4.21.-)

Earlier names: Submandibular proteinase, form D. Abbreviation: EGFbp.

Distribution: Submandibular salivary glands (equivalent to submaxillary glands) of mouse, especially adult male, and probably other species.

Source: Submandibular glands of adult male mouse.

Action: Hydrolyzes Bz-Arg-OEt and some other substrates of trypsin. May be responsible for the "processing" of a higher mol.wt. precursor of epidermal growth factor.

Requirements: Optimal pH 7.8-8.5 for Bz-Arg-OEt.

Substrate, usual: (No specific substrate has been described. The protein is usually detected by electrophoretic or immunological methods.)

Substrate, special: Ac-Met-1-ONap with a diazonium salt to stain for activity [1].

Inhibitors: Aprotinin. (Not soybean trypsin inhibitor.)

Molecular properties: Mol.wt. about 29,000, pI 5.6, A280,1% 15.6. Contains glucosamine and galactosamine, but no free thiol. The amino acid composition has been reported [2].

Comment: Epidermal growth factor binding protein occurs as a component of the high mol.wt. EGF complex obtained from the submaxillary glands of adult male mice. The complex (62,000 mol.wt.) seems to consist of two molecules each of the growth factor itself (mol.wt. 6045) and the binding protein. It has been shown that the C-terminal Arg of the growth factor is necessary for the interaction with the binding protein, and it has been suggested that this is bound at the active site of the proteinase [3], in the same way as was postulated for the nerve growth factor (see #1.51).

References
[1] Schaller & von Deimling Anal. Biochem. 93: 251-256, 1979.
[2] Taylor et al. J. Biol. Chem. 249: 2188-2194, 1974.
[3] Bothwell et al. J. Biol. Chem. 254: 7287-7294, 1979.

Bibliography

1970

Taylor, J. M., Cohen, S. & Mitchell, W. M. Epidermal growth factor: high and low molecular weight forms. Proc. Natl. Acad. Sci. U.S.A. 67: 164–171.

1972

Byyny, R., Orth, D. N. & Cohen, S. Radioimmunoassay of epidermal growth factors. Endocrinology 90: 1261–1266.

1974

Schenkein, I., Levy, M., Bueker, E. D. & Wilson, J. Immunological and enzymatic evidence for the absence of an esteroproteolytic enzyme (protease "D") in the submandibular gland of the Tfm mouse. Endocrinology 94: 840–844.

Taylor, J. M., Mitchell, W. M. & Cohen, S. Characterization of the binding protein for epidermal growth factor. J. Biol. Chem. 249: 2188–2194.

1976

Lembach, K. J. Induction of human fibroblast proliferation by epidermal growth factor: enhancement by an EGF-binding arginine esterase and by ascorbate. Proc. Natl. Acad. Sci. U.S.A. 73: 183–187.

Server, A. C., Sutter, A. & Shooter, E. M. Modification of the epidermal growth factor affecting the stability of its high molecular weight complex. J. Biol. Chem. 251: 1188–1194.

1977

Schenkein, I., Levy, M., Franklin, E. C. & Frangione, B. Proteolytic enzymes from the mouse submaxillary gland. Specificity restricted to arginine residues. Arch. Biochem. Biophys. 182: 64–70.

1979

Bothwell, M. A., Wilson, W. H. & Shooter, E. M. The relationship between glandular kallikrein and growth factor-processing proteases of mouse submaxillary gland. J. Biol. Chem. 254: 7287–7294.

Schaller, E. & von Deimling, O. Methionine-α-naphthyl ester, a useful chromogenic substrate for esteroproteases of the mouse submandibular gland. Anal. Biochem. 93: 251–256.

NERVE GROWTH FACTOR GAMMA SUBUNIT

Summary

EC Number: (3.4.21.-)

Earlier names: -

Distribution: Mouse submandibular salivary gland (equivalent to submaxillary gland) as the γ subunit of 7S nerve growth factor (NGF), espcially in the adult male. Occurs elsewhere in much lower concentration.

Source: Male mouse submandibular gland.

Action: Has general proteolytic activity, and hydrolyzes synthetic substrates of trypsin.

Requirements: pH 7.5-8.0. Activation by a chelating agent (see below).

Substrate, usual: $Bz-Arg-NH_2$, $Bz-Arg-NPhNO_2$.

Inhibitors: Benzamidine. Aprotinin. (Not soybean or lima bean trypsin inhibitors, ovomucoid, human $α_1$-proteinase inhibitor, antithrombin III or Cl inhibitor). NGF β-subunit.

Molecular properties: 7S NGF (mol.wt. 140,000) is a multisubunit protein, $α_2βγ_2$, in which β represents βNGF, itself a dimeric unit with growth factor activity. The complex is greatly stabilized by Zn^{2+}. A280,1% of the γ subunit is 15.6.

Comment: The γ subunit (mol.wt. about 25,000) probably cleaves an arginyl bond in a precursor of the β chain, and then remains bound to the new C-terminal Arg through the active site, enzymically inactive. Equilibrium dissociation of the subunits, favoured by chelation of the Zn^{2+}, liberates active free γ subunit. Aprotinin and $Bz-Arg-NPhNO_2$ occupy the active site, inhibiting reassociation [1].

References
[1] Bothwell & Shooter   J. Biol. Chem. 253: 8458-8464, 1978.

Bibliography

1968

Greene, L. A., Shooter, E. M. & Varon, S. Enzymatic activities of mouse nerve growth factor and its subunits. Proc. Natl. Acad.

Sci. U.S.A. 60: 1383-1388.

1969

Greene, L. A., Shooter, E. M. & Varon, S. Subunit interaction and enzymatic activity of mouse 7S nerve growth factor. Biochemistry 8: 3735-3741.

Smith, A. P., Greene, L. A., Fisk, H. R., Varon, S. & Shooter, E. M. Subunit equilibria of the 7S nerve growth factor protein. Biochemistry 8: 4918-4926.

1974

Moore, J. B., Jr, Mobley, W. C. & Shooter, E. M. Proteolytic modification of the β nerve growth factor protein. Biochemistry 13: 833-840.

1975

Pattison, S. E. & Dunn, M. F. On the relationship of zinc ion to the structure and function of the 7S nerve growth factor protein. Biochemistry 14: 2733-2739.

1976

Pattison, S. E. & Dunn, M. F. On the mechanism of divalent metal ion chelator induced activation of the 7S nerve growth factor esteropeptidase. Thermodynamics and kinetics of activation. Biochemistry 15: 3696-3703.

1977

Au, A.-M. J. & Dunn, M. F. Reaction of the basic trypsin inhibitor from bovine pancreas with the chelator-activated 7S nerve growth factor esteropeptidase. Biochemistry 16: 3958-3966.

Berger, E. A. & Shooter, E. M. Evidence for pro-β-nerve growth factor, a biosynthetic precursor to β-nerve growth factor. Proc. Natl. Acad. Sci. U.S.A. 74: 3647-3651.

Pantazis, N. J., Blanchard, M. H., Arnason, B. G. W. & Young, M. Molecular properties of the nerve growth factor secreted by L cells. Proc. Natl. Acad. Sci. U.S.A. 74: 1492-1496.

Server, A. C. & Shooter, E. M. Nerve growth factor. Adv. Protein Chem. 31: 339-409.

1978

Bothwell, M. A. & Shooter, E. M. Thermodynamics of interaction of the subunits of 7S nerve growth factor. The mechanism of activation of the esteropeptidase activity by chelators. J. Biol. Chem. 253: 8458-8464.

Orenstein, N. S., Dvorak, H. F., Blanchard, M. H. & Young, M. Nerve growth factor - protease that can activate plasminogen. Proc. Natl. Acad. Sci. U.S.A. 75: 5497-5500.

1979

Bothwell, M. A., Wilson, W. H. & Shooter, E. M. The relationship between glandular kallikrein and growth factor-processing proteases of mouse submaxillary gland. J. Biol. Chem. 254: 7287-7294.

Calissano, P. & Levi-Montalcini, R. Is NGF an enzyme? Nature 280: 359 only.

Schaller, E. & von Deimling, O. Methionine-α-naphthyl ester, a useful chromogenic substrate for esteroproteases of the mouse submandibular gland. Anal. Biochem. 93: 251-256.

Young, M. Proteolytic activity of nerve growth factor: a case of autocatalytic activation. Biochemistry 18: 3050-3055.

Entry 1.52

TISSUE KALLIKREIN

Summary

EC Number: 3.4.21.8 (plasma kallikrein inappropriately bears the same number).

Earlier names: Glandular kallikrein. (Sometimes also called kininogenase, although this term means any enzyme releasing a peptide with kinin-like activity from kininogen, and thus includes trypsin and many other proteinases). The mouse submandibular kallikrein seems to be identical with βNGF endopeptidase [1]. Precursor: tissue prokallikrein (or prekallikrein).

Distribution: Principally pancreas, pancreatic juice, salivary glands, saliva, and urine of all mammals investigated. The precursor has been detected only in pancreas.

Source: Pig pancreas, submandibular glands and urine; human, rat and horse urine.

Action: Liberates kallidin (lysyl-bradykinin) from kininogens, by cleavage of -Met-Lys- and -Phe-Arg≠Ser- bonds in bovine kininogens. Very little general proteolytic activity, but cleaves polypeptides rich in basic amino acids. Hydrolyzes arginine esters. Most active against Phe-(or Leu)-Arg≠X bonds.

Requirements: Optimal pH 7-8.

Substrate, usual: Bz-Arg-OEt, Tos-Arg-OMe, plasma kininogens.

Substrate, special: Ac-Phe-Arg-OEt, Pro-Phe-Arg-NMec, D-Val-Leu-Arg-NPhNO$_2$ (S-2266) for greater sensitivity and specificity. Z-Ser-Pro-Phe-Arg-2-NNapOMe with a diazonium salt to stain for activity. Gbz-NPhNO$_2$ for active site titration.

Inhibitors: Dip-F, peptidyl-arginyl chloromethanes, leupeptin, antipain, aprotinin (but see Comment). (Not Tos-Lys-CH$_2$Cl, soybean trypsin inhibitor or α-macroglobulin, in contrast to plasma kallikrein; less sensitive to Zn^{2+} than NGF γ-subunit and EGF-binding protein.)

Molecular properties: Pig: obtained from autolyzed pancreas, mainly as β-kallikrein, a two-chain form apparently arising from cleavage of a single-chain form during isolation. Occurs as β-kallikreins A (mol.wt. 27,100; 5.6% carbohydrate) and B (mol.wt. 28,900; 11.5% carbohydrate) of similar amino acid

composition. The sequence of the 232 amino acids is strongly
homologous with that of other pancreatic serine proteinases.
Submandibular and urinary kallikreins are both single-chain
α-kallikreins and occur as the major B form (apparent mol.wt.
39,600) and an A form (apparent mol.wt. 35,900). They show
immunological identity with pancreatic β-kallikrein, and have
the same N-terminal sequence, but contain more carbohydrate. The
slight differences from pancreatic β-kallikrein in amino acid
composition (about 3 residues for submandibular kallikrein) and
in catalytic properties may be due to the intrachain split (but
see Comment). The multiple forms (pI 3.2-4.4) are decreased in
number by treatment with neuraminidase.

Man: For the enzyme from urine, mol.wts. from 25,000 to
48,000 have been reported.

Comment: Tissue prokallikrein, activatable by trypsin, exists in
pancreas and (human) urine, but has not been detected in
salivary glands. In the salivary glands of rats and mice
(especially adult males) tissue kallikrein exists in association
with several closely related trypsin-like enzymes[1].

Treating tissue kallikrein as a single enzyme may be an
oversimplification. On the basis of the available data, a case
could be made for recognizing distinct, but very closely related
forms, i.e. pancreatic kallikrein, renal (and urinary)
kallikrein, and submandibular kallikrein. These would have
highly homologous sequences, and would show a high degree of
immunological identity, but would differ in degree of
glycosylation, and number of peptide chains. There are some
differences in catalytic activity, and renal kallikrein is
distinctly less sensitive to inhibition by aprotinin than the
other types. It is possible that such differences could be
accounted for by the post-translational modifications of a
single polypeptide chain, however.

References
[1] Bothwell et al. J. Biol. Chem. 254: 7287-7294, 1979.

Bibliography

1961

Trautschold, I. & Werle, E. Spektrophotometrische Bestimmung des
Kallikreins und seiner Inaktivatoren. Hoppe-Seyler's Z.
Physiol. Chem. 325: 48-59.
Webster, M. E. & Pierce, J. V. Action of kallikreins on synthetic
ester substrates. Proc. Soc. Exp. Biol. Med. 107: 186-191.
Werle, E., Trautschold, I. & Leysath, G. Isolierung und Struktur
des Kallidins. Hoppe-Seyler's Z. Physiol. Chem. 326: 174-176.

1962

Habermann, E. Trennung und Reinigung von Pankreaskallikreinen.
Hoppe-Seyler's Z. Physiol. Chem. 328: 15-23.

Prado, E. S., Prado, J. L. & Brandi, C. M. W. Further purification
and some properties of horse urinary kallikrein. Arch. Int.
Pharmacodyn. 87: 358–374.

1963

Moriya, H., Pierce, J. V. & Webster, M. E. Purification and some
properties of three kallikreins. Ann. N. Y. Acad. Sci. 104:
172–185.
Prado, J. L., Prado, E. S., Brandi, C. M. W. & Katchburian, A. C.
Some properties of highly purified horse urinary kallikrein.
Ann. N. Y. Acad. Sci. 104: 186–189.
Werle, E. & Trautschold, I. Kallikrein, kallidin, kallikrein
inhibitors. Ann. N. Y. Acad. Sci. 104: 117–129.

1964

Habermann, E. & Blennemann, G. Über substrate und
Reaktionsprodukte der kininbildenden Enzyme Trypsin, Serum- und
Pankreas-kallikrein sowie von Crotalusgift. Naunyn-
Schmiedeberg's Arch. Pharmacol. 249: 357–373.

1965

Brandi, C. M. W., Mendes, J., Paiva, A. C. M. & Prado, E. S.
Proteolysis of salmine by horse urinary kallikrein. Biochem.
Pharmacol. 14: 1665–1671.
Diniz, C. R., Pereira, A. A., Barroso, J. & Mares-Guia, M. On the
specificity of urinary kallikreins. Biochem. Biophys. Res.
Commun. 21: 448–453.
Moriya, H., Yamazaki, K., Fukushima, H. & Moriwaki, C. Biochemical
studies on kallikreins and their related substances. II. An
improved method of purification of hog pancreatic kallikrein and
its some biological properties. J. Biochem. 58: 208–213.

1966

Habermann, E. Strukturaufklärung kininliefernder Peptide aus
Rinderserum-Kininogen. Naunyn-Schmiedeberg's Arch. Pharmacol.
253: 474–483.
Habermann, E. & Müller, B. Zur enzymatischen Spaltung peptischer
kininliefernder Fragmente (PKF) sowie von Rinderserum-Kininogen.
Naunyn-Schmiedeberg's Arch. Pharmacol. 253: 464–473.
Kramer, C., Prado, E. S. & Prado, J. L. Inhibition by ions of the
esterolytic activity of horse urinary kallikrein. Med.
Pharmacol. Exp. 15: 389–398.
Prado, E. S., Stella, R. C. R., Roncada, M. J. & Prado, J. L.
Action of horse urinary kallikrein on arginine and lysine-
peptides. In: International Symposium on Vaso-active
Polypeptides. Bradykinin and Related Kinins (Rocha e Silva, M. &
Rothschild, H. A. eds), pp. 141–147, Edart, Sao Paulo.

1967

Ekfors, T. O., Malmiharju, T., Riekkinen, P. J. & Hopsu-Havu, V.
K. The depressor activity of trypsin-like enzymes purified from
rat submandibular gland. Biochem. Pharmacol. 16: 1634–1636.

Ekfors, T. O., Riekkinen, P. J., Malmiharju, T. & Hopsu-Havu, V. K. Four isozymic forms of a peptidase resembling kallikrein purified from the rat submandibular gland. Hoppe-Seyler's Z. Physiol. Chem. 348: 111-118.

Fiedler, F. & Werle, E. Vorkommen zweier Kallikreinogene im Schweinepankreas und Automation der Kallikrein- und Kallikreinogenbestimmung. Hoppe-Seyler's Z. Physiol. Chem. 348: 1087-1089.

Fritz, H., Eckert, I. & Werle, E. Isolierung und Charakterisierung von sialinsäurehaltigem und sialinsäurefreiem kallikrein aus Schweinepancreas. Hoppe-Seyler's Z. Physiol. Chem. 348: 1120-1132.

Hochstrasser, K. & Werle, E. Über kininliefernde Peptide aus pepsinverdauten Rinderplasmaproteinen. Hoppe-Seyler's Z. Physiol. Chem. 348: 177-182.

Mares-Guia, M. & Diniz, C. R. Studies on the mechanism of rat urinary kallikrein catalysis, and its relation to catalysis by trypsin. Arch. Biochem. Biophys. 121: 750-756.

1968

Babel, I., Stella, R. C. R. & Prado, E. S. Action of horse urinary kallikrein on synthetic derivatives of bradykinin. Biochem. Pharmacol. 17: 2232-2234.

Fiedler, F. & Werle, E. Activation, inhibition, and pH-dependence of the hydrolysis of α-N-benzoyl-L-arginine ethyl ester catalyzed by kallikrein from porcine pancreas. Eur. J. Biochem. 7: 27-33.

Frey, E. K., Kraut, H., Werle, E., Vogel, R., Zickgraf-Rüdel, G. & Trautschold, I. Das Kallikrein-Kinin-System und seine Inhibitoren. Ferdinand Enke Verlag, Stuttgart.

Paskhina, T. S., Zukova, V. P., Egorova, T. P., Narticova, V. F., Karpova, A. V. & Guseva, M. V. [Reaction of N-α-tosyl-L-lysyl-chloromethane with kallikrein from blood plasma of rabbit and with kallikreins from urine and saliva of man.] Biokhimia 33: 745-752.

1969

Fiedler, F., Müller, B. & Werle, E. The inhibition of porcine pancreas kallikrein by di-isopropyl-fluorophosphate. Kinetics, stoichiometry, and nature of the group phosphorylated. Eur. J. Biochem. 10: 419-425.

Geratz, J. D. Synthetic inhibitors and ester substrates of pancreatic kallikrein. Experientia 25: 483-484.

Moriya, H., Kato, A. & Fukushima, H. Further study on purification of hog pancreatic kallikrein. Biochem. Pharmacol. 18: 549-552.

Schröder, E. & Lehmann, M. Über Peptidsynthesen. XLVI. Synthese kininliefernder Kininogensequenzen. Experientia 25: 1126-1127.

Takami, T. [Purification of hog pancreatic kallikreins and their proteolytic activities.] Seikagaku 41: 777-785.

Takami, T. Action of hog pancreatic kallikrein on synthetic substrates and its some other enzymatic properties. J. Biochem.

66: 651-658.

1970

Fiedler, F., Hirschauer, C. & Werle, E. [Purification of prekallikrein B from pig pancreas and properties of different forms of pancreas kallikrein.] Hoppe-Seyler's Z. Physiol. Chem. 351: 225-238.

Fiedler, F., Müller, B. & Werle, E. [The characterization of different porcine kallikreins by means of diisopropyl fluorophosphate.] Hoppe-Seyler's Z. Physiol. Chem. 351: 1002-1006.

Fräki, J. E., Jansén, C. T. & Hopsu-Havu, V. K. Human sweat kallikrein. Acta Derm. Venereol. (Stockh.) 50: 321-326.

Kato, H. & Suzuki, T. Substrate specificities of kinin releasing enzymes: hog pancreatic kallikrein and snake venom kininogenase. J. Biochem. 68: 9-17.

Nustad, K. The relationship between kidney and urinary kininogenase. Br. J. Pharmacol. 39: 73-86.

Pierce, J. V. Purification of mammalian kallikreins, kininogens, and kinins. In: Bradykinin, Kallidin and Kallikrein. Handbook of Experimental Pharmacology (Erdös, E. G. ed.), vol. 25, pp. 21-51, Springer Verlag, Berlin.

Trautschold, I. Assay methods in the kinin system. In: Bradykinin, Kallidin and Kallikrein. Handbook of Experimental Pharmacology (Erdös, E. G. ed.), vol. 25, pp. 52-81, Springer Verlag, Berlin.

Vogel, R. & Werle, E. Kallikrein inhibitors. In: Bradykinin, Kallidin and Kallikrein. Handbook of Experimental Pharmacology (Erdös, E. G. ed.), vol. 25, pp. 213-249, Springer Verlag, Berlin.

Webster, M. E. Kallikreins in glandular tissues. In: Bradykinin, Kallidin and Kallikrein. Handbook of Experimental Pharmacology (Erdös, E. G. ed.), vol. 25, pp. 131-155, Springer Verlag, Berlin.

Webster, M. E. & Prado, E. S. Glandular kallikreins from horse and human urine and from hog pancreas. Methods Enzymol. 19: 681-699.

Werle, E. Discovery of the most important kallikreins and kallikrein inhibitors. In: Bradykinin, Kallidin and Kallikrein. Handbook of Experimental Pharmacology (Erdös, E. G. ed.), vol. 25, pp. 1-6, Springer Verlag, Berlin.

1971

Beaven, V. H., Pierce, J. V. & Pisano, J. J. A sensitive isotopic procedure for the assay of esterase activity: measurement of human urinary kallikrein. Clin. Chim. Acta 32: 67-73.

Fiedler, F., Müller, B. & Werle, E. Die Aminosäuresequenz im Bereich des Serins im aktiven Zentrum des Kallikreins aus Schweinepankreas. Hoppe-Seyler's Z. Physiol. Chem. 352: 1463-1464.

Markwardt, F., Walsmann, P., Richter, M., Klöcking, H.-P.,

Drawert, J. & Landmann, H. Aminoalkylbenzolsulfofluoride als
Fermentinhibitoren. Die Pharmazie 26: 401-404.
Moriwaki, C. & Schachter, M. Kininogenase of the guinea-pigs's
coagulating gland and the release of bradykinin. J. Physiol.
219: 341-353.
Moriwaki, C., Inoue, N., Hojima, Y. & Moriya, H. [A new assay
method for the esterolytic activity of kallikreins with
chromotropic acid.] Yakugaku Zasshi 91: 413-416.
Moriya, H., Todoki, N., Moriwaki, C. & Hojima, Y. A new sensitive
assay method of kallikrein-like arginine-esterases. J. Biochem.
69: 815-817.
Porcelli, G. & Croxatto, H. Purification of kininogenase from rat
urine. Ital. J. Biochem. 20: 66-88.
Prado, E. S., Webster, M. E. & Prado, J. L. Kallidin (lysyl-
bradykinin), the kinin formed from horse plasma by horse urinary
kallikrein. Biochem. Pharmacol. 20: 2009-2015.

                                1972

Ekfors, T. O., Suominen, J. & Hopsu-Havu, V. K. Increased
vascular permeability and activation of plasminogen by the
trypsin-like esterases from the mouse submandibular gland.
Biochem. Pharmacol. 21: 1370-1373.
Fiedler, F., Müller, B. & Werle, E. Active site titration of pig
pancreatic kallikrein with p-nitrophenyl p-guanidinobenzoate.
FEBS Lett. 24: 41-44.
Fiedler, F., Müller, B. & Werle, E. The reaction of pig
pancreatic kallikrein with cinnamoyl and indoleacryloyl
imidazoles. FEBS Lett. 22: 1-4.
Fritz, H. & Förg-Brey, B. Zur Isolierung von Organ- und
Harnkallikreinen durch Affinitätschromatography. Spezifische
Bindung an wasserunlösliche Inhibitorderivate und Dissoziation
der Komplex mit kompetitiven Hemmstoffen (Benzamidin). Hoppe-
Seyler's Z. Physiol. Chem. 353: 901-905.
Fujimoto, Y., Moriya, H., Yamaguchi, K. & Moriwaki, C. Detection
of aginine esterase of various kallikrein preparations on
gellified electrophoretic media. J. Biochem. 71: 751-754.
Geratz, J. D. Kinetic aspects of the irreversible inhibition of
trypsin and related enzymes by p-[m(m-fluorosulfonyl-
phenylureido)phenoxyethoxy]benzamidine. FEBS Lett. 20: 294-296.
Kutzbach, C. & Schmidt-Kastner, G. Kallikrein from pig pancreas.
Purification, separation of components A and B, and
crystallization. Hoppe-Seyler's Z. Physiol. Chem. 353: 1099-
1106.
Seki, T., Nakajima, T. & Erdös, E. G. Colon kallikrein, its
relation to the plasma enzyme. Biochem. Pharmacol. 21: 1227-
1235.

                                1973

Angeletti, M., Sbaraglia, G., DeCampora, E. & Frati, L.
Biological properties of esteroprotease isolated from the
submaxillary gland of mice. Life Sci. 12: 297-305.

Arens, A. & Haberland, G. L. Determination of kallikrein activity in animal tissue using biochemical methods. In: Kininogenases. Kallikrein. 1st Symposium on Physiological Properties and Pharmacological Rational (Haberland, G. L. & Rohen, J. W. eds), pp. 43-53, Schattauer Verlag, Stuttgart.

Fiedler, F., Leysath, G. & Werle, E. Hydrolysis of amino-acid esters by pig-pancreatic kallikrein. Eur. J. Biochem. 36: 152-159.

Fritz, H., Förg-Brey, B. & Umezawa, H. Leupeptin and antipain. Strong competitive inhibitors of sperm acrosomal proteinase (boar acrosin) and kallikreins from porcine organs (pancreas, submand. glands, urine). Hoppe-Seyler's Z. Physiol. Chem. 354: 1304-1306.

Fujimoto, Y., Moriya, H. & Moriwaki, C. Studies on human salivary kallikrein. I. Isolation of human salivary kallikrein. J. Biochem. 74: 239-246.

Fujimoto, Y., Moriwaki, C. & Moriya, H. Studies on human salivary kallikrein. II. Properties of purified salivary kallikrein. J. Biochem. 74: 247-252.

Geratz, J. D. & Fox, L. L. B. On the irreversible inhibition of α-chymotrypsin, trypsin, pancreatic kallikrein, thrombin and elastase by N-(βpyridylmethyl)-3,4-dichlorophenoxyacetamide-p-fluorosulfonylacetanilide bromide. Biochim. Biophys. Acta 321: 296-302.

Geratz, J. D., Whitmore, A. C., Cheng, M. C.-F. & Piantadosi, C. Diamidino-α,ω-diphenoxyalkanes. Structure-activity relationships for the inhibition of thrombin, pancreatic kallikrein, and trypsin. J. Med. Chem. 16: 970-975.

Kutzbach, C. & Schmidt-Kastner, G. Properties of highly purified hog pancreatic kallikrein. In: Kininogenases. Kallikrein. 1st Symposium on Physiological Properties and Pharmacological Rational (Haberland, G. L. & Rohen, J. W. eds), pp. 23-35, Schattauer Verlag, Stuttgart.

Moriya, H., Matsuda, Y., Fujimoto, Y., Hojima, Y. & Moriwaki, C. Some aspects of multiple components on the kallikrein. Human urinary kallikrein (HUK). In: Kininogenases. Kallikrein. 1st Symposium on Physiological Properties and Pharmacological Rational (Haberland, G. L. & Rohen, J. W. eds), pp. 37-42, Schattauer Verlag, Stuttgart.

1974

Hial, V., Diniz, C. R. & Mares-Guia, M. Purification and properties of a human urinary kallikrein (kininogenase). Biochemistry 13: 4311-4318.

Moriwaki, C., Hojima, Y. & Moriya, H. A proposal - use of combined assays of kallikrein measurement. Chem. Pharm. Bull. 22: 975-983.

Moriwaki, C., Watanuki, N., Fujimoto, Y. & Moriya, H. Further purification and properties of kininogenase from the guinea pig's coagulating gland. Chem. Pharm. Bull. 22: 628-633.

Mutt, V. & Said, S. I. Structure of the porcine vasoactive

intestinal octacosapeptide. The amino-acid sequence. Use of kallikrein in its determination. Eur. J. Biochem. 42: 581-589.

Nustad, K. & Pierce, J. V. Purification of rat urinary kallikreins and their specific antibody. Biochemistry 13: 2312-2319.

Porcelli, G., Marini-Bettòlo, G. B., Croxatto, H. R. & DiIorio, M. Purification and chemical studies on human urinary kallikrein. Ital. J. Biochem. 23: 44-55.

Porcelli, G., Marini-Bettòlo, G. B., Croxatto, H. R. & DiIorio, M. Purification and chemical studies on rabbit urinary kallikrein. Ital. J. Biochem. 23: 154-164.

Silva, E., Diniz, C. R. & Mares-Guia, M. Rat urinary kallikrein. Purification and properties. Biochemistry 13: 4304-4310.

Worthington, K. J. & Cuschieri, A. A study on the kinetics of pancreatic kallikrein with the substrate benzoyl arginine ethyl ester using a spectrophotometric assay. Clin. Chim. Acta 52: 129-136.

Zuber, M. & Sache, E. Isolation and characterization of porcine pancreatic kallikrein. Biochemistry 13: 3098-3110.

1975

Fiedler, F., Hirschauer, C. & Werle, E. Characterization of pig pancreatic kallikreins A and B. Hoppe-Seyler's Z. Physiol. Chem. 356: 1879-1891.

Geratz, J. D., Cheng, M. C.-F. & Tidwell, R. R. New aromatic diamidines with central α-oxyalkane or α,ω-dioxyalkane chains. Structure-activity relationships for the inhibition of trypsin, pancreatic kallikrein, and thrombin and for the inhibition of the overall coagulation process. J. Med. Chem. 18: 477-481.

Hojima, Y., Matsuda, Y., Moriwaki, C. & Moriya, H. Homogeneity and multiplicity forms of glandular kallikreins. Chem. Pharm. Bull. 23: 1120-1127.

Hojima, Y., Moriwaki, C. & Moriya, H. Isolation of dog, rat, and hog pancreatic kallikreins by preparative disc electrophoresis and their properties. Chem. Pharm. Bull. 23: 1128-1136.

Kaizu, T. & Margolis, H. S. Studies on rat renal cortical cell kallikrein. 1. Separation and measurement. Biochim. Biophys. Acta 411: 305-315.

Loeffler, L. J., Mar, E.-C., Geratz, J. D. & Fox, L. B. Synthesis of isosteres of p-amidinophenylpyruvic acid. Inhibitors of trypsin, thrombin and pancreatic kallikrein. J. Med. Chem. 18: 287-292.

Moriwaki, C., Kizuki, K. & Moriya, H. Kininogenase from the coagulating gland. In: Kininogenases. Kallikrein. 2nd Symposium on Physiological Properties and Pharmacological Rationale - Reproduction (Haberland, G. L., Rohen, J. W., Schirren, C. & Huber, P. eds), pp. 23-31, Schattauer Verlag, Stuttgart.

Moriya, H., Hojima, Y., Yamashita, Y. & Moriwaki, C. Some studies on dog and rat pancreatic kallikreins. In: Kininogenases. Kallikreins. 3rd Symposium on Physiological Properties and Pharmacological Rationale - Proliferation, Reparation, Function (Haberland, G. L., Rohen, J. W., Blümel, G. & Huber, P. eds), pp. 19–26, Schattauer Verlag, Stuttgart.

Nustad, K., Pierce, J. V. & Vaaje, K. Synthesis of kallikreins by rat kidney slices. Br. J. Pharmacol. 53: 229–234.

Pisano, J. J. Chemistry and biology of the kallikrein-kinin system. In: Proteases and Biological Control (Reich, E., Rifkin, D. B. & Shaw, E. eds), pp. 199–222, Cold Spring Harbor Laboratory, New York.

Porcelli, G., Marini-Bettòlo, G. B., Croxatto, H. R. & DiIorio, M. Purification and chemical studies on rat urinary kallikrein. Ital. J. Biochem. 24: 175–187.

Sampaio, C. A. M. & Prado, E. S. Action of proteolytic enzymes upon horse urinary kallikrein. Biochem. Pharmacol. 24: 1438–1439.

Ward, P. E., Gedney, C. D., Dowben, R. M. & Erdös, E. G. Isolation of membrane-bound kallikrein and kininase. Biochem. J. 151: 755–758.

1976

Brandtzaeg, P., Gautvik, K. M., Nustad, K. & Pierce, J. V. Rat submandibular gland kallikreins: purification and cellular localization. Br. J. Pharmacol. 56: 155–167.

Bretzel, G. & Hochstrasser, K. Liberation of an acid stable proteinase inhibitor from the human inter-α-trypsin inhibitor by the action of kallikrein. Hoppe-Seyler's Z. Physiol. Chem. 357: 487–489.

Carretero, O. A., Oza, N. B., Piwonska, A., Ocholik, T. & Scicli, A. Measurement of urinary kallikrein by kinin radioimmunoassay. Biochem. Pharmacol. 25: 2265–2270.

Fiedler, F. Pig pancreatic kallikrein: structure and catalytic properties. In: Chemistry and Biology of the Kallikrein-Kinin System in Health and Disease, DHEW Publication No. (NIH) 76-791 (Pisano, J. J. & Austen, K. F. eds), pp. 93–95, U.S. Government Printing Office, Washington.

Fiedler, F. Pig pancreatic kallikreins A and B. Methods Enzymol. 45: 289–303.

Geratz, J. D., Cheng, M. C.-F. & Tidwell, R. R. Novel bis(benzamidine) compounds with an aromatic central link. Inhibitors of thrombin, pancreatic kallikrein, trypsin and complement. J. Med. Chem. 19: 634–639.

Han, Y. N., Kato, S. & Iwanaga, S. Identification of Ser-Leu-Met-Lys-bradykin isolated from chemically modified high-molecular weight bovine kininogen. FEBS Lett. 71: 45–48.

Imanari, T., Kaizu, T., Yoshida, H., Yates, K., Pierce, J. V. & Pisano, J. J. Radiochemical assays for human urinary, salivary, and plasma kallikreins. In: Chemistry and Biology of the Kallikrein-Kinin System in Health and Disease. DHEW Publication No. (NIH) 76-791 (Pisano, J. J. & Austen, K. F., eds), pp. 205–213, U.S. Government Printing Office, Washington.

Kato, H., Han, Y. N., Iwanaga, S., Suzuki, T. & Komiya, M. Bovine plasma HMW and LMW kininogens: isolation and characterization of the polypeptide fragments produced by plasma and tissue kallikreins. In: Kinins - Pharmacodynamics and Biological Roles (Sicuteri, F., Back, N. & Haberland, G. L. eds), pp. 135-150, Plenum Press, New York.

Mares-Guia, M., Diniz, C. R., Silva, E. & Hial, V. A comparative study of urinary kallikreins isolated from man and the rat. In: Chemistry and Biology of the Kallikrein-Kinin System in Health and Disease, DHEW Publication No. (NIH) 76-791 (Pisano, J. J. & Austen, K. F. eds), pp. 97-101, U.S. Government Printing Office, Washington.

Matsuda, Y., Miyazaki, K., Moriya, H., Fujimoto, Y., Hojima, Y. & Moriwaki, C. Studies on urinary kallikreins. I. Purification and characterization of human urinary kallikreins. J. Biochem. 80: 671-679.

Matsuda, Y., Moriya, H., Moriwaki, C., Fujimoto, Y. & Matsuda, M. Fluorometric method for the assay of kallikrein-like arginine esterases. J. Biochem. 79: 1197-1200.

Moriwaki, C., Hojima, Y. & Schachter, M. Purification of kallikrein from cat submaxillary gland. In: Kinins - Pharmacodynamics and Biological Roles (Sicuteri, F., Back, N. & Haberland, G. L. eds), pp. 151-156, Plenum Press, New York.

Moriwaki, C., Miyazaki, K., Matsuda, Y., Moriya, H., Fujimoto, Y. & Ueki, H. Dog renal kallikrein: purification and some properties. J. Biochem. 80: 1277-1285.

Nustad, K., Gautvik, K. M. & Pierce, J. V. Glandular kallikreins: purification, characterization and biosynthesis. In: Chemistry and Biology of the Kallikrein-Kinin System in Health and Disease, DHEW Publication No. (NIH) 76-791 (Pisano, J. J. & Austen, K. F. eds), pp. 77-92, U.S. Government Printing Office, Washington.

Oza, N. B., Amin, V. M., McGregor, R. K., Scicli, A. G. & Carretero, O. A. Isolation of rat urinary kallikrein and properties of its antibodies. Biochem. Pharmacol. 25: 1607-1612.

Porcelli, G., Cozzari, C., DiIorio, M., Croxatto, R. H. & Angeletti, P. Isolation and partial characterization of a kallikrein from mouse submaxillary glands. Ital. J. Biochem. 25: 337-348.

Prado, E. S., Araújo-Viel, M. S., Sampaio, M. U., Prado, J. L. & Paiva, A. C. M. Enzymatic specificity of horse urinary kallikrein. In: Chemistry and Biology of the Kallikrein-Kinin System in Health and Disease, DHEW Publication No. (NIH) 76-791 (Pisano, J. J. & Austen, K. F. eds), pp. 103-105, U.S. Government Printing Office, Washington.

Sampaio, C. A. M. & Prado, E. S. Active-site labelling of kallikreins by chloromethylketone derivatives. Gen. Pharmac. 7: 163-166.

Sampaio, M. U., Galembeck, F., Paiva, A. C. M. & Prado, E. S. Kinetics of the hydrolysis of synthetic substrates by horse urinary kallikrein and trypsin. Gen. Pharmac. 7: 167-171.

Tidwell, R. R., Fox, L. L. & Geratz, J. D. Aromatic tris-amidines.

A new class of highly active inhibitors of trypsin-like
proteases. Biochim. Biophys. Acta 445: 729-738.
Tschesche, H., Ehret, W., Godec, G., Hirschauer, C., Kutzbach, C.,
Schmidt-Kastner, G. & Fiedler, F. The primary structure of pig
pancreatic kallikrein B. In: Kinins - Pharmacodynamics and
Biological Roles (Sicuteri, F., Back, N. & Haberland, G. L.
eds), pp. 123-133, Plenum Press, New York.
Tschesche, H., Mair, G., Förg-Brey, B. & Fritz, H. Isolation of
porcine urinary kallikrein. In: Kinins - Pharmacodynamics and
Biological Roles (Sicuteri, F., Back, N. & Haberland, G. L.
eds), pp. 119-122, Plenum Press, New York.
Vahtera, E. & Hamberg, U. Absence of binding of pancreatic and
urinary kallikreins to α_2-macroglobulin. Biochem. J. 157: 521-
524.
Ward, P. E., Erdös, E. G., Gedney, C. D., Dowben, R. M. &
Reynolds, R. C. Isolation of membrane-bound renal enzymes that
metabolize kinins and angiotensins. Biochem. J. 157: 643-650.

                                  1977

Béchet, J.-J., Dupaix, A., Roucous, C. & Bonamy, A.-M.
Inactivation of proteases and esterases by halomethylated
derivates of dihydrocoumarins. Biochemie 59: 241-246.
Claeson, G., Aurell, L., Karlsson, G. & Friberger, P. Substrate
structure and activity relationship. In: New Methods for the
Analysis of Coagulation using Chromogenic Substrates (Witt, I.
ed.), pp. 37-54, Walter de Gruyter, Berlin.
Eeckhout, Y. & Vaes, G. Further studies on the activation of
procollagenase, the latent precursor of bone collagenase.
Effects of lysosomal cathepsin B, plasmin and kallikrein, and
spontaneous activation. Biochem. J. 166: 21-31.
Fiedler, F., Ehret, W., Godec, G., Hirschauer, C., Kutzbach, C.,
Schmidt-Kastner, G. & Tschesche, H. The primary structure of
pig pancreatic kallikrein B. In: Kininogenases. Kallikrein. 4th
Symposium on Physiological Properties and Pharmacological
Rationale (Haberland, G. L., Rohen, J. W. & Suzuki, T. eds), pp.
7-14, Schattauer Verlag, Stuttgart.
Fiedler, F., Hirschauer, C. & Fritz, H. Inhibition of three
porcine glandular kallikreins by chloromethyl ketones. Hoppe-
Seyler's Z. Physiol. Chem. 358: 447-451.
Fink, E. Development of a radioimmunoassay for hog pancreatic
kallikrein and its application in pharmacological studies. In:
Kininogenases. Kallikrein. 4th Symposium on Physiological
Properties and Pharmacological Rationale (Haberland, G. L.,
Rohen, J. W. & Suzuki, T. eds), pp. 47-54, Schattauer Verlag,
Stuttgart.
Fritz, H., Fiedler, F., Dietl, T., Warwas, M., Truscheit, E.,
Kolb, H. J., Mair, G. & Tschesche, H. On the relationships
between porcine pancreatic, submandibular, and urinary
kallikreins. In: Kininogenases. Kallikrein. 4th Symposium on
Physiological Properties and Pharmacological Rationale
(Haberland, G. L., Rohen, J. W. & Suzuki, T. eds), pp. 15-28,

Schattauer Verlag, Stuttgart.

Geiger, R., Mann, K. & Bettels, T. Isolation of human urinary kallikrein by affinity chromatography. J. Clin. Chem. Clin. Biochem. 15: 479-483.

Halabi, M. & Back, N. Kinetic studies with serine proteases and protease inhibitors utilizing a synthetic nitro-anilide chromogenic substrate. In: Kininogenases. Kallikrein. 4th Symposium on Physiological Properties and Pharmacological Rationale. (Haberland, G. L., Rohen, J. W. & Suzuki, T. eds), pp. 35-46, Schattauer Verlag, Stuttgart.

Hojima, Y., Isobe, M. & Moriya, H. Kallikrein inhibitors in rat plasma. J. Biochem. 81: 37-46.

Hojima, Y., Yamashita, M., Ochi, N., Moriwaki, C. & Moriya, H. Isolation and some properties of dog and rat pancreatic kallikreins. J. Biochem. 81: 599-610.

Mann, K. & Geiger, R. Radioimmunoassay of human urinary kallikrein. In: Kininogenases. Kallikrein. 4th Symposium on Physiological Properties and Pharmacological Rationale (Haberland, G. L., Rohen, J. W. & Suzuki, T. eds), pp. 55-61, Schattauer Verlag, Stuttgart.

Modéer, T. Characterization of kallikrein from human saliva isolated by use of affinity chromatography. Acta Odont. Scand. 35: 31-39.

Modéer. T. Human saliva kallikrein. Biological properties of saliva kallikrein isolated by use of affinity chromatography. Acta Odont. Scand. 35: 23-29.

Morita, T., Kato, H., Iwanaga, S., Takada, K., Kimura, T. & Sakakibara, S. New fluorogenic substrates for α-thrombin, factor X, kallikreins and urokinase. J. Biochem. 82: 1495-1498.

Moriwaki, C., Fujimori, H. & Toyono, Y. Intestinal kallikrein and the influence of kinin on the intestinal transport. In: Kininogenases. Kallikrein. 4th Symposium on Physiological Properties and Pharmacological Rationale (Haberland, G. L., Rohen, J. W. & Suzuki, T. eds), pp. 283-290, Schattauer Verlag, Stuttgart.

Moriya, H., Matsuda, Y., Mijazaki, K., Moriwaki, C. & Hojima, Y. Some aspects of urinary and renal, kallikreins. In: Kininogenases. Kallikrein. Physiological Properties and Pharmacological Rationale (Haberland, G. L., Rohen, J. W. & Suzuki, T. eds), pp. 29-34, Schattauer Verlag, Stuttgart.

ole-MoiYoi, O., Austen, K. F. & Spragg, J. Kinin-generating and esterolytic activity of purified human urinary kallikrein (urokallikrein). Biochem. Pharmacol. 26: 1893-1900.

ole-MoiYoi, O., Spragg, J., Halbert, S. P. & Austen, K. F. Immunologic reactivity of purified human urinary kallikrein (urokallikrein) with antiserum directed against human pancreas. J. Immunol. 118: 667-672.

Oza, N. B. A direct assay for urinary kallikrein. Biochem. J. 167: 305-307.

Paskhina, T. S., Makevnina, L. G., Egorova, T. P. & Levina, G. O.

[Identification of kinins releasing by pig pancreatic kallikrein from rabbit low molecular weight kininogen.] Biokhimiya 42: 1144-1147.

Proud, D., Bailey, G. S., Nustad, K. & Gautvik, K. M. The immunological similarity of rat glandular kallikreins. Biochem. J. 167: 835-838.

Proud, D., Bailey, G. S., Ørstavik, T. B. & Nustad, K. The location of prekallikrein in the rat pancreas. Biochem. Soc. Trans. 5: 1402-1404.

Sampaio, C. A. M., Saad, A. D. & Grisolia, D. Hydrolysis of histones by horse urinary kallikreins. Experientia 33: 292-294.

1978

Carretero, O. A., Amin, V. M., Ocholik, T., Scicli, A. G. & Koch, J. Urinary kallikrein in rats bred for their susceptibility and resistance to the hypertensive effect of salt. A new radioimmunoassay for its direct determination. Circ. Res. 42: 727-731.

Chao, J. Activity and structural changes of purified kallikrein with detergent treatment. Arch. Biochem. Biophys. 185: 549-556.

Dietl, T., Kruck, J. & Fritz, H. Localization of kallikrein in porcine pancreas and submandibular gland as revealed by the indirect immunofluorescence technique. Hoppe-Seyler's Z. Physiol. Chem. 359: 499-505.

Fiedler, F., Geiger, R., Hirschauer, C. & Leysath, G. Peptide esters and nitroanilides as substrates for the assay of human urinary kallikrein. Hoppe-Seyler's Z. Physiol. Chem. 359: 1667-1673.

Figarella, C., Sampaio, M. U. & Greene, L. J. Comparison of the kininogenase activity of human pancreatic trypsins and porcine kallikrein on Met-Lys-bradykinin and human plasma kininogen. Hoppe-Seyler's Z. Physiol. Chem. 359: 1225-1228.

Fink, E. & Güttel, C. Development of a radioimmunoassay for pig pancreatic kallikrein. J. Clin. Chem. Clin. Biochem. 16: 381-385.

Geratz, J. D. & Tidwell, R. R. Current concepts on action of synthetic thrombin inhibitors. Haemostasis 7: 170-176.

Han, Y. N., Kato, H., Iwanaga, S. & Komiya, M. Actions of urinary kallikrein, plasmin, and other kininogenases on bovine plasma high-molecular-weight kininogen. J. Biochem. 83: 223-235.

Kraut, H. 50 Years Ago: hormone research. Kallikrein and its inhibitors. Trends Biochem. Sci. 3: 283-284.

Miller, J. A., Clappison, B. H. & Johnston, C. I. Kallikrein and plasmin as activators of inactive renin. Lancet 1: 1375-1376.

Moriya, H., Fukuoka, Y., Hojima, Y. & Moriwaki, C. Measurement of the sugar contents and their effect on the electrophoretical behaviors of multiple forms of hog pancreatic kallikrein. Chem. Pharm. Bull. 26: 3178-3185.

Nustad, K., Ørstavik, T. B., Gautvik, K. M. & Pierce, J. V. Glandular kallikreins. Gen. Pharmac. 9: 1-9.

Ørstavik, T. B. An immunohistochemical study of kallikrein in the rat parotid and exorbital lacrimal glands. Arch. Oral Biol. 23: 1023-1025.

Ørstavik, T. B. The distribution and secretion of kallikrein in some exocrine organs of the rat. Acta Physiol. Scand. 104: 431-442.

Ørstavik, T. B. & Glenner, G. G. Localization of kallikrein and its relation to other trypsin-like esterases in the rat pancreas. A comparison with the submandibular gland. Acta Physiol. Scand. 103: 384-393.

Odya, C. E., Levin, Y., Erdös, E. G. & Robinson, C. J. G. Soluble dextran complexes of kallikrein, bradykinin and enzyme inhibitors. Biochem. Pharmacol. 27: 173-179.

ole-MoiYoi, O., Spragg, J. & Austen, K. F. Inhibition of human urinary kallikrein (urokallikrein) by anti-enzyme Fab. J. Immunol. 121: 66-71.

Oza, N. B. & Ryan, J. W. A simple high-yield procedure for isolation of human urinary kallikrein. Biochem. J. 171: 285-288.

Porcelli, G., Marini-Bettolo, G. B., Croxatto, H. R., DiIorio, M. & Micotti, G. Purification of renal rat kallikrein and chemical relations with urinary rat kallikrein. Ital. J. Biochem. 27: 201-210.

Schachter, M., Maranda, B. & Moriwaki, C. Localization of kallikrein in the coagulating and submandibular glands of the guinea pig. J. Histochem. Cytochem. 26: 318-321.

Sealey, J. E., Atlas, S. A., Laragh, J. H., Oza, N. B. & Ryan, J. W. Human urinary kallikrein converts inactive to active renin and is a possible physiological activator of renin. Nature 275: 144-145.

Stocker, M. & Hornung, J. Application of bradykinin radioimmunoassay for the measurement of urinary kallikrein activity in rats. Klin. Wochenschr. 56: suppl. I, 127-129.

Tyler, D. W. Localization of renal kallikrein in the dog. Experientia 34: 621-622.

Uddin, M. & Tyler, D. W. Demilune cells of the cat submandibular gland, an unlikely source of kallikrein. Experientia 34: 609-610.

Uddin, M. & Tyler, D. W. Localization of kallikrein in the guinea-pig submandibular gland. Life Sci. 22: 1603-1606.

1979

Abe, K., Kato, H., Sakurai, Y., Itoh, T., Saito, K., Haruyama, T., Otsuka, Y. & Yoshinaga, K. Estimation of urinary kininogenase activity using bovine serum low molecular weight kininogen. In: Kinins-II. Biochemistry, Pathophysiology, and Clinical Aspects (Fujii, S., Moriya, H. & Suzuki, T. eds), pp. 105-114, Plenum Press, New York.

Amouric, M. & Figarella, C. Characterization and purification of a kallikrein from human pancreatic juice and immunological comparison with other kallikreins. Hoppe-Seyler's Z. Physiol. Chem. 360: 457-465.

Amundsen, E., Putter, J., Friberger, O., Knos, M., Larsbraten, M. & Claeson, G. Methods for the determination of glandular kallikrein by means of a chromogenic tripeptide substrate. In: Kinins-II. Biochemistry, Pathophysiology, and Clinical Aspects (Fujii, S., Moriya, H. & Suzuki, T. eds), pp. 83-95, Plenum Press, New York.

Ayling, C. M., Caygill, C. P. J., Hartley, R. E., Sutcliffe, N. & Tsakok, T. I. The evaluation of hog pancreatic kininogenase and methods of measurement. In: Current Concepts in Kinin Research (Haberland, G. L. & Hamberg, U. eds), pp. 173-181, Pergamon Press, Oxford.

Bhoola, K. D., Higginson, J., Lemon, M. J. C., Matthews, R. W. & Whicher, J. T. Immunogenicity of guinea-pig submandibular and coagulating gland kininogenases. In: Kinins-II. Systemic Proteases and Cellular Function (Fujii, S., Moriya, H. & Suzuki, T. eds), pp. 255-259, Plenum Press, New York.

Bhoola, K. D., Lemon, M. & Matthews, R. Kallikrein in exocrine glands. In: Bradykinin, Kallidin and Kallikrein. Handbook of Experimental Pharmacology (Erdös, E. G. ed.), vol. 25 suppl., pp. 489-523, Springer-Verlag, Berlin.

Bothwell, M. A., Wilson, W. H. & Shooter, E. M. The relationship between glandular kallikrein and growth factor-processing proteases of mouse submaxillary gland. J. Biol. Chem. 254: 7287-7294.

Chao, J. & Margolius, H. S. Isozymes of rat urinary kallikrein. Biochem. Pharmacol. 28: 2071-2079.

Chung, A., Ryan, J. W., Pena, G. & Oza, N. B. A simple radioassay for human urinary kallikrein. In: Kinins-II. Biochemistry, Pathophysiology and Clinical Aspects (Fujii, S., Moriya, H. & Suzuki, T. eds), pp. 115-125, Plenum Press, New York.

Claeson, G., Aurell, L., Karlsson, G., Gustavsson, S., Friberger, P., Arielly, A. & Simonsson, R. Design of chromogenic peptide substrates. In: Chromogenic Peptide Substrates: Chemistry and Clinical Usage (Scully, M. F. & Kakkar, V. V. eds), pp. 20-31, Churchill Livingstone, Edinburgh.

Corthorn, J., Imanari, T., Yoshida, H., Kaizu, T., Pierce, J. V. & Pisano, J. J. Isolation of prokallikrein from human urine. In: Kinin-II. Systemic Proteases and Cellular Function (Fujii, S., Moriya, H. & Suzuki, T. eds), pp. 575-579, Plenum Press, New York.

Dietl, T., Huber, C., Geiger, R., Iwanaga, S. & Fritz, H. Inhibition of porcine glandular kallikreins by structurally homologous proteinase inhibitors of the Kunitz (Trasylol) type. Significance of the basic nature of amino acid residues in subside positions for kallikrein inhibition. Hoppe-Seyler's Z. Physiol. Chem. 360: 67-71.

Dittmann, B. & Wimmer, R. Comparison of kinin release and blood pressure activity of porcine pancreatic, submandibular and urinary kallikrein. In: Current Concepts in Kinin Research (Haberland, G. L. & Hamberg, U. eds), pp. 121-126, Pergamon

Press, Oxford.

Fiedler, F. Enzymology of glandular kallikreins. In: Bradykinin, Kallidin and Kallikrein. Handbook of Experimental Pharmacology (Erdös, E. G. ed.), vol. 25 suppl., pp. 103-161, Springer-Verlag, Berlin.

Fiedler, F. & Leysath, G. Substrate specificity of porcine pancreatic kallikrein. In: Kinin-II. Biochemistry, Pathophysiology, and Clinical Aspects (Fujii, S., Moriya, H. & Suzuki, T. eds), pp. 261-271, Plenum Press, New York.

Figueiredo, A. F. S. & Mares-Guia, M. A simple, large scale process for purifying human urinary kallikrein, based on reverse osmosis and ammonium sulfate precipitation. In: Kinins-II. Biochemistry, Pathophysiology, and Clinical Aspects (Fujii, S., Moriya, H. & Suzuki, T. eds), pp. 305-312, Plenum Press, New York, London.

Figueiredo, A. F. S. & Mares-Guia, M. Some properties of human urinary kallikrein obtained through scaled-up purification. In: Current Concepts in Kinin Research (Haberland, G. L. & Hamberg, U. eds), pp. 133-138, Pergamon Press, Oxford.

Fink, E., Dietl, T., Seifert, J. & Fritz, H. Studies on the biological function of glandular kallikrein. In: Kinins-II. Systemic Proteases and Cellular Function (Fujii, S., Moriya, H. & Suzuki, T. eds), pp. 261-273, Plenum Press, New York.

Fukuoka, Y., Hojima, Y., Miyaura, S. & Moriwaki, C. Purification of cat submaxillary kallikrein. J. Biochem. 85: 549-557.

Geiger, R., Fink, E., Stuckstedte, U. & Förg-Brey, B. Isolation and determination of human urinary kallikrein. In: Current Concepts in Kinin Research (Haberland, G. L. & Hamberg, U. eds), pp. 127-132, Pergamon Press, Oxford.

Geiger. R., Stuckstedte, U., Förg-Brey, B. & Fink, E. Human urinary kallikrein - biochemical and physiological aspects. In: Kinins-II. Biochemistry, Pathophysiology, and Clinical Aspects (Fujii, S., Moriya, H. & Suzuki, T. eds), pp. 235-244, Plenum Press, New York.

Inagami, T., Yokosawa, N., Takahashi, N. & Takii, Y. Partial purification of prorenin and activation by kallikreins: a possible link betwen renin and kallikrein systems. In: Kinins-II. Systemic Proteases and Cellular Function (Fujii, S., Moriya, H. & Suzuki, T. eds), pp. 415-428, Plenum Press, New York.

Iwanaga, S., Morita, T., Kato, H., Harda, T., Adachi, N., Sugo, T., Maruyama, I., Takada, K., Kimura, T. & Sakakibara, S. Fluorogenic peptide substrates for proteases in blood coagulation, kallikrein-kinin and fibrinolysis systems. In: Kinins-II. Biochemistry, Pathophysiology, and Clinical Aspects (Fujii, S., Moriya, H. & Suzuki, T. eds), pp. 147-163, Plenum Press, New York.

Kettner, C. & Shaw, E. The susceptibility of urokinase to affinity labeling by peptides of arginine chloromethyl ketone. Biochim. Biophys. Acta 569: 31-40.

Kira, H., Hiraku, S. & Terashima, H. Purification and

characterization of hog pancreatic kallikrein I, II and III.
In: Kinins-II. Biochemistry, Pathophysiology, and Clinical
Aspects (Fujii, S., Moriya, H. & Suzuki, T. eds), pp. 273-290,
Plenum Press, New York.

Lemon, M., Fiedler, F., Förg-Brey, B., Hirschauer, C., Leysath, G.
& Fritz, H. The isolation and properties of pig submandibular
kallikrein. Biochem. J. 177: 159-168 (see also correction:
Biochem. J. 181: page preceding 257).

Levinsky, N. G., ole-MoiYoi, O., Austen, K. F. & Spragg, J.
Measurement of human urinary kallikrein and evidence for non-
kallikrein urinary TAMe esterases by direct immunoassay and by
affinity chromatography. Biochem. Pharmacol. 28: 2491-2495.

Lottspeich, F., Geiger, R., Henschen, A. & Kutzback, C.
N-Terminal amino acid sequence of human urinary kallikrein.
Homology with other serine proteases. Hoppe-Seyler's Z.
Physiol. Chem. 360: 1947-1950.

Makevnina, L. G., Levina, G. O., Sherstjuk, S. F. & Paskhina, T.
S. Fragmentation of rabbit low molecular weight kininogen by
pig pancreatic kallikrein. In: Current Concepts in Kinin
Research (Haberland, G. L. & Hamberg, U. eds), pp. 151-159,
Pergamon Press, Oxford.

Mills, I. H. Renal kallikrein and regulation of blood pressure in
man. In: Bradykinin, Kallidin and Kallikrein. Handbook of
Experimental Pharmacology, (Erdös, E. G. ed.), vol. 25 suppl.,
pp. 549-567, Springer-Verlag, Berlin.

Morris, D. H. & Stewart, J. M. A new assay for urinary kallikrein.
In: Current Concepts in Kinin Research (Haberland, G. L. &
Hamberg, U. eds), pp. 139-143, Pergamon Press, Oxford.

Nustad, K., Gautvik, K. & Ørstavik, T. Radioimmunoassay of rat
submandibular gland kallikrein and the detection of
immunoreactive antigen in blood. In: Kinins-II. Biochemistry,
Pathophysiology, and Clinical Aspects (Fujii, S., Moriya, H. &
Suzuki, T. eds), pp. 225-234, Plenum Press, New York.

Ørstavik, T. B., Nustad, K. & Gautvik, K. M. Localization of
glandular kallikreins and secretion of kallikrein from the major
salivary glands of the rat. In: Kinins-II. Biochemistry,
Pathophysiology, and Clinical Aspects (Fujii, S., Moriya, H. &
Suzuki, T. eds), pp. 439-450, Plenum Press, New York.

ole-MoiYoi, O., Pinkus, G. S., Spragg, J. & Austen, K. F.
Identification of human glandular kallikrein in the beta cell of
the pancreas. N. Engl. J. Med. 300: 1289-1294.

ole-MoiYoi, O., Seldin, D. C., Spragg, J., Pinkus, G. S. & Austen,
K. F. Sequential cleavage of proinsulin by human pancreatic
kallikrein and a human pancreatic kininase. Proc. Natl. Acad.
Sci. U.S.A. 76: 3612-3616.

ole-MoiYoi, O., Spragg, J. & Austen, K. F. Structural studies of
human urinary kallikrein (urokallikrein). Proc. Natl. Acad.
Sci. U.S.A. 76: 3121-3125.

Oza, N. & Ryan, J. W. A direct radioimmunoassay for human urinary
kallikrein. In: Kinins-II. Biochemistry, Pathophysiology, and

Clinical Aspects (Fujii, S., Moriya, H. & Suzuki, T. eds), pp.
97-103, Plenum Press, New York.
Paskhina, T. S. & Levitskiy, A. P. Bradykinin, kallidin, and
kallikreins (survey of the Russian literature). In: Bradykinin,
Kallidin and Kallikrein. Handbook of Experimental Pharmacology
(Erdös, E. G. ed.), vol. 25 suppl., pp. 609-656, Springer-Verlag,
Berlin.
Porcelli, G., Marini-Bettòlo, G. B., Croxatto, H. R. & DiIorio, M.
Chemical relations between renal and urinary kallikrein of rat.
In: Kinins-II. Biochemistry, Pathophysiology, and Clinical
Aspects (Fujii, S., Moriya, H. & Suzuki, T. eds), pp. 221-224,
Plenum Press, New York.
Porcelli, G., Marini-Bettòlo, G. B., Croxatto, H. R. & DiIorio, M.
Purification of horse renal kallikrein and chemical relations
with horse urinary kallikrein. In: Kinins-II. Biochemistry,
Pathophysiology, and Clinical Aspects (Fujii, S., Moriya, H. &
Suzuki, T. eds), pp. 325-333, Plenum Press, New York.
Rabito, S. F., Amin, V., Scicli, A. G. & Carretero, O. A.
Glandular kallikrein in plasma and urine: evaluation of a direct
RIA for its determination. In: Kinins-II. Biochemistry,
Pathophysiology, and Clinical Aspects (Fujii, S., Moriya, H. &
Suzuki, T. eds), pp. 127-142, Plenum Press, New York.
Ryan, J. W., Oza, N. B., Martin, L. B. & Pena, G. A. Components
of the kallikrein-kinin system in urine. In: Kinins-II.
Biochemistry, Pathophysiology, and Clinical Aspects (Fujii, S.,
Moriya, H. & Suzuki, T. eds), pp. 313-323, Plenum Press, New
York.
Saito, H., Goldsmith, G. H., Moroi, M. & Aoki, N. Inhibitory
spectrum of alpha-2-plasmin inhibitor. Proc. Natl. Acad. Sci.
U.S.A. 76: 2013-2017.
Schachter, M. Biological and physiological considerations arising
from cellular localization and other recent observations on
kallikreins. In: Kinins-II. Systemic Proteases and Cellular
Function (Fujii, S., Moriya, H. & Suzuki, T. eds), pp. 249-254,
Plenum Press, New York.
Shimamoto, K., Margolius, H. S., Chao, J. & Crosswell, A. R. A
direct radioimmunoassay of rat urinary kallikrein and comparison
with other measures of urinary kallikrein activity. J. Lab.
Clin. Med. 94: 172-179.
Stewart, J. M. & Morris, D. H. A system for the study of
kallikreins. In: Kinins-II. Biochemistry, Pathophysiology, and
Clinical Aspects (Fujii, S., Moriya, H. & Suzuki, T. eds), pp.
213-218, Plenum Press, New York.
Tschesche, H., Mair, G., Godec, G., Fiedler, F., Ehret, W.,
Hirschauer, C., Lemon, M., Fritz, H., Schmidt-Kastner, G. &
Kutzbach, G. The primary structure of porcine glandular
kallikreins. In: Kinins-II. Biochemistry, Pathophysiology, and
Clinical Aspects (Fujii, S., Moriya, H, & Suzuki, T. eds), pp.
245-260, Plenum Press, New York.
Uchida, K., Yokoshima, A., Niinobe, M., Kato, H. & Fujii, S. Rat

stomach kallikrein: its purification and properties. In:
Kinins-II. Biochemistry, Pathophysiology, and Clinical Aspects
(Fujii, S., Moriya, H. & Suzuki, T. eds), pp. 291-303, Plenum
Press, New York.
Vogel, R. Kallikrein inhibitors. In: Bradykinin, Kallidin and
Kallikrein. Handbook of Experimental Pharmacology (Erdös, E. G.
ed.), vol. 25 suppl., pp. 163-225, Springer-Verlag, Berlin.
Ward, P. E. & Margolius, H. S. Renal and urinary kallikreins.
In: Bradykinin, Kallidin and Kallikrein. Handbook of Experimental
Pharmacology (Erdös, E. G. ed.), vol. 25 suppl., pp. 525-548,
Springer-Verlag, Berlin.
Yokosawa, N., Takahashi, N. & Inagami, T. Isolation of completely
inactive plasma prorenin and its activation by kallikreins: a
possible new link between renin and kallikrein. Biochim.
Biophys. Acta 569: 211-219.
Zimmermann, A., Geiger, R. & Kortmann, H. Similarity between a
kininogenase (kallikrein) from human large intestine and human
urinary kallikrein. Hoppe-Seyler's Z. Physiol. Chem. 360: 1767-
1773.

Entry 1.53

UTERINE TRYPSIN-LIKE PROTEINASE

Summary

EC Number: (3.4.21.-)

Earlier names: -

Distribution: Rat and human uterus, in organelles sedimenting at 12,000 x g.

Source: Rat uterus.

Action: Cleaves arginyl bonds in proteins and synthetic substrates. Activates plasminogen and perhaps latent collagenase.

Requirements: pH 8-9.

Substrate, usual: Bz-Arg-NPhNO$_2$, protamine.

Inhibitors: Dip-F, Tos-Lys-CH$_2$Cl (but see below), antipain, leupeptin. Not soybean trypsin inhibitor or Tos-Phe-CH$_2$Cl.

Molecular properties: Mol.wt. apparently high. Solubilized by 1.5 M KCl.

Comment: The uterine trypsin-like enzyme could well be identical with the chromatin-associated proteinases of thymus (#1.58) and/or liver (#1.59). The bovine uterine activator of latent collagenase [1] was not inhibited by latent collagenase.

References
[1] Woessner   Biochem. J. 161: 535-542, 1977.

Bibliography

1973

Notides, A. C., Hamilton, D. E. & Rudolph, J. H.   The action of a human uterine protease on the estrogen receptor.   Endocrinology 93: 210-216.

1974

Levitz, M., Katz, J., Krone, P., Prochoroff, N. N. & Troll, W. Hormonal influences on histones and template activity in the rat uterus.   Endocrinology 94: 633-640.
Rybák, M., Rybáková, B. & Koutský, J.   Trypsin-like enzymes of human uterus wall and their relationship to plasminogen and kinin-system activator.   Physiol. Bohemoslov. 23: 467-473.

1975

Troll, W., Rossman, T., Katz, J., Levitz, M. & Sugimura, T.
Proteinases in tumor production and hormone action. In:
Proteases and Biological Control (Reich, E., Rifkin, D. B. &
Shaw, E. eds), pp. 977-987, Cold Spring Harbor Laboratory, New
York.

1976

Katz, J., Troll, W., Levy, M., Filkins, K., Russo, J. & Levitz, M.
Estrogen-dependent trypsin-like activity in the rat uterus.
Localization of activity in the 12,000g pellet and nucleus.
Arch. Biochem. Biophys. 173: 347-354.

1977

Katz, J., Troll, W., Adler, S. W. & Levitz, M. Antipain and
leupeptin restrict uterine DNA synthesis and function in mice.
Proc. Natl. Acad. Sci. U.S.A. 74: 3754-3757.

Woessner, J. F., Jr. A latent form of collagenase in the
involuting rat uterus and its activation by a serine proteinase.
Biochem. J. 161: 535-542.

Entry 1.54

MUSCLE TRYPSIN-LIKE PROTEINASE

Summary

EC Number: (3.4.21.-)

Earlier names: -

Distribution: Rat intestinal smooth muscle, apparently particle-bound.

Source: Intestinal muscle of rats on low-protein diet.

Action: Hydrolyzes Bz-Arg-NPhNO$_2$, Tos-Arg-OMe, azo-casein, histones, with specific activity similar to trypsin, but much more active in the limited proteolysis of native enzymes including malate dehydrogenase.

Requirements: pH 7.5-8.5. Unstable to acid and alkali.

Substrate, usual: Malate dehydrogenase (inactivation assay).

Substrate, special: Tos-Gly-Pro-Arg-NPhNO$_2$.

Inhibitors: Dip-F, Tos-Lys-CH$_2$Cl, leupeptin, benzamidine, proflavine. Soybean and lima bean trypsin inhibitors. Endogenous inhibitor in source tissue.

Molecular properties: Mol.wt. 33,000 (by gel chromatography).

Comment: Seems quite distinct from the chymotrypsin-like inactivator of apo-forms of pyridoxal enzymes from the same tissue, now thought to be mast cell chymase II (#1.43).

Bibliography

1977

Beynon, R. J. & Kay, J.   A neutral protease from rat intestinal muscle. A possible role in the degradation of native enzymes. Acta Biol. Med. Ger. 36: 1625-1635.

Carney, I. T. & Kay, J.   The effects of ligands on the proteolysis of native phosphorylases a and b by a neutral proteinase from rat intestinal muscle. Biochem. Soc. Trans. 5: 1335-1337.

Kay, J., Carney, I. T. & Beynon, R. J.   The susceptibility of glycogen phosphorylase to inactivation by endogenous and exogenous proteases. Acta Biol. Med. Ger. 36: 1637-1644.

1978

Beynon, R. J. & Kay, J.   The inactivation of native enzymes by a
neutral proteinase from rat intestinal muscle. Biochem. J. 173:
291–298.
Carney, I. T., Beynon, R. J., Kay, J. & Birket, N.   The
susceptibility of muscle phosphorylases a and b to digestion by
a neutral proteinase from rat intestinal muscle. Biochem. J.
175: 105–113.

1980

Carney, I. T., Curtis, C. G., Kay, J. & Birket, N.   A low-
molecular-weight inhibitor of the neutral proteinase from rat
intestinal muscle. Biochem. J. 185: 423–433.

Entry 1.55

PROLYL ENDOPEPTIDASE

Summary

EC Number: 3.4.21.26

Earlier names: Post-proline cleaving enzyme, post-proline endopeptidase. Possibly identical with brain kininase B [1].

Distribution: Human uterus, lamb kidney and other tissues.

Source: Lamb kidney. Rabbit brain (see Comments).

Action: Specific for cleavage of -Pro-X- bonds, excluding -Pro-Pro-. Inactivates oxytocin, vasopressin, bradykinin and other biologically active peptides [1].

Requirements: pH 7.5-8.0. EDTA (1mM) and dithiothreitol (0.5mM) have been found to enhance activity.

Substrate, usual: Z-Gly-Pro⧸Leu-Gly (Km 6.0×10^{-5} M), Z-Gly-Pro-OPhNO$_2$, Z-Gly-Pro-NNap, Z-Gly-Pro-NMec [2], Z-Gly-Pro-sulphamethoxazole [1].

Substrate, special: ^{14}C-Labelled oxytocin.

Inhibitors: Dip-F, Z-Gly-Pro-CH$_2$Cl.

Molecular properties: Mol.wt. 115,000, composed of two similar subunits. pI 4.8.

Comment: Suggested to be responsible for prolyl endopeptidase activity sometimes associated with dipeptidyl aminopeptidase IV [3]. Flavobacterium meningosepticum produces an enzyme with similar substrate specificity [2].

   The properties summarized above are mainly those of the lamb kidney enzyme: the rabbit and rat brain enzymes are somewhat different, being more sensitive to thiol-blocking reagents, and of mol.wt. about 70,000 [1,4].

References
   [1] Orlowski et al. J. Neurochem. 33: 461-469, 1979.
   [2] Yoshimoto & Tsuru Agr. Biol. Chem. 42: 2417-2419, 1978.
   [3] Yoshimoto & Walter Biochim. Biophys. Acta 485: 391-401, 1977.
   [4] Rupnow & Taylor Biochemistry 18: 1206-1212, 1979.

Bibliography

1971

Walter, R., Shlank, H., Glass, J. D., Schwartz, I. L. & Kerenyi, T. D. Leucylglycinamide release from oxytocin by human uterine enzyme. Science 173: 827–829.

1972

Shlank, H. & Walter, R. Enzymic cleavage of post–proline peptide bonds: degradation of arginine–vasopressin and angiotensin II. Proc. Soc. Exp. Biol. Med. 141: 452–455.

1976

Koida, M. & Walter, R. Post–proline cleaving enzyme. Purification of this endopeptidase by affinity chromatography. J. Biol. Chem. 251: 7593–7599.

Oliveira, E. B., Martins, A. R. & Camargo, A. C. M. Isolation of brain endopeptidases: influence of size and sequence of substrates structurally related to bradykinin. Biochemistry 15: 1967–1974.

Walter, R. Partial purification and characterization of post–proline cleaving enzyme: enzymatic inactivation of neurohypophyseal hormones by kidney preparations of various species. Biochim. Biophys. Acta 422: 138–158.

1977

Yoshimoto, T. & Walter, R. Post–proline dipeptidyl aminopeptidase (dipeptidyl aminopeptidase IV) from lamb kidney. Biochim. Biophys. Acta 485: 391–401.

Yoshimoto, T., Orlowski, R. C. & Walter, R. Postproline cleaving enzyme: identification as serine protease using active site specific inhibitors. Biochemistry 16: 2942–2948.

1978

Walter, R. & Yoshimoto, T. Postproline cleaving enzyme: kinetic studies of size and stereospecificity of its active site. Biochemistry 17: 4139–4144.

Yoshimoto, T., Fischl, M., Orlowski, R. C. & Walter, R. Post–proline cleaving enzyme and post–proline dipeptidyl aminopeptidase. Comparison of two peptidases with high specificity for proline residues. J. Biol. Chem. 253: 3708–3716.

1979

Knisatschek, H. & Bauer, K. Characterization of "thyroliberin–deamidating enzyme" as a post–proline–cleaving enzyme. Partial purification and analysis of the enzyme from anterior pituitary tissue. J. Biol. Chem. 254: 10936–10943.

Orlowski, M., Wilk, E., Pearce, S. & Wilk, S. Purification and properties of a prolyl endopeptidase from rabbit brain. J. Neurochem. 33: 461–469.

Rupnow, J. H., Taylor, W. L. & Dixon, J. E. Purification and
  characterization of a thyrotropin-releasing hormone deamidase
  from rat brain. Biochemistry 18: 1206-1212.
Wilk, S. & Orlowski, M. Degradation of bradykinin by isolated
  neutral endopeptidases of brain and pituitary Biochem. Biophys.
  Res. Commun. 90: 1-6.
Yoshimoto, T., Ogita, K., Walter, R., Koida, M. & Tsuru, D. Post-
  proline cleaving enzyme. Synthesis of a new fluorogenic
  substrate and distribution of the endopeptidase in rat tissues
  and body fluids of man. Biochim. Biophys. Acta 569: 184-192.

Entry 1.56

RIBOSOMAL SERINE PROTEINASE

Summary

EC Number: (3.4.21.-)

Earlier names: The name "cathepsin R" has been used in verbal communications, and may become the usual name.

Distribution: Associated with ribosomes (40S subunit) and polysomes.

Source: Ribosomes of rat liver and Hela cells.

Action: Hydrolyzes casein and ribosomal proteins producing high mol.wt. products.

Requirements: pH 7-8.

Substrate, usual: Ribosomal proteins.

Substrate, special: Radiolabelled protein (e.g. ^{125}I-labelled casein) bound to Sepharose.

Inhibitors: Dip-F, leupeptin. Aprotinin, soybean trypsin inhibitor, α_1-proteinase inhibitor. (Not Tos-Lys-CH$_2$Cl of Tos-Phe-CH$_2$Cl.)

Molecular properties: -

Comment: Largely solubilized when the ribosomal subunits are separated with puromycin.

Bibliography

1970

Paik, W. K. & Lee, H. W.   Enzymatic hydrolysis of histones in rat kidney microsomes. Biochem. Biophys. Res. Commun. 38: 333-340.

1975

Levyant, M. I., Bilinkina, V. S., Trudolyubova, M. G. & Orekhovich, V. N. [Proteolytic activity of ribosomes.] Biochemistry U.S.S.R. (Engl. Transl.) 40: 1110-1111.

1976

Levyant, M. I., Bilinkina, V. S., Trudolyubova, M. G. & Orekhovich, V. N. Presence of a proteinase in polyribosomes of rat liver. Molecular Biology U.S.S.R. (Engl. Transl.) 10: 634-639.

1977

Korant, B. D. Protease activity associated with HeLa cell ribosomes. Biochem. Biophys. Res. Commun. 74: 926-933.

Korant, B. D. Protein cleavage in virus-infected cells. Acta Biol. Med. Ger. 36: 1565-1573.

Langner, J., Ansorge, S., Bohley, P., Welfle, H. & Bielka, H. Presence of an endopeptidase activity in rat liver ribosomes. Acta Biol. Med. Ger. 36: 1729-1733.

1978

Bylinkina, V. S., Levyant, M. I., Gorach, G. G. & Orekhovich, V. N. Localization of neutral proteinase on the ribosomes. Biochemistry U.S.S.R. (Engl. Transl.) 43: 83-85.

Levjant, M. I., Bylinkina, V. S., Spivak, V. A. & Orekhovich, V. N. [On the specificity of neutral proteinase from ribosomes.] Biokhimiya 43: 1423-1428.

Entry 1.57

MICROSOMAL MEMBRANE PROTEINASE

Summary

EC Number: (3.4.21.-)

Earlier names: May be "signalase", signal-peptide cleaving enzyme.

Distribution: Rat liver endoplasmic reticular membrane.

Source: Rat liver microsomes.

Action: Degrades histones. Signalase cleaves pre-pieces from
nascent polypeptide chains of secreted proteins as they are
transported into the cisternae of the endoplasmic reticulum. An
activity in the Golgi vesicles then cleaves the pro-peptides,
often at the C-terminal side of pairs of basic residues.

Requirements: Neutral pH.

Substrate, usual: Proalbumin, proparathormone, proinsulin, etc.

Inhibitors: -

Molecular properties: -

Comment: Clearly, this is likely to be a heterogeneous group of
enzymes, rather than a single entity. Very little
characterization has been achieved, and even the classification
as serine proteinase(s) may be premature.

Bibliography

1970

Paik, W. K. & Lee, H. W.   Enzymatic hydrolysis of histones in rat
kidney microsomes.   Biochem. Biophys. Res. Commun. 38: 333-340.

1975

Blobel, G. & Dobberstein, B.   Transfer of proteins across
membranes. 1. Presence of proteolytically processed and
unprocessed nascent immunoglobulin light chains on membrane-bound
ribosomes of murine myeloma.   J. Cell Biol. 67: 835-851.

1976

MacGregor, R. R., Chu, L. L. H. & Cohn, D. V.   Conversion of
proparathyroid hormone to parathyroid hormone by a particulate
enzyme of the parathyroid gland.   J. Biol. Chem. 251: 6711-6716.

1977

Birken, S., Smith, D. L., Canfield, R. E. & Boime, L. Partial amino acid sequence of human placental lactogen precursor and its mature hormone form produced by membrane-associated enzyme activity. Biochem. Biophys. Res. Commun. 74: 106-112.

Jackson, R. C. & Blobel, G. Post-translational cleavage of presecretory proteins with an extract of rough microsomes from dog pancreas containing signal peptidase activity. Proc. Natl. Acad. Sci. U.S.A. 74: 5598-5602.

1978

Dorner, A. J. & Kemper, B. Conversion of pre-proparathyroid hormone to proparathyroid hormone by dog pancreatic microsomes. Biochemistry 17: 5550-5555.

MacGregor, R. R., Hamilton, J. W. & Cohn, D. V. The mode of conversion of proparathormone to parathormone by a particulate converting enzymic activity of the parathyroid gland. J. Biol. Chem. 253: 2012-2017.

Sogawa, K. & Takahashi, K. Evidence for the presence of a serine proteinase(s) associated with the microsomal membranes of rat liver. J. Biochem. 84: 763-770.

1979

Erickson, A. H. & Blobel, G. Early events in the biosynthesis of the lysosomal enzyme cathepsin D. J. Biol. Chem. 254: 11771-11774.

Sogawa, K. & Takahashi, K. A neutral proteinase of monkey liver microsomes. Solubilization, partial purification, and properties. J. Biochem. 86: 1313-1322.

Strauss, A. W., Zimmerman, M., Boime, I., Ashe, B., Mumford, R. A. & Alberts, A. W. Characterization of an endopeptidase involved in pre-protein processing. Proc. Natl. Acad. Sci. U.S.A. 76: 4225-4229.

Entry 1.58

THYMUS CHROMATIN PROTEINASE

Summary

EC Number: (3.4.21.-)

Earlier names: -

Distribution: Calf thymus nuclei.

Source: Calf thymus chromatin.

Action: Cleaves -Lys-X- and -Arg-X- bonds in histones and ribosomal basic proteins. Much less active against hemoglobin and casein, and almost inactive against ribonuclease.

Requirements: pH optimum 8.5.

Substrate, usual: Histones.

Inhibitors: Dip-F, Tos-Lys-CH$_2$Cl, Gbz-OPhNO$_2$, Pms-F, (not Tos-Phe-CH$_2$Cl). Soybean trypsin inhibitor, aprotinin.

Molecular properties: Mol.wt. 15,000 (by gel chromatography).

Comment: The enzyme may well be identical with those of uterus (#1.53) and/or liver nuclei (#1.59).

Bibliography

1967

Furlan, M. & Jericijo, M. Protein catabolism in thymus nuclei. 1. Hydrolysis of nucleoproteins by proteases present in calf-thymus nuclei. Biochim. Biophys. Acta 147: 135-144.

Furlan, M. & Jericijo, M. Protein catabolism in thymus nuclei. II. Binding of histone-splitting nuclear proteases to deoxyribonucleic acid. Biochim. Biophys. Acta 147: 145-153.

1968

Furlan, M., Jericijo, M. & Suhar, A. Purification and properties of a neutral protease from calf thymus nuclei. Biochim. Biophys. Acta 167: 154-160.

1970

Bartley, J. & Chalkley, R. Further studies of a thymus nucleohistone-associated protease. J. Biol. Chem. 245: 4286-4292.

1973

Fraki, J., Ruuskanen, O., Hopsu-Havu, V. K. & Kouvalainen, K.
Thymus proteases: extraction, distribution, comparison to lymph
node proteases and species variation. Hoppe-Seyler's Z.
Physiol. Chem. 354: 933-943.

1974

Kurecki, T. & Toczko, K. Purification and partial characterization
of protease from calf thymus chromatin. Acta Biochim. Polon.
21: 225-233.

1975

Destree, O. H. J., D'Adelhart-Toorop, H. A. & Charles, R.
Cytoplasmic origin of the so-called nuclear neutral histone
protease. Biochim. Biophys. Acta 378: 450-458.
Kurecki, T., Kowalska-Loth, B., Toczko, K. & Chmielewska, I.
Evidence that neutral protease from calf thymus chromatin is a
serine type enzyme. FEBS Lett. 53: 313-315.

1976

Kowalska-Loth, B., Zieleński, J., Toczko, K. & Chmielewska, I.
Similarity in active site arrangement of neutral protease from
calf thymus chromatin and trypsin. Acta Biochim. Pol. 23: 139-
144.

1978

Pestaña, A. DNA-bound proteases. Partial characterization of
proteolytic activities in commercial sources of DNA. Biochim.
Biophys. Acta 521: 547-556.

Entry 1.59
RAT LIVER CHROMATIN PROTEINASE

Summary

EC Number: (3.4.21.-)

Earlier names: -

Distribution: Rat liver nuclei.

Source: Rat liver nuclei.

Action: Degrades histones.

Requirements: Optimal pH 7.5.

Substrate, usual: Histone.

Inhibitors: Dip-F, o-methylphenylethylboronic acid, EDTA (then reactivated with divalent cations).

Molecular properties: Mol.wt. values from 25,000 to 200,000 have been reported.

Comment: Rat liver nuclei are likely to be contaminated by mast cell granules which are very rich in chymase I (#1.42), so care is needed in positively identifying a chymotrypsin-like enzyme as nuclear.

Bibliography

1972

Garrells, J. I., Elgin, S. C. R. & Bonner, J. A histone protease of rat liver chromatin. Biochem. Biophys. Res. Commun. 46: 545-551.

1974

Chae, C.-B. & Carter, D. B. Degradation of chromosomal proteins during dissociation and reconstitution of chromatin. Biochem. Biophys. Res. Commun. 57: 740-746.

Chong, M. T., Garrard, W. T. & Bonner, J. Purification and properties of a neutral protease from rat liver chromatin. Biochemistry 13: 5128-5134.

1975

Chae, C.-B., Gadski, R. A., Carter, D. B. & Efird, P. H. Integrity of proteins in reconstituted chromatin. Biochem.

Biophys. Res. Commun. 67: 1459-1465.

Destree, O. H. J., D'Adelhart-Toorop, H. A. & Charles, R. Cytoplasmic origin of the so-clled nuclear neutral histone protease. Biochim. Biophys. Acta 378: 450-458.

1976

Carter, D. B. & Chae, C.-B. Chromatin-bound protease: degradation of chromosomal proteins under chromatin dissociation conditions. Biochemistry 15: 180-185.

Carter, D. B., Efird, P. H. & Chae, C.-B. Chromatin-bound protease: [^{3}H]diisopropyl fluorophosphate labeling patterns of chromatin. Biochemistry 15: 2603-2607.

1977

Anachkova, B. & Russev, G. Endogenous proteolytic activity of chromatin. Acta Biol. Med. Ger. 36: 1945-1950.

Carter, D. B., Ross, D. A., Ishaq, K. S., Suarez, G. M. & Chae, C.-B. The inhibition of rat liver chromatin protease by congeners of the phenylboronic acids. Biochim. Biophys. Acta 484: 103-108.

Schmidt, G. & Lindigkeit, R. Degradation of phosphorylated chromosomal nonhistone proteins. Acta Biol. Med. Ger. 36: 1951-1954.

1978

Ramponi, G., Nassi, P., Liguri, G., Cappugi, G. & Grisolia, S. Purification and properties of a histone-specific protease from rat liver chromatin. Effect on acylated histones. FEBS Lett. 90: 228-232.

1979

Wyrick, S., Kim, Y.-J., Ishaq, K. & Carter, C.-B. Synthesis of active site-directed organometallic irreversible protease inhibitors. Biochim. Biophys. Acta 568: 11-18.

Section 2

CYSTEINE PROTEINASES

Entry 2.01

CATHEPSIN B

Summary

EC Number: 3.4.22.1

Earlier names: Cathepsin Bl; cathepsin B_1; cathepsin B'.

Distribution: Seems generally distributed in lysosomes of mammalian cells.

Source: Spleen or liver, usually bovine or human.

Action: Endopeptidase action on proteins, and cleaves N-terminal peptides of collagen, but also liberates C-terminal dipeptides sequentially from glucagon and aldolase (causing inactivation). Cleaves synthetic esters and amides, especially of arginine.

Requirements: pH optima for protein substrates vary in the range 3.5–6.0, but maximal activity is close to pH 6.0 with synthetic substrates. A thiol activator (e.g. 2 mM cysteine) is required, and sometimes a chelator (1 mM EDTA).

Substrate, usual: Bz-Arg-2-NNap, Bz-Arg-NPhNO$_2$. (Both of these substrates are sensitive to cathepsin H as well as to cathepsin B, and Bz-Arg-NH$_2$, sometimes used in the past, is even more non-specific, also being cleaved by lysosomal carboxypeptidase B and by cathepsin L, #2.03.)

Substrate, special: Z-Arg-Arg-2-NNap or Z-Phe-Arg-NMec for assays of maximal sensitivity free from interference by cathepsin H. Z-Arg-2-NNapOMe with a diazonium salt to stain gels for activity [1], or Z-Ala-Arg-Arg-2-NNapOMe with a diazonium salt for tissue sections [2].

Inhibitors: Thiol-blocking reagents, Z-Phe-Phe-CHN$_2$, L-trans-epoxysuccinyl-Leu-agmatine (E-64)[2], leupeptin. Egg white papain inhibitor, α_2-macroglobulin.

Molecular properties: Mol.wt. about 25,000. pI 4.5–5.5 (human; multiple forms). Leu- N-terminus. A280,1% 20. Contains little (or no?) hexosamine, and has no affinity for concanavalin A-Sepharose, unlike other lysosomal endopeptidases. Unstable above pH 7.

Comment: Cathepsin B is thought to play a part in proteolysis in the lysosomal system.

References

[1] Barrett Biochem. J. 131: 809–822, 1973.

[2] Towatari et al. J. Biochem. 84: 659–671, 1978.

[2] Smith & Van Frank In: Lysosomes in Biology and Pathology (Dingle, J. T. & Dean, R. T. eds), vol. 4, pp. 193–249, North-Holland Publishing Co., Amsterdam, 1975.

Bibliography

1946

Greenstein, J. P. & Leuthardt, F. M. Enzymatic hydrolysis of benzoylarginineamide in normal and neoplastic tissues. J. Nat. Cancer Inst. 6: 203–206.

1947

Greenstein, J. P. & Leuthardt, F. M. Note on benzoylarginineamide activity in extracts of rat liver and hepatoma. J. Nat. Cancer Inst. 8: 77–78.

1951

Maver, M. E. & Greco, A. E. The intracellular distribution of cathepsin, benzoylarginine amidase and leucine amidase activities in normal rat tissues and primary rat hepatoma. J. Nat. Cancer Inst. 12: 37–48.

1957

Greenbaum, L. M. & Fruton, J. S. Purification and properties of beef spleen cathepsin B. J. Biol. Chem. 226: 173–180.

1958

Fujii, S. & Fruton, J. S. Transamidation reactions catalyzed by cathepsins. J. Biol. Chem. 230: 1–11.

1959

Greenbaum, L. M., Hirshkowitz, A. & Shoichet, I. The activation of trypsinogen by cathepsin B. J. Biol. Chem. 234: 2885–2890.

1960

Barnhart, M. L. Partial activation of prothrombin and generation of prothrombin-R with cathepsin B. Am. J. Physiol. 198: 899–905.

1962

Blackwood, C. E., Mateyko, G. M. & Mandl, I. Changes in proteolytic enzyme systems of rat tissues in response to heterologous growth of human ovarian tumors. Cancer Res. 22: 993–997.

1964

Bouma, J. M. W. & Gruber, M. The distribution of cathepsins B and C in rat tissues. Biochim. Biophys. Acta. 89: 545–547.

1966

Bouma, J. M. W. & Gruber, M. Intracellular distribution of cathepsin B and cathepsin C in rat liver. Biochim. Biophys. Acta 113: 350–358.

1967

Ali, S. Y. The presence of cathepsin B in cartilage . Biochem. J. 102: 10c–11c.

Otto, K. Über ein neues Kathepsin. Reinigung aus Rindermilz, Eigenschaften, sowie Vergleich mit Kathepsin B. Hoppe-Seyler's Z. Physiol. Chem. 348: 1449–1460.

Otto, K. & Schepers, P. Über die Katheptische inaktivierung einiger Enzyme der Rattenleber, insbesondere der Glucokinase. Hoppe-Seyler's Z. Physiol. Chem. 348: 482–490.

Snellman, O. & Sylvén, B. Haptoglobin acting as a natural inhibitor of cathepsin B activity. Nature 216: 1033 only.

1968

Sylvén, B. Cellular detachment by purified lysosomal cathepsin B. Eur. J. Cancer 4: 559–562.

1969

Keilová, H. & Keil, B. Isolation and specificity of cathepsin B. FEBS Lett. 4: 295–298.

Otto, K. & Bhakdi, S. [Studies on cathepsin B': specificity and properties.] Hoppe-Seyler's Z. Physiol. Chem. 350: 1577–1588.

Snellman, O. Cathepsin B, the lysosomal thiol proteinase of calf liver. Biochem. J. 114: 673–678.

1970

Keilová, H. & Turková, J. Analogy between active sites of cathepsin B1 and papain. FEBS Lett. 11: 287–288.

McDonald, J. K., Zeitman, B. B. & Ellis, S. 'Leucine naphthylamide: an inappropriate substrate for the histochemical detection of cathepsin B and B'. Nature 225: 1048–1049.

1971

Barrett, A. J. Terminology of the tissue proteinases. In: Tissue Proteinases (Barrett, A. J. & Dingle, J. T., eds), pp. ix–x, North-Holland Publishing Co., Amsterdam.

Keilová, H. On the specificity and inhibition of cathepsins D and B. In: Tissue Proteinases (Barrett, A. J. and Dingle, J. T. eds), pp. 45–68, North-Holland Publishing Co., Amsterdam.

Otto, K. Cathepsins B1 and B2. In: Tissue Proteinases (Barrett, A. J. & Dingle, J. T. eds), pp. 1–28, North-Holland Publishing Co., Amsterdam.

Snellman, O. A study on the reactivity of the thiol group of cathepsin B. In: Tissue Proteinases (Barrett, A. J. & Dingle, J. T. eds), pp. 29–43, North-Holland Publishing Co., Amsterdam..

Suominen, J. & Hopsu-Havu, V. K. Cathepsin B' in the thyroid gland. Acta Chem. Scand. B 25: 2531–2540.

1972

Barrett, A. J.   A new assay for cathepsin B1 and other thiol proteinases. Anal. Biochem. 47: 280-293.

De Lumen, B. O. & Tappel, A. L.   α-N-Benzoylarginine-β-naphthylamide amidohydrolase of rat liver lysosomes. J. Biol. Chem. 247: 3552-3557.

De Lumen, B. O. & Tappel, A. L.   Continuous fluorometric assay for cathepsin B-like activity in lysosomes. Anal. Biochem. 48: 378-385.

Distelmaier, P., Hübner, H. & Otto, K.   Cathepsins B1 and B2 in various organs of the rat. Enzymologia 42: 363-375.

Franklin, S. G. & Metrione, R. M.   Chromatographic evidence for the existence of multiple forms of cathepsin B1. Biochem. J. 127: 207-213.

Smith, R. E., Smithwick, E. L., Jr. & Allen, C. H.   A new assay for cathepsin B1 using CBZ-Ala-Arg-Arg-4MeOβNA. J. Histochem. Cytochem. 20: 843 only.

1973

Barrett, A. J.   Human cathepsin B1. Purification and some properties of the enzyme. Biochem. J. 131: 809-822.

Keilová, H. & Tomášek, V.   On the isozymes of cathepsin B1. FEBS Lett. 29: 335-338.

Starkey, P. M. & Barrett, A. J.   Human cathepsin B1. Inhibition by α$_2$-macroglobulin and other serum proteins. Biochem. J. 131: 823-831.

1974

Burleigh, M. C., Barrett, A. J. & Lazarus, G. S.   Cathepsin B1. A lysosomal enzyme that degrades native collagen. Biochem. J. 137: 387-398.

Davies, P., Allison, A. C. & Hylton, W. J.   The identification, properties and subcellular distribution of cathepsins B1 and C (dipeptidyl aminopeptidase I) in human peripheral-blood leucocytes. Biochem. Soc. Trans. 2: 432-434.

Etherington, D. J.   The purification of bovine cathepsin B1 and its mode of action on bovine collagens. Biochem. J. 137: 547-557.

Huisman, W., Lanting, L., Doddema, H. J., Bouma, J. M. W. & Gruber, M.   Role of individual cathepsins in lysosomal protein digestion as tested by specific inhibitors. Biochim. Biophys. Acta 370: 297-307.

Keilová, H. & Tomášek, V.   Effect of papain inhibitor from chicken egg white on cathepsin B1. Biochim. Biophys. Acta 334: 179-186.

Nakashima, K. & Ogino, K.   Regulation of rabbit liver fructose-1,6-diphosphatase. II. Modification by lysosomal cathepsin B$_1$ from the same cell. J. Biochem. 75: 355-365.

Ogino, K. & Nakashima, K.   Purification of rabbit liver cathepsin B1. J. Biochem. 75: 723-730.

Swanson, A. A., Martin, B. J. & Spicer, S. S.   Human placental cathepsin B1. Isolation and some physical properties. Biochem.

J. 137: 223-228.
Sylvén, B. Biochemical factors involved in the cellular
detachment from tumors. Schweiz. Med. Wochenschr. 104: 258-261.
Sylvén, B. & Snellman, O. The immunofluorescent demonstration of
cathepsin Bl in tissue sections. Histochemistry 38: 35-41.
Sylvén, B., Snellman, O. & Sträuli, P. The immunofluorescent
demonstration of cathepsin Bl in tissue sections. Virchows
Arch. B Cell Pathol. 17, 97-112.

1975

Bajkowski, A. S. & Frankfater, A. Specific spectrophotometric
assays for cathepsin B_1. Anal. Biochem. 68: 119-127.
Barrett, A. J. Lysosomal and related proteinases. In: Proteases
and Biological Control (Reich, E., Rifkin, D. B. & Shaw, E.
eds), pp. 467-482, Cold Spring Harbor Laboratory, New York.
Davidson, E. & Poole, B. Fractionation of the rat liver enzymes
that hydrolyze benzoyl-arginine-2-naphthylamide. Biochim.
Biophys. Acta 397: 437-442.
Järvinen, M. & Hopsu-Havu, V. K. α-N-Benzoylarginine-2-
naphthylamide hydrolase (cathepsin Bl?) from rat skin. I.
Preliminary experiments with skin extract. Acta Chem. Scand. B
29: 671-676.
Keilová, H. & Tomásek, V. Inhibition of cathepsin C by papain
inhibitor from chicken egg white and by complex of this
inhibitor with cathepsin B_1. Coll. Czech. Chem. Commun. 40:
218-224.
McDonald, J. K. & Ellis, S. On the substrate specificity of
cathepsin Bl and B2 including a new fluorogenic substrate for
cathepsin Bl. Life Sciences 17: 1269-1276.
Otto, K. & Riesenkönig, H. Improved purification of cathepsin Bl
and cathepsin B2. Biochim. Biophys. Acta 379: 462-475.

1976

Bajkowski, A. S. & Frankfater, A. Effects of added nucleophiles
on the cathepsin Bl catalyzed hydrolysis of CBZ-L-lysine
p-nitrophenyl ester. Evidence for an acyl enzyme intermediate
and leaving group specificity. In: Proteolysis and Physiological
Regulation (Ribbons, D. W. & Brew, K. eds), p. 392 only,
Academic Press, New York.
Barrett, A. J. An improved color reagent for use in Barrett's
assay of cathepsin B. Anal. Biochem. 76: 374-376.
Barrett, A. J. Some properties of human cathepsin Bl, the binding
of the enzyme by $α_2$-macroglobulin, and the role of $α_2$-
macroglobulin in the physiological control of proteolytic
activity. In: Intracellular Protein Catabolism (Hanson, H. &
Bohley, P. eds), pp. 227-236, J. A. Barth, Leipzig.
Dean, R. T. The roles of cathepsins Bl and D in the digestion of
cytoplasmic proteins in vitro by lysosomal extracts. Biochem.
Biophys. Res. Commun. 68: 518-523.
Etherington, D. J. Bovine spleen cathepsin Bl and collagenolytic
cathepsin. A comparative study of the properties of the two

enzymes in the degradation of native collagen. Biochem. J. 153: 199-209.

Etherington, D. J. & Evans, P. The action of cathepsin B1 and collagenolytic cathepsin in the degradation of collagen. In: Intracellular Protein Catabolism (Hanson, H. & Bohley, P. eds), pp. 25-31, J. A. Barth, Leipzig.

Fräki, J. E. Human skin proteases. Separation and characterization of two acid proteases resembling cathepsin B1 and cathepsin D and of an inhibitor of cathepsin B1. Arch. Derm. Res. 255: 317-330.

Gráf, L. & Kenessey, A. Specific cleavage of a single peptide bond (residues 77-78) in β-lipotropin by a pituary endopeptidase. FEBS Lett. 69: 255-260.

Husain, S. S. Intracellular cysteine proteases from bovine liver. In: Proteolysis and Physiological Regulation (Ribbons, D. W. & Brew, K. eds), p. 397 only, Academic Press, New York.

Husain, S. S. & Baqai, J. Evidence for two types of sulfhydryl groups in cathepsins B_1 and B_1A. Fed. Proc. Fed. Am. Soc. Exp. Biol. 35: 1460 only.

Keilová, H. & Tomášek, V. Contribution to studies on active site of cathepsin B_1 and D. In: Intracellular Protein Catabolism (Hanson, H. & Bohley, P. eds), pp. 237-251, J. A. Barth, Leipzig.

Kobayashi, K., Endo, Y., Matsuda, K., Tanaka, K. & Misaka, E. A role of cathepsin B1 in polymorphonuclear leukocytes chemotaxis. Life Sci. 19: 1199-1210.

Kruze, D., Fehr, K. & Böni, A. Effect of antirheumatic drugs on cathepsin B_1 from bovine spleen. Zeit. Rheumatol. 35: 95-102.

Menninger, H., Fehr, K., Böni, A. & Otto, K. Digestion of human immunoglobulin G by bovine cathepsin B1. Immunochemistry 13: 633-637.

Szego, C. M., Seeler, B. J. & Smith, R. E. Lysosomal cathepsin B1: partial characterization in rat preputial gland and recompartmentation in response to estradiol-17β. Eur. J. Biochem. 69: 463-474.

Towatari, T., Tanaka, K., Yoshikawa, D. & Katunuma, N. Separation of a new protease from cathepsin B1 of rat liver lysosomes. FEBS Lett. 67: 284-288.

Warwas, M. & Dobryszycka, W. Cathepsin B1 from human fetal membranes. Biochim. Biophys. Acta 429: 573-580.

1977

Ansorge, S., Kirschke, H. & Friedrich, K. Conversion of proinsulin into insulin by cathepsins B and L from rat liver lysosomes. Acta Biol. Med. Ger. 36: 1723-1727.

Barrett, A. J. Cathepsin B and other thiol proteinases. In: Proteinases in Mammalian Cells and Tissues (Barrett, A. J. ed.), pp. 181-207. North-Holland Publishing Co., Amsterdam.

Bird, J. W. C., Schwartz, W. N. & Spanier, A. M. Degradation of myofibrillar proteins by cathepsins B and D. Acta Biol. Med. Ger. 36: 1587-1604.

DeMartino, G. N., Doebber, T. W. & Miller, L. L. Pepstatin-insensitive proteolytic activity of rat liver lysosomes. Isolation and purification of a new pepstatin-insensitive proteolytic enzyme from rat lysosomes. J. Biol. Chem. 252: 7511-7516.

Eeckhout, Y. & Vaes, G. Further studies on the activation of procollagenase, the latent precursor of bone collagenase. Effects of lysosomal cathepsin B, plasmin and kallikrein, and spontaneous activation. Biochem. J. 166: 21-31.

Etherington, D. J. The dissolution of insoluble bovine collagens by cathepsin B1, collagenolytic cathepsin and pepsin. The influence of collagen type, age and chemical purity on susceptibility. Connect. Tissue Res. 5: 135-145.

Etherington, D. J. & Evans, P. J. The action of cathepsin B and collagenolytic cathepsin in the degradation of collagen. Acta Biol. Med. Ger. 36: 1555-1563.

Judah, J. D. & Quinn, P. S. Ca^{2+} dependent vesicle fusion: a general mechanism for the conversion of precursor proteins. Nature 271: 384-385.

Kar, N. C. & Pearson, C. M. Early elevation of cathepsin B_1 in human muscle disease. Biochem. Med. 18: 126-129.

Leary, R. & Shaw, E. Inactivation of cathepsin B_1 by diazomethyl ketones. Biochem. Biophys. Res. Commun. 79: 926-931.

Nakagawa, H., Aihara, K. & Tsurufuji, S. The role of cathepsin B1 in the collagen breakdown of carrageenin granuloma in rats. J. Biochem. 81: 801-804.

Pierart-Gallois, M., Trouet, A. & Tulkens, P. Production of rabbit antibodies against active rat cathepsin B. Acta Biol. Med. Ger. 36: 1887-1891.

Pietras, R. J., Szego, C. E., Mangan, B. J., Seeler, M. M., Burtnett, M. M. & Orevi, M. Elevated serum cathepsin B1 activity with vaginal adenosis and adenocarcinoma in young women exposed in utero to diethylstilbestrol. Fed. Proc. Fed. Am. Soc. Exp. Biol. 36: 387 only.

Roughley, P. J. The degradation of cartilage proteolycans by tissue proteinases. Proteoglycan heterogeneity and the pathway of proteolytic degradation. Biochem. J. 167: 639-646.

Roughley, P. J. & Barrett, A. J. The degradation of cartilage proteoglycans by tissue proteinases. Proteoglycan structure and its susceptibility to proteolysis. Biochem. J. 167: 629-637.

Schwartz, W. N. & Bird, J. W. C. Degradation of myofibrillar proteins by cathepsins B and D. Biochem. J. 167: 811-820.

1978

Aronson, N. N., Jr. & Barrett, A. J. The specificity of cathepsin B. Hydrolysis of glucagon at the C-terminus by a peptidyldipeptidase mechanism. Biochem. J. 171: 759-765.

Baggiolini, M., Bretz, U. & Dewald, B. Subcellular localization of granulocyte enzymes. In: Neutral Proteases of Human Polymorphonuclear Leukocytes (Havemann, K. & Janoff, A. eds), pp.3-17, Urban & Schwartzenberg, Baltimore.

Bayliss, M. T. & Ali, S. Y. Studies on cathepsin B in human articular cartilage. Biochem. J. 171: 149-154.

Bird, J. W. C., Spanier, A. M. & Schwartz, W. N. Cathepsins B and D: proteolytic activity and ultrastructural localization in skeletal muscle. In: Protein Turnover and Lysosome Function (Segal, H. L. & Doyle, D. J. eds), pp. 589-604, Academic Press, New York.

Davies, M., Barrett, A. J., Travis, J., Sanders, J. & Coles, G. A. The degradation of human glomerular basement membrane with purified lysosomal proteinases: evidence for the pathogenic role of the polymorphonuclear leucocyte in glomerulonephritis. Clin. Sci. Mol. Med. 54: 233-240.

Evans, P. & Etherington, D. J. Characterisation of cathepsin B and collagenolytic cathepsin from human placenta. Eur. J. Biochem. 83: 87-97.

Hanada, K., Tamai, M., Yamagishi, M., Ohmura, S., Sawada, J. & Tanaka, I. Isolation and characterization of E-64, a new thiol protease inhibitor. Agr. Biol. Chem. (Tokyo) 42: 523-528.

Hayasaka, S., Hara, S., Takaku, Y. & Mizuno, K. Distribution and some properties of cathepsin B in the bovine eyes. Exp. Eye Res. 26: 57-63.

MacGregor, R. R., Hamilton, J. W. & Cohn, D. V. Isolation and properties of porcine parathyroid cathepsin B1. Fed. Proc. Fed. Am. Soc. Exp. Biol. 37: 1331 only.

Nakai, N., Wada, K., Kobashi, K. & Hase, J. The limited proteolysis of rabbit muscle aldolase by cathepsin B1. Biochem. Biophys. Res. Commun. 83: 881-885.

Pietras, R. J., Szego, C. M., Mangan, C. E., Seelers, B. J., Burtnett, M. M. & Orevi, M. Elevated serum cathepsin B1 and vaginal pathology after prenatal DES exposure. Obstetrics Gynecology 52: 321-327.

Poole, A. R., Tiltman, K. J., Recklies, A. D. & Stoker, T. A. M. Differences in secretion of the proteinase cathepsin B at the edges of human breast carcinomas and fibroadenomas. Nature 273: 545-547.

Quinn, P. S. & Judah, J. D. Calcium-dependent golgi-vesicle fusion and cathepsin B in the conversion of proalbumin into albumin in rat liver. Biochem. J. 172: 301-309.

Roughley, P. J., Murphy, G. & Barrett, A. J. Proteinase inhibitors of bovine nasal cartilage. Biochem. J. 169: 721-724.

Singh, H. & Kalnitsky, G. Separation of a new α-N-benzoylarginine-β-naphthylamide hydrolase from cathepsin B1. Purification, characterization, and properties of both enzymes from rabbit lung. J. Biol. Chem. 253: 4319-4326.

Singh, H., Kuo, T. & Kalnitsky, G. Collagenolytic activity of lung BANA hydrolase and cathepsin B1. In: Protein Turnover and Lysosome Function (Segal, H. L. & Doyle, D. J. eds), pp. 315-331, Academic Press, New York.

Towatari, T. & Katunuma, N. Crystallization and amino acid composition of cathepsin B from rat liver lysosomes. Biochem.

Biophys. Res. Commun. 83: 513-520.
Towatari, T., Tanaka, K., Yoshikawa, D. & Katunuma, N.
Purification and properties of a new cathepsin from rat liver.
J. Biochem. 84: 659-671.

1979

Graf, M., Leeman, U., Ruch, F. & Sträuli, P. The fluorescence and
bright field microscope demonstration of cathepsin B in human
fibroblasts. Histochemistry 64: 319-322.
Hardy, M. F. & Pennington, R. J. T. Separation of cathepsin B1
and related enzymes from rat skeletal muscle. Biochim. Biophys.
Acta 577: 253-266.
Hayasaka, S. & Hayasaka, I. Cathepsin B and collagenolytic
cathepsin in the aqueous humor of patients with Behçet's
disease. Albrecht von Graefes Arch. Ophthalmol. 210: 103-107.
MacGregor, R. R., Hamilton, J. W., Kent, G. N., Shofstall, R. E. &
Cohn, D. V. The degradation of proparathormone and parathormone
by parathyroid and liver cathepsin B. J. Biol. Chem. 254: 4428-
4433.
MacGregor, R. R., Hamilton, J. W., Shofstall, R. E. & Cohn, D. V.
Isolation and characterization of porcine parathyroid cathepsin
B. J. Biol. Chem. 254: 4423-4427.
Pietras, R. J., Szego, C. M., Mangan, C. E., Seeler, B. J. &
Burtnett, M. M. Elevated serum cathepsin B1-like activity in
women with neoplastic disease. Gynecologic Oncology 7: 1-17.
Suhar, A. & Marks, N. Purification and properties of brain
cathepsin B. Evidence for cleavage of pituitary lipotropins.
Eur. J. Biochem. 101: 23-30.
Takahashi, K., Isemura, M. & Ikenaka, T. Isolation and
characterization of three forms of cathepsin B from porcine
liver. J. Biochem. 85: 1053-1060.
Watanabe, H., Green, G. D. J. & Shaw, E. A comparison of the
behaviour of chymotrypsin and cathepsin B towards peptidyl
diazomethyl ketones. Biochem. Biophys. Res. Commun. 89: 1354-
1360.

1980

Barrett, A. J. Thiol proteinases of human lysosomes. In: Enzyme
Regulation and Mechanism of Action (Mildner, P. & Ries, B. eds),
pp. 307-315, Pergamon, Oxford.
Takahashi, K., Isemura, M., Ono, T. & Ikenaka, T. Location of the
essential thiol of porcine liver cathepsin B. J. Biochem. 87:
347-350.
Turk, V., Kregar, I., Gubensek, T., Popovic, P., Locnicar, P. &
Lah, T. Carboxyl and thiol intracellular proteinases. In:
Enzyme Regulation and Mechanism of Action (Mildner, P. & Ries,
B. eds), pp. 317-330, Pergamon, Oxford.

Entry 2.02

CATHEPSIN H

Summary

EC Number: (3.4.22.-)

Earlier names: Cathepsin B_3, cathepsin B_1a, BANA-hydrolase (see Comments).

Distribution: Rat liver lysosomes, rabbit lung.

Source: As above.

Action: Degrades proteins. Hydrolyzes blocked synthetic substrates of endopeptidases, but acts equally well as an aminopeptidase on the substrates with free N-termini. Almost inactive on Z-Arg-Arg-2-NNap and Z-Phe-Arg-NMec, excellent substates of cathepsin B.

Requirements: pH 6.0, 1-5mM glutathione and EDTA.

Substrate, usual: Bz-Arg-NNap (but this is non-specific).

Substrate, special: Arg-NNap, Leu-NNap or Arg-NMec for discrimination from cathepsin B.

Inhibitors: Leu-CH_2Cl; thiol-blocking reagents. Leupeptin (but rather insensitive: K_i 5 x 10^{-6} M).

Molecular properties: Mol.wt. 28,000. pI 7.1. Unstable to alkali, but the rat enzyme is rather stable to heat at slightly acid pH values.

Comment: Designated an "endoaminopeptidase" because of the combination of endopeptidase with aminopeptidase activity. This enzyme may be the acid leucine aminopeptidase, cathepsin III [1], and the lysosomal leucine naphthylamidase revealed histochemically [2] (although aminopeptidases more sensitive to puromycin also contribute activity against Leu-NNap).

    The rabbit lung "BANA-hydrolase" [3] differs from rat and human cathepsin H in having greater activity against collagen, but is similar in many other respects.

References
    [1] Fruton et al. J. Biol. Chem. 138: 249-262, 1941.
    [2] Sylvén Histochemie 15: 150-159, 1968.
    [3] Singh & Kalnitsky J. Biol. Chem. 253: 4319-4326, 1978.

Bibliography

1968

Sylvén, B. Studies on the histochemical "leucine aminopeptidase" reaction. VI. The selective demonstration of cathepsin B activity by means of the naphthylamide reaction. Histochemie 15: 150-159.

1971

Sandström, B. Some observations on a histochemically demonstrable leucyl-naphthylamidase activity (cathepsin B?) in chicken liver. Histochemie 26: 35-39.

1972

Kirschke, H., Langner, J., Wiederanders, B., Ansorge, S. & Bohley, P. Intrazellulärer Proteinabbau. IV. Isolierung und Charakterisierung von peptidasen aus Rattenleberlysosomen. Acta Biol. Med. Ger. 28: 305-322.

1975

Davidson, E. & Poole, B. Fractionation of the rat liver enzymes that hydrolyze benzoyl-arginine-2-naphthylamide. Biochim. Biophys. Acta 397: 437-442.

Järvinen, M. & Hopsu-Havu, V. K. α-N-Benzoylarginine-2-naphthylamide hydrolase (cathepsin B1?) from rat skin. II. Purification of the enzyme and demonstration of two inhibitors in the skin. Acta Chem. Scand. B 29: 772-780.

1976

Husain, S. S. Intracellular cysteine proteases from bovine liver. In: Proteolysis and Physiological Regulation (Ribbons, D. W. & Brew, K. eds), p. 397 only, Academic Press, New York.

Husain, S. S. & Baqai, J. Evidence for two types of sulfhydryl groups in cathepsins B_1 and B_1A. Fed. Proc. Fed. Am. Soc. Exp. Biol. 35: 1460 only.

Järvinen, M. α-N-Benzoylarginine-2-naphthylamide hydrolase (cathepsin B1?) from rat skin. III. Substrate specificity, modifier characteristics, and transformation of the enzyme at acidic pH. Acta Chem. Scand. B 30: 53-60.

Kirschke, H., Langner, J., Wiederanders, B., Ansorge, S., Bohley, P. & Broghammer, U. Intrazellulärer Proteinabbau. VII. Kathepsin L und H: zwei neue Proteinasen aus Rattenleberlysosomen. Acta Biol. Med. Ger. 35: 285-299.

Kirschke, H., Langner, J., Wiederanders, B., Ansorge, S., Bohley, P. & Hanson, H. Cathepsin L and proteinases with cathepsin B1-like activity from rat liver lysosomes. In: Intracellular Protein Catabolism (Hanson, H. & Bohley, P. eds), pp. 210-217, J. A. Barth, Leipzig.

1977

Kirschke, H. Cathepsin H: an endoaminopeptidase. Acta Biol. Med.

Ger. 36: 1547–1548.

Kirschke, H., Langner, J., Wiederanders, B., Ansorge, S., Bohley, P. & Hanson, H. Cathepsin H: an endoaminopeptidase from rat liver lysosomes. Acta Biol. Med. Ger. 36: 185–199.

Kirschke, H., Langner, J., Wiederanders, B., Ansorge, S., Bohley, P. & Hanson, H. Cathepsin L and cathepsin B₃ from rat liver lysosomes. In: Intracellular Protein Catabolism II (Turk, V., & Marks, N. eds), pp. 299–303, Plenum Press, New York.

1978

Singh, H. & Kalnitzky, G. Separation of a new α-N-benzoylargine-β-naphthylamide hydrolase from cathepsin Bl. Purification, characterization, and properties of both enzymes from rabbit lung. J. Biol. Chem. 253: 4319–4326.

Singh, H., Kuo, T. & Kalnitsky, G. Collagenolytic activity of lung BANA hydrolase and cathepsin Bl. In: Protein Turnover and Lysosome Function (Segal, H. L. & Doyle, D. J. eds), pp. 315–331, Academic Press, New York.

1980

Barrett, A. J. Thiol proteinases of human lysosomes. In: Enzyme Regulation and Mechanism of Action (Mildner, P. & Ries, B. eds), pp. 307–315, Pergamon, Oxford.

Singh, H. & Kalnitsky, G. α-N-Benzoylarginine-β-naphthylamide hydrolase, an aminoendopeptidase from rabbit lung. J. Biol. Chem. 255: 369–374.

Turk, V., Kregar, I., Gubensek, T., Popovic, P., Locnicar, P. & Lah, T. Carboxyl and thiol intracellular proteinases. In: Enzyme Regulation and Mechanism of Action (Mildner, P. & Ries, B. eds), pp. 317–330, Pergamon, Oxford.

Entry 2.03

CATHEPSIN L

Summary

EC Number: 3.4.22.15

Earlier names: Initially termed L20 C5 [1].

Distribution: In rat liver lysosomes, and probably lysosomes generally.

Source: Rat liver lysosomes.

Action: Acts on proteins, and also hydrolyzes Bz-Arg-NH$_2$, but has very little action on other synthetic substrates of cathepsin B.

Requirements: pH optimum 5.0-6.0. Thiol activator; sometimes a chelator.

Substrate, usual: Azo-casein.

Substrate, special: Labelled cytosol proteins, Z-Lys-OPhNO$_2$ [3].

Inhibitors: Thiol-blocking reagents, chloromethylketones, Z-Phe-Phe-CHN$_2$, L-trans-epoxysuccinyl-Leu-agmatine (E-64)[2], leupeptin.

Molecular properties: Molecular weight about 24,000. pI 5.8-6.1 (multiple forms).

Comment: Resembles cathepsin N (#2.04) in many respects. K$_m$ for Bz-Arg-NH$_2$ (3 mM) seems 10-fold lower than that of cathepsin B.

References
[1] Bohley et al. In: Tissue Proteinases (Barrett, A. J. & Dingle, J. T. eds), pp.187-219, North-Holland Publishing Co., Amsterdam, 1971.
[2] Towatari et al. J. Biochem. 84: 659-671, 1978.
[3] Strewler & Manganiello J. Biol. Chem. 254: 11891-11898, 1979.

Bibliography

1971

Bohley, P., Kirschke, H., Langner, J., Ansorge, S., Wiederanders, B. & Hanson, H. Intracellular protein breakdown. In: Tissue Proteinases (Barrett, A. J. & Dingle, J. T. eds), pp. 187-219, North-Holland Publishing Co., Amsterdam.

1972

Kirschke, H., Langner, J., Wiederanders, B., Ansorge, S. & Bohley, P. Intrazellulärer Proteinabbau. IV. Isolierung und Charakterisierung von peptidasen aus Rattenleberlysosomen. Acta Biol. Med. Ger. 28: 305-322.

1976

Kirschke, H., Langner, J., Wiederanders, B., Ansorge, S., Bohley, P. & Broghammer, U. Intrazellulärer Proteinabbau. VII. Kathepsin L und H: zwei neue Proteinasen aus Rattenleberlysosomen. Acta Biol. Med. Ger. 35: 285-299.

Kirschke, H., Langner, J., Wiederanders, B., Ansorge, S., Bohley, P. & Hanson, H. Cathepsin L and proteinases with cathepsin B1-like activity from rat liver lysosomes. In: Intracellular Protein Catabolism (Hanson, H. & Bohley, P. eds), pp. 210-217, J. A. Barth, Leipzig.

Towatari, T., Tanaka, K., Yoshikawa, D. & Katunuma, N. Separation of a new protease from cathepsin B1 of rat liver lysosomes. FEBS Lett. 67: 284-288.

1977

Ansorge, S., Kirschke, H. & Friedrich, K. Conversion of proinsulin into insulin by cathepsins B and L from rat liver lysosomes. Acta Biol. Med. Ger. 36: 1723-1727.

DeMartino, G. N., Doebber, T. W. & Miller, L. L. Pepstatin-insensitive proteolytic activity of rat liver lysosomes. Isolation and purification of a new pepstatin-insensitive proteolytic enzyme from rat lysosomes. J. Biol. Chem. 252: 7511-7516.

Katunuma, N. New intracellular proteases and their role in intracellular degradation. Trends Biochem. Sci. 2: 122-125.

Kirschke, H., Langner, J., Wiederanders, B., Ansorge, S. & Bohley, P. Cathepsin L. A new proteinase from rat-liver lysosomes. Eur. J. Biochem. 74: 293-301.

Kirschke, H., Langner, J., Wiederanders, B., Ansorge, S., Bohley, P. & Hanson, H. Cathepsin L and cathepsin B3 from rat liver lysosomes. In: Intracellular Protein Catabolism II (Turk, V., & Marks, N. eds), pp. 299-303, Plenum Press, New York.

1978

Strewler, G. J., Manganiello, V. C. & Vaughan, M. Phosphodiesterase activator from rat kidney cortex. J. Biol. Chem. 253: 390-394.

Towatari, T., Tanaka, K., Yoshikawa, D. & Katunuma, N. Purification and properties of a new cathepsin from rat liver. J. Biochem. 84: 659-671.

1979

Pinkett, M. O., Strewler, G. J. & Anderson, W. B. Stimulation of adenylate cyclase activity by a protease which activates cyclic nucleotide phosphodiesterase. Biochem. Biophys. Res. Commun.

90: 1159–1165.
Strewler, G. J. & Manganiello, V. C. Purification and characterization of phosphodiesterase activator from kidney. A lysosomal protease. J. Biol. Chem. 254: 11891–11898.

1980

Barrett, A. J. Thiol proteinases of human lysosomes. In: Enzyme Regulation and Mechanism of Action (Mildner, P. & Ries, B. eds), pp. 307–315, Pergamon, Oxford.

Entry 2.04

CATHEPSIN N

Summary

EC Number: (3.4.22.-)

Earlier names: Collagenolytic cathepsin.

Distribution: Detected in rat granulomata, bovine spleen, rabbit neutrophil leucocytes and human placenta. Lysosomal.

Source: Bovine spleen, rat granuloma, etc.

Action: Attacks the N-terminal peptides of native collagen, thus eliminating the cross-links, and forming modified tropocollagen very much as cathepsin B and leukocyte elastase do, but the specific activity of the enzyme is higher than that of cathepsin B. Cleaves the insulin B chain at several points. Little action on azocasein, but probably degrades proteoglycan and gelatin. Little or no action on synthetic substrates of cathepsin B.

Requirements: pH optimum about 3.5; thiol-activated.

Substrate, usual: Collagen (but there is no known substrate specific for this enzyme in crude samples).

Inhibitors: Thiol-blocking reagents, Tos-Lys-CH$_2$Cl, leupeptin.

Molecular properties: Bovine: mol.wt. about 20,000, pI 6.4. Human: mol.wt. 34,600, pI 5.1.

Comment: In some studies, thiol-dependent collagenolytic activity has been attributed to "collagenolytic cathepsin" without proper consideration of the possible contribution of cathepsins B and L. Cathepsin N is not specific for collagen, but its low activity on azocasein seems to distinguish it from cathepsin L.

Bibliography

1966

Bazin, S. & Delaunay, A.   Caractères de cathepsines collagénolytiques présente dans les tissus enflammés du rat. Ann. Inst. Pasteur Paris 110: 192-204.

Bazin, S. & Delaunay, A.   Influence exercée par une hypervitaminose A sur la constitution de la trame conjonctive et l'activité collagénolytique de tissus normaux ou enflammes.   Ann. Inst. Pasteur Paris, 110: 487-492.

Bazin, S. & Delaunay, A.   Variations de l'activité collagénolytique dans des foyers inflammatoires en évolution.   Ann. Inst. Pasteur Paris 110: 347-354.

1971

Bazin, S. & Delaunay, A.   Biochimie de l'inflammation. IX. Étude chimique et physiologique de deux cathepsines isolées de foyers inflammatoires.   Ann. Inst. Pasteur Paris 120: 50-61.

1972

Etherington, D. J.   The nature of the collagenolytic cathepsin of rat liver and its distribution in other rat tissues.   Biochem. J. 127: 685-692.

1973

Etherington, D. J.   Collagenolytic-cathepsin and acid-proteinase activities in the rat uterus during post partum involution.   Eur. J. Biochem. 32: 126-128.

1974

Etherington, D. J.   The purification of bovine cathepsin B1 and its mode of action on bovine collagens.   Biochem. J. 137: 547-557.
Stančikova, M. & Trnavský, K.   Collagenolytic enzymes and acid-proteinase activity in granuloma tissue.   Experientia 30: 594-595.

1976

Etherington, D. J.   Bovine spleen cathepsin B1 and collagenolytic cathepsin. A comparative study of the properties of the two enzymes in the degradation of native collagen.   Biochem. J. 153: 199-209.
Etherington, D. J. & Evans, P.   The action of cathepsin B1 and collagenolytic cathepsin in the degradation of collagen.   In: Intracellular Protein Catabolism (Hanson, H. & Bohley, P. eds), pp. 25-31, J. A. Barth, Leipzig.
Gibson, W. T., Milsom, D. W., Steven, F. S. & Lowe, J. S. Collagenolytic cathepsin activity in rabbit peritoneal polymorphonuclear leucocytes.   Biochem. Soc. Trans. 4: 627-628.
Gibson, W. T., Milsom, D. W., Steven, F. S. & Lowe, J. S.   The subcellular location of a collagenolytic cathepsin in rabbit peritoneal polymorphonuclear leucocytes.   Biochem. Soc. Trans. 4: 628-630.

1977

Burleigh, M. C.   Degradation of collagen by non-specific proteinases.   In: Proteinases in Mammalian Cells and Tissues (Barrett, A. J. ed.), pp. 285-309, North-Holland Publishing Co., Amsterdam.
Etherington, D. J.   The dissolution of insoluble bovine collagens by cathepsin B1, collagenolytic cathepsin and pepsin. The influence of collagen type, age and chemical purity on

susceptibility. Connect. Tissue Res. 5: 135-145.

Etherington, D. J. & Evans, P. J. The action of cathepsin B and collagenolytic cathepsin in the degradation of collagen. Acta Biol. Med. Ger. 36: 1555-1563.

Stražiščar, S., Zvonar, T. & Turk, V. Cathepsins from experimental granuloma and their action on collagen. In: Intracellular Protein Catabolism II (Turk, V. & Marks, N. eds), pp. 224-229, Plenum Press, New York.

1978

Ducastaing, A. & Etherington, D. J. Purification of bovine spleen collagenolytic cathepsin (cathepsin N). Biochem. Soc. Trans. 6: 938-940.

Evans, P. & Etherington, D. J. Characterisation of cathepsin B and collagenolytic cathepsin from human placenta. Eur. J. Biochem. 83: 87-97.

Gibson, W. T., Milsom, D. W., Steven, F. S. & Lowe, J. S. Collagenolytic cathepsin activity in rabbit peritoneal polymorphonuclear leucocyte granules. Biochem. J. 172: 83-89.

Hayasaka, S. & Hayasaka, I. The distribution and some properties of collagenolytic cathepsin in the bovine eye. Albrecht von Graefes Arch. Ophthalmol. 206: 163-168.

Hayasaka, S. & Hayasaka, I. The presence of collagenolytic cathepsin in uveal lysosomes of bovine eye. Albrecht von Graefes Arch. Ophthalmol. 206: 25-32.

1979

Evans, P. & Etherington, D. J. Action of cathepsin N on the oxidized B-chain of bovine insulin. FEBS Lett. 99: 55-58.

Hayasaka, S. & Hayasaka, I. Cathepsin B and collagenolytic cathepsin in the aqueous humor of patients with Behçet's disease. Albrecht von Graefes Arch. Ophthalmol. 210: 103-107.

Entry 2.05

CATHEPSIN S

Summary

EC Number: (3.4.22.-)

Earlier names: -

Distribution: Bovine lymph nodes, spleen.

Source: Bovine lymph nodes, spleen.

Action: Digests hemoglobin and other proteins. No activity against Bz-Arg-NNap.

Requirements: pH optimum about 3.5.

Substrate, usual: Hemoglobin.

Inhibitors: 4-Chloromercuribenzoate, leupeptin.

Molecular properties: Mol.wt. 25,000 (SDS electrophoresis, gel chromatography). pI 6.3-6.9 (multiple forms).

Comment: May well be identical with cathepsin L.

Bibliography

1975

Turnšek, T., Kregar, I. & Lebez, D. Acid sulphydryl protease from calf lymph nodes. Biochim. Biophys. Acta 403: 514-520.

1978

Turk, V., Kregar, I., Gubensek, F. & Locnikar, P. Bovine spleen cathepsins D and S: purification, characterization and structural studies. In: Protein Turnover and Lysosome Function (Segal, H. L. & Doyle, D. J. eds), pp. 353-361, Academic Press, New York.

1980

Turk, V., Kregar, I., Gubensek, T., Popovic, P., Locnicar, P. & Lah, T. Carboxyl and thiol intracellular proteinases. In: Enzyme Regulation and Mechanism of Action (Mildner, P. & Ries, B. eds), pp. 317-330, Pergamon, Oxford.

Entry 2.06

LYSOSOMAL INSULIN-GLUCAGON PROTEINASE

Summary

EC Number: (3.4.22.-)

Earlier names: Insulinase, insulin-specific protease.

Distribution: Detected in lysosomal fractions from rat liver and kidney.

Source: Rat liver or kidney.

Action: Degrades insulin, glucagon, and possibly other proteins.

Requirements: pH 4, and a thiol compound.

Substrate, usual: ^{125}I-labelled insulin.

Inhibitors: Thiol-blocking reagents.

Molecular properties: -

Comment: Requires higher substrate concentrations than the cytosol insulin-glucagon proteinase. The activity is not well characterized, and may well be due to cathepsins B, H and L.

Bibliography

1976

Duckworth, W. C.   Insulin and glucagon degradation by the kidney. I. Subcellular distribution under different assay conditions. Biochim. Biophys. Acta 437: 518-530.

Grisolia, S. & Wallace, R.   Insulin degradation by lysosomal extracts from rat liver: model for a role of lysosomes in hormone degradation. Biochem. Biophys. Res. Commun. 70: 22-27.

Entry 2.07

CALCIUM-DEPENDENT PROTEINASE I

Summary

EC Number: (3.4.22.-)

Earlier names: -

Distribution: Rabbit skeletal muscle.

Source: Same.

Action: Endopeptidase.

Requirements: Ca^{2+}, thiol activation, pH 7.5.

Substrate, usual: Denatured serum albumin.

Substrate, special: Succinylated albumin, casein or protamine.

Inhibitors: ECTA.

Molecular properties: Eluted from DEAE-cellulose before proteinase II (#2.08) in a gradient of increasing salt concentration (hence numbering).

Comment: Not active in removing Z-lines of myofibrils, unlike proteinase II (#2.08), and also differs in relative activities with protein substrates. Has not been studied in any detail.

Bibliography

1975

Reddy, M. K., Etlinger, J. D., Rabinowitz, M., Fischman, D. A. & Zak, R. Removal of Z-lines and alpha-actinin from isolated myofibrils by a calcium-activated neutral protease. J. Biol. Chem. 250: 4278-4284.

1979

Mellgren, R. L., Aylward, J. H., Killilea, S. D. & Lee, E. Y. C. The activation and dissociation of a native high molecular weight form of rabbit skeletal muscle phosphorylase phosphatase by endogenous Ca^{2+}-dependent proteases. J. Biol. Chem. 254: 648-652.

Entry 2.08

CALCIUM-DEPENDENT PROTEINASE II

Summary

EC Number: (3.4.22.-)

Earlier names: Calcium-activated factor (CAF), calcium-activated sarcoplasmic factor (CASF), kinase-activating factor (KAF, but note that KAF is also an abbreviation used for C3b inactivator), calcium-activated neutral protease (CANP), receptor transforming factor.

Distribution: Detected in rat and rabbit brain, skeletal muscle of rat, rabbit, pig and man, bovine cardiac muscle and uterus, rat liver and human platelets. Seems to be predominantly cytoplasmic, but immunofluorescence shows a small proportion of the enzyme bound to Z-bands in chicken muscle [3].

Source: Sarcoplasmic fraction of rat, rabbit or pig muscle.

Action: Removes Z-disks (and M-lines) from myofibrils. Limited cleavage of troponin-I, tropomyosin and C-protein from myofibrils. Degrades casein. Activates phosphorylase kinase, presumably by limited proteolysis.

Requirements: pH 7.5 (labile above pH 8). 1.0 mM Ca^{2+} optimal (but unstable to storage in the presence of Ca^{2+}), 2 mM 2-mercaptoethanol, 25° (autolyzes at 37°).

Substrate, usual: Casein.

Substrate, special: Myofibrils.

Inhibitors: EDTA, EGTA (but reversible by Ca^{2+}), leupeptin, iodoacetate, tetrathionate. L-trans-epoxysuccinyl-Leu-agmatine (E-64) for the chicken enzyme [4]. Natural inhibitor (270,000 mol.wt.) in muscle, liver and brain [1,2], and α_2-macroglobulin.

Molecular properties: Molecular weight about 110,000, comprised of chains of about 80,000 and 30,000, or may simply be 80,000 mol.wt. single chain. 5.9S. Molecular dimensions calculated as 4 x 23 nM.

Comment: Although the similarities between Ca^{2+}-dependent cysteine proteinases of various tissues are striking, it clearly cannot yet be decided with certainty whether the enzymes are identical.

References
[1] Waxman & Krebs   J. Biol. Chem. 253: 5888-5891, 1978.

[2] Nishiura et al. Experientia 35: 1006-1007, 1979.
[3] Ishiura et al. J. Biochem. 87: 343-346, 1980.
[4] Sugita et al. J. Biochem. 87: 339-341, 1980.

Bibliography

1964

Guroff, G. A neutral, calcium-activated proteinase from the soluble fraction of rat brain. J. Biol. Chem. 239: 149-155.

Meyer, W. L., Fischer, E. H. & Krebs, E. G. Activation of skeletal muscle phosphorylase b kinase by Ca^{2+}. Biochemistry 3: 1033-1039.

1965

Belocopitow, E., Appleman, M. M. & Torres, H. N. Factors affecting the activity of muscle glycogen synthetase. II. The regulation by Ca^{++}. J. Biol. Chem. 240: 3473-3478.

1966

Drummond, G. I. & Duncan, L. The action of calcium ion on cardiac phosphorylase b kinase. J. Biol. Chem. 241: 3097-3103.

1968

Drummond, G. I. & Duncan, L. On the mechanism of activation of phosphorylase b kinase by calcium. J. Biol. Chem. 243: 5532-5538.

Huston, R. B. & Krebs, E. G. Activation of skeletal muscle phosphorylase kinase by Ca^{2+}. II. Identification of the kinase activating factor as a proteolytic enzyme. Biochemistry 7: 2116-2122.

1969

Kohn, R. R. A proteolytic system involving myofibrils and a soluble factor from normal and atrophying muscle. Lab. Invest. 20: 202-206.

1972

Busch, W. A., Stromer, M. H., Goll, D. E. & Suzuki, A. Ca^{2+}-specific removal of Z-lines from rabbit skeletal muscle. J. Cell Biol. 52: 367-381.

1974

Okitani, A., Otsuka, Y., Sugitani, M. & Fujimaki, M. Some properties of neutral proteolytic system in rabbit skeletal muscle. Agr. Biol. Chem. 38: 573-579.

Suzuki, A. & Goll, D. E. Quantitative assay for CASF (Ca^{2+}-activated sarcoplasmic factor) activity, and effect of CASF treatment on ATPase activities of rabbit myofibrils. Agr. Biol. Chem. 38: 2167-2175.

1975

Dayton, W. R., Goll, D. E., Stromer, M. H., Reville, W. J., Zeece, M. G. & Robson, R. M. Some properties of a Ca^{2+}-activated protease that may be involved in myofibrillar protein turnover. In: Proteases and Biological Control (Reich, E., Rifkin, D. B. & Shaw, E. eds), pp. 551-577, Cold Spring Harbor Laboratory, New York.

Reddy, M. K., Etlinger, J. D., Rabinowitz, M., Fischman, D. A. & Zak, R. Removal of Z-lines and α-actinin from isolated myofibrils by a calcium-activated neutral protease. J. Biol. Chem. 250: 4278-4284.

1976

Dayton, W. R., Goll, D. E., Zeece, M. G., Robson, R. M. & Reville, W. J. A Ca^{2+}-activated protease possibly involved in myofibrillar protein turnover. Purification from porcine muscle. Biochemistry 15: 2150-2158.

Dayton, W. R., Reville, W. J., Goll, D. E. & Stromer, M. H. A Ca^{2+}-activated protease possibly involved in myofibrillar protein turnover. Partial characterization of the purified enzyme. Biochemistry 15: 2159-2167.

Kar, N. C. & Pearson, C. M. A calcium-activated neutral protease in normal and dystrophic human muscle. Clin. Chim. Acta 73: 293-297.

Reville, W. J., Goll, D. E., Stromer, M. H., Robson, R. M. & Dayton, W. R. A Ca^{2+}-activated protease possibly involved in myofibrillar protein turnover. Subcellular localization of the protease in porcine skeletal muscle. J. Cell Biol. 70: 1-8.

1977

Hamon, M., Bourgoin, S., Artaud, F. & Héry, F. Rat brain stem tryptophan hydroxylase: mechanism of activation by calcium. J. Neurochem. 28: 811-818.

Inoue, M., Kishimoto, A., Takai, Y. & Nishizuka, Y. Studies on a cyclic nucleotide-independent protein kinase and its proenzyme in mammalian tissues. J. Biol. Chem. 252: 7610-7616.

Phillips, D. R. & Jakábová, M. Ca^{2+}-dependent protease in human platelets. Specific cleavage of platelet polypeptides in the presence of added Ca^{2+}. J. Biol. Chem. 252: 5602-5605.

Puca, G. A., Nola, E., Sica, V. & Bresciani, F. Estrogen binding proteins of calf uterus. Molecular and functional characterization of the receptor transforming factor: a Ca^{2+}-activated protease. J. Biol. Chem. 252: 1358-1366.

Takai, Y., Yamamoto, M., Inoue, M., Kishimoto, A. & Nishizuka, Y. A proenzyme of cyclic nucleotide-independent protein kinase and its activation by calcium-dependent neutral protease from rat liver. Biochem. Biophys. Res. Commun. 77: 542-550.

1978

Davies, P. J. A., Wallach, D., Willingham, M. C., Pastan, I., Yamaguchi, M. & Robson, R. M. Filamin-actin interaction.

Dissociation of binding from gelation by Ca^{2+}-activated proteolysis. J. Biol. Chem. 253: 4036-4042.

Goll, D. E., Okitani, A., Dayton, W. R. & Reville, W. J. A Ca^{2+}-activated muscle protease in myofibrillar protein turnover. In: Protein Turnover and Lysosome Function (Segal, H. L. & Doyle, D. J. eds), pp. 587-588, Academic Press, New York.

Ishiura, S., Murofushi, H., Suzuki, K. & Imahori, K. Studies of a calcium-activated neutral protease from chicken skeletal muscle. 1. Purification and characterization. J. Biochem. 84: 225-230.

Nishiura, I., Tanaka, K., Yamato, S. & Murachi, T. The occurrence of an inhibitor of Ca^{2+}-dependent neutral protease in rat liver. J. Biochem. 84: 1657-1659.

Sandoval, I. V. & Weber, K. Calcium-induced inactivation of microtubule formation in brain extracts. Eur. J. Biochem. 92: 463-470.

Waxman, L. Characterization of two unique protease inhibitors and a calcium-activated protease from bovine cardiac muscle. In: Protein Turnover and Lysosome Function (Segal, H. L. & Doyle, D. J. eds), pp. 363-377, Academic Press, New York.

Waxman, L. & Krebs, E. G. Identification of two protease inhibitors from bovine cardiac muscle. J. Biol. Chem. 253: 5888-5891.

1979

Azanza, J.-L., Raymond, J., Robin, J.-M., Cottin, P. & Ducastaing, A. Purification and some physico-chemical and enzymic properties of a calcium ion-activated neutral proteinase from rabbit skeletal muscle. Biochem. J. 183: 339-347.

Dayton, W. R., Schollmeyer, J. V., Chan, A. C. & Allen, C. E. Elevated levels of a calcium-activated muscle protease in rapidly atrophying muscles from vitamin E-deficient rabbits. Biochim. Biophys. Acta 584: 216-230.

Hamon, M. & Bourgoin, S. Characterization of the Ca^{2+}-induced proteolytic activation of tryptophan hydroxylase from the rat brain stem. J. Neurochem. 32: 1837-1844.

Mellgren, R. L., Aylward, J. H., Killilea, S. D. & Lee, E. Y. C. The activation and dissociation of a native high molecular weight form of rabbit skeletal muscle phosphorylase phosphatase by endogenous Ca^{2+}-dependent proteases. J. Biol. Chem. 254: 648-652.

Neerunjun, J. S. & Dubowitz, V. Increased calcium-activated neutral protease activity in muscles of dystrophic hamsters and mice. J. Neurol. Sci. 40: 105-111.

Nishiura, I., Tanaka, K. & Murachi, T. A high molecular weight inhibitor of Ca^{2+}-dependent neutral protease in rat brain. Experientia 35: 1006-1007.

Sandoval, I. V. & Weber, K. Calcium-induced inactivation of microtubule formation in brain extracts. Presence of a calcium-dependent protease acting on polymerization-stimulating microtubule-associated proteins. Eur. J. Biochem. 92: 463-470.

Suzuki, K., Ishiura, S., Tsuji, S., Katamoto, T., Sugita, H. &

Imahori, K.   Calcium activated neutral protease from human
skeletal muscle. FEBS Lett. 104: 355-358.
Toyo-Oka, T. & Masaki, T.   Calcium-activated neutral protease from
bovine ventricular muscle: isolation and some of its properties.
J. Mol. Cell. Cardiol. 11: 769-786.

1980

Mellgren, R. L.   Canine cardiac calcium-dependent proteases:
resolution of two forms with different requirements for calcium.
FEBS Lett. 109: 129-133.

Entry 2.09
LEUCOEGRESIN-GENERATING PROTEINASE

Summary

EC Number: (3.4.22.-)

Earlier names: SH-Dependent protease II (protease I being the higher mol.wt. form - see below).

Distribution: Found in rabbit skin Arthus reaction and burn sites.

Source: As above.

Action: Liberates a chemotactic factor for polymorphonuclear leucocytes by limited cleavage of rabbit or human IgG.

Requirements: Optimal pH 7.1 for casein. Activated by cysteine (1mM).

Substrate, usual: Casein.

Substrate, special: Immunoglobulin G.

Inhibitors: 4-Chloromercuribenzoate, O_2 (then reactivated by cysteine or glutathione). Tissue inhibitor (12,000 mol.wt. heat-stable protein) and plasma inhibitor (72,000 mol.wt. glycoprotein).

Molecular properties: Mol.wt. 14,000 (by gel chromatography). Also occurs as a 200,000 mol.wt. aggregate.

Bibliography

1965

Hayashi, H., Udaka, K., Miyoshi, H. & Kudo, S. Further study of correlative behaviour between specific protease and its inhibitor in cutaneous Arthus reactions. Lab. Invest. 14: 665-673.

Udaka, K. & Hayashi, H. Further purification of a protease inhibitor from rabbit skin with healing inflammation. Biochim. Biophys. Acta 97: 251-261.

Udaka, K. & Hayashi, H. Molcular weight determination of a protease inhibitor from rabbit skin with healing inflammation. Biochim. Biophys. Acta 104: 600-603.

1968

Koono, M., Muto, M. & Hayashi, H. Proteases associated with

Arthus skin lesions: their purification and biological
significance. Tohoku J. Exp. Med. **94**: 231-235.

1971

Kouno, T. The role of neutral SH-dependent protease of PMN
leukocyte lysosomes in inflammation. Kumamoto Med. J. **24**: 135-
150.

Yamamoto, S., Yoshinaga, M. & Hayashi, H. The natural mediator
for PMN emigration in inflammation. II. Common antigenicity of
leucoegresin with immunoglobulin G. Immunology 20: 803-808.

Yoshinaga, M., Yamamoto, S., Maeda, S. & Hayashi, H. The natural
mediator for PMN emigration in inflammation. III. In vitro
production of a chemotactic factor by inflammatory SH-dependent
protease from serum immunoglobulin G. Immunology 20: 809-815.

Yoshinaga, M., Yoshida, K., Tashiro, A. & Hayashi, H. The natural
mediator for PMN emigration in inflammation. I. Purification and
characterization of leucoegresin from Arthus skin site.
Immunology **21**: 281-298.

1973

Yamamoto, S., Nishiura, M. & Hayashi, H. The natural mediator for
PMN emigration in inflammation. V. The site of structural change
in the chemotactic generation of immunoglobulin G by inflammatory
SH-dependent protease. Immunology 24: 791-801.

1974

Hayashi, H. Chemotaxis in immune tissue injury. Int. Cong. Ser.
No. 323. Allergology: Proc. 8th Congress Int. Assoc. Allergology,
pp. 252-260, Excerpta Medica, Amsterdam.

Nishiura, M., Yamamoto, S. & Hayashi, H. The natural mediator for
PMN emigration in inflammation. VI. Production of leucoegressin
by inflammatory SH-dependent protease in active Arthus reactions
in complement-depleted rabbits. Immunology 27: 1023-1031.

Yamamoto, S., Nishiura, M. & Matsumura, K. In vitro production of
a chemotactic factor by inflammatory SH-dependent protease from
rabbit and human immunoglobulin G subclasses. Tohoku J. Exp.
Med. 114: 49-54.

1975

Hayashi, H. The intracellular neutral SH-dependent protease
associated with inflammatory reactions. Int. Rev. Cytol. 40:
101-151.

Entry 2.10
ATP-DEPENDENT PROTEINASE

Summary

EC Number: (3.4.22.-)

Earlier names: -

Distribution: Rabbit reticulocytes, rat liver soluble fraction.

Source: Rat liver cytosol.

Action: Degrades hemoglobin and globin.

Requirements: Optimal pH 7.5-9.5. Stimulated by dithiothreitol, by ATP (5 mM) and other pyrophosphates, and also (additively) by Triton X-100.

Substrate, usual: [^{14}C]Methyl-globin or ^{125}I-labelled hemoglobin.

Inhibitors: Iodoacetate (partial).

Molecular properties: Apparent mol.wt. 550,000 (gel chromatography).

Comment: The role played by the ATP is unclear, and it is not known to be consumed. Pyrophosphates chelate various cations, but the specificity for activators of this enzyme suggests that this may not be merely a cation-sensitive proteinase. Activation of cathepsin D by ATP has also been reported [1].

Dependence of a proteinase on ATP is of special interest because of the physiological relationship between protein breakdown and metabolic activity.

References
[1] Watabe et al. Biochem. Biophys. Res. Commun. 89: 1161-1167, 1979.

Bibliography

1977

Etlinger, J. D. & Goldberg, A. L.   A soluble ATP-dependent proteolytic system responsible for the degradation of abnormal proteins in reticulocytes. Proc. Natl. Acad. Sci. U.S.A. 74: 54-58.

1978

Hershko, A., Heller, H., Ganoth, D. & Ciehanover, A.   Mode of degradation of abnormal globin chains in rabbit reticulocytes. In: Protein Turnover and Lysosome Function (Segal, H. L. & Doyle, D. J. eds), pp. 149-169, Academic Press, New York.

1979

DeMartino, G. N. & Goldberg, A. L. Identification and partial purification of an ATP-stimulated alkaline protease in rat liver. J. Biol. Chem. 254: 3712-3715.

Rose, I. A., Warms, J. V. B. & Hershko, A. A high molecular weight protease in liver cytosol. J. Biol. Chem. 254: 8135-8138.

Entry 2.11

MITOCHONDRIAL CYSTEINE PROTEINASE

Summary

EC Number: (3.4.22.-)

Earlier names: -

Distribution: Rat liver mitochondrial membrane fraction.

Source: Same.

Action: Degrades insulin.

Requirements: Optimal pH 8. Stimulated by EDTA.

Substrate, usual: ^{125}I-Labelled insulin-Sepharose.

Inhibitors: MalNEt. (Not Pms-F, soybean trypsin inhibitor, 1,10-phenanthroline.)

Molecular properties: -

Comment: Chymase I (#1.42) was extracted from the starting material at high ionic strength, and lysosomal acid phosphatase was removed in a digitonin treatment, before the enzyme was solubilized with Triton X-100.

Bibliography

1978

Hare, J. F. A novel proteinase associated with mitochondrial membranes. Biochem. Biophys. Res. Commun. 83: 1206-1215.

Entry 2.12

CYTOSOL INSULIN-GLUCAGON PROTEINASE

Summary

EC Number: (3.4.22.-, previously 3.4.22.11 and 3.4.99.10).

Earlier names: Insulinase, insulin-specific protease.

Distribution: Known from various rat tissues.

Source: Rat skeletal muscle, kidney, liver.

Action: Seems highly specific for insulin (K_m 2 x 10^{-8} M) and glucagon (K_m 3 x 10^{-6} M). Cleaves insulin at $-Tyr_{16}-Leu_{17}-$ in the B chain. Inactive against proinsulin and the separated A and B chains of insulin.

Requirements: pH about 7.5. (Usually no thiol activation.)

Substrate, usual: ^{125}I-Labelled glucagon or insulin.

Inhibitors: 1,10-Phenanthroline, EDTA, Nbs_2, MalNEt, 4-chloromercuribenzoate. (Zn^{2+}, Mn^{2+} and Co^{2+}, but not Ca^{2+}, may prevent inactivation by 1,10-phenanthroline or EDTA.) Non-supressible insulin-like factor of serum, proinsulin (K_i 4 x 10^{-8} M), ACTH.

Molecular properties: Mol.wt. 80,000.

Comment: This enzyme is effective in insulin degradation at far lower substrate concentrations than the lysosomal insulin-glucagon proteinase (#2.06). It seems to be a member of the rather vaguely defined group of metal-dependent cysteine proteinases. Other enzymes apparently dependent on both a thiol group and a metal include the calcium-dependent cysteine proteinases (#2.07, #2.08), collagenase-like peptidase (#5.01), lens neutral proteinase (#5.02) and possibly a myofibril-associated proteinase.

Bibliography

1957

Mirsky, I. A. Insulinase, insulinase-inhibitors, and diabetes mellitus. Recent Prog. Hormone Res. 13: 429-471.

1971

Brush, J. S. Purification and characterization of a protease with specificity for insulin from rat muscle. Diabetes 20: 140–145.

1972

Burghen, G. A., Kitabchi, A. E. & Brush, J. S. Characterization
of a rat liver protease with specificity for insulin.
Endocrinology 91: 633–642.
Duckworth, W. C., Heinemann, M. A. & Kitabchi, A. E. Purification
of insulin-specific protease by affinity chromatography. Proc.
Natl. Acad. Sci. U.S.A. 69: 3698–3702.
Kitabchi, A. E. & Stentz, F. B. Degradation of insulin and
proinsulin by various organ homogenates of rat. Diabetes 21:
1091–1101.

1973

Baskin, F. K. & Kitabchi, A. E. Substrate studies of insulin-
specific protease. Eur. J. Biochem. 37: 489–496.
Brush, J. S. Purification and characterization of inhibitors of
insulin specific protease in human serum. Biochem. Biophys.
Res. Commun. 53: 894–903.

1974

Duckworth, W. C. & Kitabchi, A. E. Insulin and glucagon
degradation by the same enzyme. Diabetes 23: 536–543.

1975

Baskin, F. K., Duckworth, W. C. & Kitabchi, A. E. Sites of
cleavage of glucagon by insulin-glucagon protease. Biochem.
Biophys. Res. Commun. 67: 163–169.
Duckworth, W. C., Heinemann, M. & Kitabchi, A. E. Proteolytic
degradation of insulin and glucagon. Biochim. Biophys. Acta
377: 421–430.

1976

Burghen, G. A., Duckworth, W. C., Kitabchi, A. E., Solomon, S. S.
& Poffenbarger, P. L. Inhibition of insulin degradation by non-
suppressible insulin-like activity. J. Clin. Invest. 57: 1089–
1092.
Duckworth, W. C. Insulin and glucagon degradation by the kidney.
I. Subcellular distribution under different assay conditions.
Biochim. Biophys. Acta 437: 518–530.
Duckworth, W. C. Insulin and glucagon degradation by the kidney.
II. Characterization of the mechanisms at neutral pH. Biochim.
Biophys. Acta 437: 531–542.
Kahn, C. R., Megyesi, K. & Roth, J. Non suppressible insulin-like
activity of human serum. A potent inhibitor of insulin
degradation. J. Clin. Invest. 57: 526–529.

1977

Ansorge, S., Kirschke, H., Köhler, E., Langner, J., Römer, I. &
Wiederanders, B. Insulin and glucagon degradation in livers of
sand rats (Psammomys obesus). Acta Biol. Med. Ger. 36: 1713–
1727.

Ansorge, S., Kirschke, H., Langner, J., Wiederanders, B., Bohley, P. & Hanson, H. Proteolytic degradation of insulin and glucagon by enzymes of rat liver. In: Intracellular Protein Catabolism II (Turk, V. & Marks, N. eds), pp. 163-166, Plenum Press, New York.

Brush, J. S. Studies on inhibitors of the insulin protease of rat liver. Biochem. Pharmacol. 26: 2349-2354.

1978

Duckworth, W. C. Insulin and glucagon binding and degradation by kidney cell membrane. Endocrinology 102: 1766-1774.

1979

Duckworth, W. C. Insulin degradation by liver cell membranes. Endocrinology 104: 1758-1764.

Duckworth, W. C., Stentz, F. B., Heinemann, M. & Kitabchi, A. E. Initial site of insulin cleavage by insulin protease. Proc. Natl. Acad. Sci. U.S.A. 76: 635-639.

Runyan, K., Duckworth, W. C., Kitabchi, A. E., & Huff, G. The effect of age on insulin degrading activity in rat tissue. Diabetes 28: 324-325.

Section 3

ASPARTIC PROTEINASES

Entry 3.01

PEPSIN A

Summary

EC Number: 3.4.23.1

Earlier names: Funduspepsin, PII, PI. Porcine pepsin D is the un-
phosphorylated form.

Distribution: Pepsins seem to occur in all vertebrates. Precursor
(pepsinogen) in gastric mucosa, active enzyme in stomach.
Pepsin(ogen) A represents the major component, secreted mainly
in the fundus region of the stomach.

Source: Usually pig gastric mucosa (for pepsinogen) or gastric
juice for pepsin.

Action: Preferentially cleaves bonds of Phe-, Met-, Leu- or Trp-
to another hydrophobic residue. Tyr-X bonds are relatively
resistant.

Requirements: General proteolytic activity maximal at pH 1.8-2.2.

Substrate, usual: Ac-Phe-Tyr(I_2). Hemoglobin.

Substrate, special: Phe-Gly-His-Phe(NO_2)$\neq$Phe-Ala-Phe-OMe (scarcely
hydrolyzed by chymosin, #3.06).

Inhibitors: N_2Ac-Nle-OMe with Cu^{2+}, p-bromophenacylbromide, 1,2-
epoxy-3-(p-nitrophenoxy)propane, pepstatin. Ascaris pepsin
inhibitor.

Molecular properties: The zymogen is a single chain (without
carbohydrate) activated by an intra- or intermolecular reaction
releasing about 40 residues from the N-terminus. pI about 2.7
due to many carboxyls and a serine phosphate group, and mol.wt.
about 34,500. Active enzyme (but not zymogen) unstable at pH 6
and above. Pepsin(ogen) A corresponds to the component most
mobile in electrophoresis at pH 8.5 (as the zymogen) or at pH 5
(the enzyme), in mammals generally.

Comment: Correspondence of pepsin A (the major gastric acid
proteinase of the pig) with pepsins in other species is not
precise.

Bibliography

1939

Fruton, J. S. & Bergmann, M.   The specificity of pepsin.   J. Biol. Chem. 127: 627-641.

1951

Baker, L. E.   New synthetic substrates for pepsin.   J. Biol. Chem. 193: 809-819.

1955

Herriott, R. M.   Swine pepsin and pepsinogen.   Methods Enzymol. 2: 3-7.

1959

Schlamowitz, M. & Peterson, L. U.   Studies on the optimum pH for the action of pepsin on "native" and denatured bovine serum albumin and bovine hemoglobin.   J. Biol. Chem. 234: 3137-3145.

Taylor, W. H.   Studies on gastric proteolysis. 3. The secretion of different pepsins by fundic and pyloric glands of the stomach. Biochem. J. 71: 384-388.

1960

Bovey, F. A. & Yanari, S. S.   Pepsin.   In: The Enzymes (Boyer, P. D., Lardy, H. & Myrbäck, K. eds), 2nd edn, vol. 4, pp. 63-92, Academic Press, New York.

Hartley, B. S.   Proteolytic enzymes.   Annu. Rev. Biochem. 29: 45-72.

1962

Taylor, W. H.   Proteinases of the stomach in health and disease. Physiol. Rev. 42: 519-553.

1963

Seijffers, M. J., Segal, H. L. & Miller, L. L.   Separation of pepsin I, pepsin IIA, pepsin IIB, and pepsin III from human gastric mucosa.   Am. J. Physiol. 205: 1099-1105.

Seijffers, M. J., Segal, H. L. & Miller, L. L.   Separation of pepsinogen I, pepsinogen II, and pepsinogen III from human gastric mucosa.   Am. J. Physiol. 205: 1106-1112.

1964

Kushner, I., Rapp, W. & Burtin, P.   Electrophoretic and immunochemical demonstration of the existence of four human pepsinogens.   J. Clin. Invest. 43: 1983-1993.

Seijffers, M. J., Miller, L. L. & Segal, H. L.   Partial characterization of human pepsin I, pepsin IIA, pepsin IIB, and pepsin III.   Biochemistry 3: 1203-1209.

1965

Delpierre, G. R. & Fruton, J. S.   Inactivation of pepsin by diphenyldiazomethane.   Proc. Natl. Acad. Sci. U.S.A. 54: 1161-

1167.
Erlanger, B. F., Vratsanos, S. M., Wassermann, N. & Cooper, A. G.
Specific and reversible inactivation of pepsin. J. Biol. Chem.
240: PC 3447-3448.
Jackson, W. T., Schlamowitz, M. & Shaw, A. Kinetics of pepsin
catalysed hydrolysis of N-acetyl-L-phenylalanyl-L-diiodotyrosine.
Biochemistry 4: 1537-1543.

1966

Delpierre, G. R. & Fruton, J. S. Specific inactivation of pepsin
by diazo ketone. Proc. Natl. Acad. Sci. U.S.A. 56: 1817-1822.
Gross, E. & Morell, J. L. Evidence for an active carboxyl group
in pepsin. J. Biol. Chem. 241: 3638-3642.
Hanley, W. B., Boyer, S. H. & Naughton, M. A. Electrophoretic and
functional heterogeneity of pepsinogen in several species.
Nature 209: 996-1002.
Rajagopalan, T. G., Moore, S. & Stein, W. H. Pepsin from
pepsinogen. Preparation and properties. J. Biol. Chem. 241:
4940-4950.
Rajagopalan, T. G., Stein, W. H. & Moore, S. The inactivation of
pepsin by diazoacetylnorleucine methyl ester. J. Biol. Chem.
241: 4295-4297.

1967

Erlanger, B. F., Vratsanos, S. M., Wasssermann, N. & Cooper, A. G.
Stereochemical investigation of the active center of pepsin
using a new inactivator. Biochem. Biophys. Res. Commun. 28:
203-208.
Etherington, D. J. & Taylor, W. H. Nomenclature of the pepsins.
Nature 216: 279-280.
Hamilton, G. A., Spona, J. & Cromwell, L. D. The inactivation of
pepsin by an equimolar amount of l-diazo-4-phenylbutanone-2.
Biochem. Biophys. Res. Commun. 26: 193-198.
Inouye, K. & Fruton, J. S. Studies on the specificity of pepsin.
Biochemistry 6: 1765-1777.
Lee, D. & Ryle, A. P. Pepsin D. A minor component of commercial
pepsin preparations. Biochem. J. 104: 742-748.
Lee, D. & Ryle, A. P. Pepsinogen D. A fourth proteolytic zymogen
from pig gastric mucosa. Biochem. J. 104: 735-741.
Mills, J. N. & Tang, J. Molecular weight and amino acid
composition of human gastricsin and pepsin. J. Biol. Chem. 242:
3093-3097.
Ong, E. B. & Perlmann, G. E. Specific inactivation of pepsin by
benzyloxycarbonyl-L-phenylalanyldiazomethane. Nature 215: 1492-
1494.
Reid, T. W. & Fahney, D. The pepsin-catalysed hydrolysis of
sulfite esters. J. Am. Chem. Soc. 89: 3941-3993.

1968

Hollands, T. R. & Fruton, J. S. Kinetics of the hydrolysis of
synthetic substrates by pepsin and by acetyl-pepsin.

Biochemistry 7: 2045-2053.
Humphreys, R. E. & Fruton, J. S.   The substrate binding site
    pepsin. Proc. Natl. Acad. Sci. U.S.A. 59: 519-525.
Knowles, J. R. & Wybrandt, G. B.   The sequence around an active-
    site aspartyl residue in pepsin. FEBS Lett. 1: 211-212.
Ong, E. B. & Perlmann, G. E.   The amino terminal sequence of
    porcine pepsinogen. J. Biol. Chem. 243: 6104-6109.
Schlamowitz, M. & Trujillo, R.   A new solubilizing group for
    synthetic pepsin substrates. Biochem. Biophys. Res. Commun. 33:
    156-159.
Schlamowitz, M., Shaw, A. & Jackson, W. T.   Nature of binding of
    inhibitors to pepsin and kinetics of inhibited peptic hydrolysis
    of N-acetyl-L-phenylalanyl-L-tyrosine. J. Biol. Chem. 243:
    2821,2828.
Stepanov, V. M. & Vaganova, T. I.   Identification of the carboxyl
    group of pepsin reacting with diazoacetamide derivatives.
    Biochem. Biophys. Res. Commun. 31: 825-830.
Taylor, W. H.   Biochemistry of pepsins. In: Handbook of
    Physiology, Section 6: Alimentary Canal (Code, C. F. ed.),
    vol.5, pp. 2567-2587, American Physiological Society, Washington.

                                   1969

Bayliss, R. S., Knowles, J. R. & Wybrandt, G. B.   An aspartic acid
    residue at the active site of pepsin. Biochem. J. 113: 377-386.
Cornish-Bowden, A. J. & Knowles, J. R.   The pH-dependence of
    pepsin-catalysed reactions. Biochem. J. 113: 353-362.
Etherington, D. J. & Taylor, W. H.   The pepsins of normal human
    gastric juice. Biochem. J. 113: 663-668.
Hollands, T. R. & Fruton, J. S.   On the mechanism of pepsin
    action. Proc. Natl. Acad. Sci. U.S.A. 62: 1116-1120.
Hollands, T. R., Voynick, I. M. & Fruton, J. S.   Action of pepsin
    on cationic synthetic substrates. Biochemistry 8: 575-585.
Knowles, J. R., Sharp, H. & Greenwell, P.   The pH-dependence of
    the binding of competitive inhibitors to pepsin. Biochem. J.
    113: 343-351.
Lundblad, R. L. & Stein, W. H.   On the reaction of diazoacetyl
    compounds with pepsin. J. Biol. Chem. 244: 154-160.
Sachdev, G. P. & Fruton, J. S.   Pyridyl esters of peptides as
    synthetic substrates of pepsin. Biochemistry 8: 4231-4238.
Trout, G. E. & Fruton, J. S.   The side-chain specificity of
    pepsin. Biochemistry 8: 4183-4190.
Trujillo, R. & Schlamowitz, M.   Pepsins - a chromotographic and
    kinetic comparison. Anal. Biochem. 31: 149-158.

                                   1970

Etherington, D. J. & Taylor, W. H.   The pepsins from human gastric
    mucosal extracts. Biochem. J. 118: 587-594.
Fruton, J. S.   The specificity and mechanism of pepsin action.
    Adv. Enzymol. 33: 401-443.
Fry, K. T., Kim, O.-K., Spona, J. & Hamilton, G. A.   Site of
    reaction of a specific diazo inactivator of pepsin. Biochemistry

9: 4624-4632.

Hubbard, C. D. & Stein, T. P.   The pepsin-catalysed hydrolysis of
bis-p-nitrophenyl sulfite. Biochem. Biophys. Res. Commun. 45:
293-296.

Kassell, B. & Meitner, P. A.   Bovine pepsinogen and pepsin.
Methods Enzymol. 19: 337-347.

Keil, B.   Chemical studies of porcine pepsin. Philos. Trans. R.
Soc. London Ser. B 257: 125-133.

Knowles, J. R.   On the mechanism of action of pepsin. Philos.
Trans. Roy. Soc. London Ser. B 257: 135-146.

Medzihradszky, K., Voynick, I. M., Medzihradszky-Schweiger, H. &
Fruton, J. S.   Effect of secondary enzyme-substrate interactions
on the cleavage of synthetic peptides by pepsin. Biochemistry
9: 1154-1162.

Ryle, A. P.   The porcine pepsins and pepsinogens. Methods
Enzymol. 19: 316-336.

Samloff, I. M. & Townes, P. L.   Pepsinogens: genetic polymorphism
in man. Science 168: 144-145.

Tanksley, T. D., Jr., Neumann, H., Lyman, C. M., Pace, C. N. &
Prescott, J. M.   Inhibition of pepsinogen activation by
Gossypol. J. Biol. Chem. 245: 6456-6461.

Turner, M. D., Mangla, J. C., Samloff, I. M., Miller, L. L. &
Segal, H. L.   Studies on the heterogeneity of human gastric
zymogens. Biochem. J. 116: 397-404.

Umezawa, H., Aoyagi, T., Morishima, H., Matsuzaki, M., Hamada, M.
& Takeuchi, T.   Pepstatin, a new pepsin inhibitor produced by
actinomycetes. J. Antibiot. (Tokyo) 33: 259-262.

Valueva, T. A. & Ginodman, L. M.   Investigation of the reactivation
of hog pepsin incubated with diazocarbonyl reagents.
Biochemistry U.S.S.R. (Engl. Transl.) 35: 732-736.

1971

Antonini, J. & Ribadeau Dumas, B.   Isolement, purification et
propriétés de deux zymogènes gastrique bovins. Propriétés des
protéases correspondentes. Biochimie 53: 321-329.

Etherington, D. J. & Taylor, W. H.   The separation and preparation
of human pepsins 3 and 5 from the gastric juice of single
individuals, and the mode of action of human and swine pepsins
on the B-chain of oxidised insulin. Biochim. Biophys. Acta 236:
92-98.

Fruton, J. S.   Pepsin. In: The Enzymes (Boyer, P. D. ed.), vol.
3, pp. 119-164, Academic Press, New York.

Fukumura, M., Satoi, S., Kuwana, N. & Murao, S.   Structure
elucidation of new pepsin inhibitor (S-P1). Agr. Biol. Chem.
(Tokyo) 35: 1310-1312.

Husain, S.S., Ferguson, J. B. & Fruton, J. S.   Bifunctional
inhibitors of pepsin. Proc. Natl. Acad. Sci. U.S.A. 68: 2765-
2768.

Kitson, T. M. & Knowles, J. R.   The inhibition of pepsin-catalysed
reactions by structural and stereochemical product analogues.
Biochem. J. 122: 241-247.

Kitson, T. M. & Knowles, J. R. The pathway of pepsin-catalysed transpeptidation. Evidence for the reactive species being the anion of the acceptor molecule. Biochem. J. 122: 249-256.

Lang, H. M. & Kassell, B. Bovine pepsinogens and pepsins. III. Composition and specificity of the pepsins. Biochemistry 10: 2296-2301.

Meitner, P. A. & Kassell, B. Bovine pepsinogens and pepsins. A series of zymogens and enzymes that differ in organic phosphate content. Biochem. J. 121: 249-256.

Nevaldine, B. & Kassell, B. Bovine pepsinogen and pepsin. IV. A new method of purification of pepsin. Biochim. Biophys. Acta 250: 207-209.

Stein, P., Reid, T. W. & Fahrney, D. A useful spectrophotometric rate assay for pepsin. Anal. Biochem. 41: 360-364.

Tang, J. Specific and irreversible inactivation of pepsin by substrate-like epoxides. J. Biol. Chem. 246: 4510-4517.

Vretblad, P. & Axén, R. Covalent fixation of pepsin to agarose derivatives. FEBS Lett. 18: 254-256.

1972

Aoyagi, T., Morishima, H., Nishizawa, R., Kunimoto, S., Takeuchi, T. & Umezawa, H. Biological activity of pepstatins, pepstanone A and partial peptides on pepsin, cathepsin D and renin. J. Antibiot. (Tokyo) 25: 689-694.

Chen, K. C. S. & Tang, J. Amino acid sequence around the epoxide-reactive residues in pepsin. J. Biol. Chem. 247: 2566-2574.

Hartsuck, J. A. & Tang, J. The carboxylate ion in the active center of pepsin. J. Biol. Chem. 247: 2575-2580.

Hunkapiller, M. W. & Richards, J. H. Studies on the catalytic mechanism of pepsin using a new synthetic substrate. Biochemistry 11: 2829-2839.

Kunimoto, S., Aoyagi, T., Morishima, H., Takeuchi, T. & Umezawa, H. Mechanism of inhibition of pepsin by pepstatin. J. Antibiot. (Tokyo) 25: 251-255.

May, S. W. & Kaiser, E. T. On the mechanism of pepsin-catalysed hydrolysis of sulfite esters. Biochemistry 11: 592-600.

Raju, E. V., Humphreys, R. E. & Fruton, J. S. Gel filtration studies on the binding of peptides to pepsin. Biochemistry 11: 3533-3536.

Roberts, N. B. & Taylor, W. H. A comparison of the properties of pure human pepsins 1, 2, 3 and 5. Biochem. J. 128: 103P only.

Sachdev, G. P., Johnston, M. A. & Fruton, J. S. Fluorescence studies on the interactions of pepsin with substrates. Biochemistry 11: 1080-1086.

Silver, M. S. & Stoddard, M. Amino-enzyme intermediates in pepsin-catalysed reactions. Biochemistry 11: 191-200.

1973

Doleschel, W. & Auerswald, W. Gel diffusion method for determination of pepsin. Clin. Chim. Acta 49: 105-108.

Matyash, L. F., Oglobina, O. G. & Stepanov, V. M. Modification of

carboxyl groups in pepsin. Eur. J. Biochem. 35: 540-545.
Pedersen, V. B. & Foltmann, B. The amino acid sequence of a
hitherto unobserved segment from porcine pepsinogen preceeding
the N-terminus of pepsin. FEBS Lett. 35: 255-256.
Perlmann, G. E. & Kerwar, G. A. Conformational change of pepsin
induced by its substrate N-acetyl-L-phenylalanyl-3,5-
diiodotyrosine. Arch. Biochem. Biophys. 157: 145-147.
Roberts, N. B. & Taylor, W. H. The inactivation by carbenoxolone
of individual human pepsinogens and pepsins. Clin. Sci. Mol.
Med. 45: 213-224.
Sachdev, G. P., Brownstein, A. D. & Fruton, J. S. N-Methyl-2-
anilinonaphthalene-6-sulfonyl peptides as fluorescent probes for
pepsin-substrate interaction. J. Biol. Chem. 248: 6292-6299.
Vesterberg, O. Isoelectric focusing of proteins. Studies on
pepsin. Acta Chem. Scand. B 27: 2415-2420.
Yonezawa, H., Terada, S. & Izumiya, N. Action of pepsin on
synthetic substrates. II. Diglycyl-L-phenylalanyl-L-phenylalanyl
(and L-tyrosyl)-glycines: sensitive synthetic substrates for
pepsin. J. Biochem. 73: 861-870.

                              1974

Abu-Erreish, G. M. & Peanasky, R. J. Pepsin inhibitors from
Ascaris lumbricoides. Isolation, purification, and some
properties. J. Biol. Chem. 249: 1558-1566.
Harboe, M., Andersen, P. M., Foltmann, B., Kay, J. & Kassell, B.
Activation of bovine pepsinogen. Sequence of the peptides
released, identification of a pepsin inhibitor. J. Biol Chem.
249: 4487-4494.
Sampath-Kumar, P. S. & Fruton, J. S. Studies on the extended
active sites of acid proteinases. Proc. Natl. Acad. Sci. U.S.A.
71: 1070-1072.
Takahashi, M., Wang, T. T. & Hofmann, T. Acyl intermediates in
pepsin and penicillopepsin catalyzed reactions. Biochem.
Biophys. Res. Commun. 57: 39-46.

                              1975

Newmark, A. K. & Knowles, J. R. Acyl- and amino-transfer routes
in pepsin-catalyzed reactions. J. Am. Chem. Soc. 97: 3557-3559.
Sepulveda, P., Marciniszyn, J., Jr., Liu, D. & Tang, J. Primary
structure of porcine pepsin. III. Amino acid sequence of a
cyanogen bromide fragment, CB2A and the complete structure of
porcine pepsin. J. Biol. Chem. 250: 5082-5088.
Takahashi, M. & Hofmann, T. Acyl intermediates in
penicillopepsin-catalysed reactions and a discussion of the
mechanism of action of pepsins. Biochem. J. 147: 549-563.

                              1976

Fruton, J. S. The mechanism of the catalytic action of pepsin and
related acid proteinases. Adv. Enzymol. 44: 1-36.
Hirsch-Marie, H., Loisillier, F., Touboul, J. P. & Burtin, P.
Immunochemical study and cellular localization of human

pepsinogens during ontogenesis and in gastric cancers. Lab. Invest. 34: 623-632.

Irvine, G. B. & Elmore, D. T. The use of a radiochemical titrant for the determination of the operational molarity of solutions of pepsin. Biochem. Soc. Trans. 4: 643-645.

Kassell, B., Wright, C. L. & Ward, P. H. Canine pepsinogen and pepsin. Methods Enzymol. 45: 452-458.

Marciniszyn, J., Jr., Huang, J. S., Hartsuck, J. A. & Tang, J. Mechanism of intramolecular activation of pepsinogen. Evidence for an intermediate δ and the involvement of the active site of pepsin in the intramolecular activation of pepsinogen. J. Biol. Chem. 251: 7095-7102.

Rapp, W. & Lehmann, H. E. Isolation of components of pepsinogens I, II and mucosal cathepsin of human gastric mucosa by preparative polyacrylamide gel electrophoresis, and preparation of specific antisera. J. Clin. Chem. Clin. Biochem. 14: 569-576.

Salesse, R. & Garnier, J. Synthetic peptides for chymosin and pepsin assays: pH effect and pepsin-independent determination in mixtures. J. Dairy Sci. 59: 1215-1221.

Zöller, M., Matzku, S. & Rapp, W. Purification of human gastric proteases by immunoadsorbents. Pepsinogen I group. Biochim. Biophys. Acta 427: 708-718.

1977

Antonov, V. K. New data on pepsin mechanism and specificity. In: Acid Proteases (Tang, J. ed.), pp. 179-198, Plenum Press, New York.

Christensen, K. A., Pedersen, V. B. & Foltmann, B. Identification of an enzymatically active intermediate in the activation of porcine pepsinogen. FEBS Lett. 76: 214-218.

Etherington, D. J. The dissolution of insoluble bovine collagens by cathepsin B1, collagenolytic cathepsin and pepsin. The influence of collagen type, age and chemical purity on susceptibility. Connect. Tissue Res. 5: 135-145.

Foltmann, B. & Pedersen, V. B. Comparison of the primary structures of acidic proteinases and of their zymogens. In: Acid Proteases (Tang, J. ed.), pp. 3-22, Plenum Press, New York.

Fox, P. F., Whitaker, J. R. & O'Leary, P. A. Isolation and characterization of sheep pepsin. Biochem. J. 161: 389-398.

Fruton, J. S. Specificity and mechanism of pepsin action on synthetic substrates. In: Acid Proteases (Tang, J. ed.), pp. 131-140, Plenum Press, New York.

Kaiser, E. T. & Nakagawa, Y. Anhydride intermediates in catalysis by pepsin: is pepsin an enzyme with two active sites? In: Acid Proteases (Tang, J. ed.), pp. 159-177, Plenum Press, New York.

Marciniszyn, J., Jr., Hartsuck, J. A. & Tang, J. Pepstatin inhibition mechanism. In: Acid Proteases (Tang, J. ed.), pp. 199-210, Plenum Press, New York.

Pedersen, U. D. Comparison of the proteolytic specificities of bovine and porcine pepsin A. Acta Chem. Scand. B 31: 149-156.

Powers J. C., Harley, A. D. & Myers, D. V.   Subsite specificity of
porcine pepsin. In: Acid Proteases (Tang, J. ed.), pp. 141-157,
Plenum Press, New York.

Rao, S. N., Koszelak, S. N. & Hartsuck, J. A.   Crystallization and
preliminary crystal data of porcine pepsinogen. J. Biol. Chem.
252: 8728-8730.

1978

Ahmad, F. & McPhie, P.   Spectrophotometric titration of phenolic
groups of pepsin. Biochim. Biophys. Acta 537: 247-254.

Andreeva, N. S., Fedorov, A. A., Gushchina, A. E., Riskulov, R.
R., Shutskever, N. E. & Safro, M. G.   X-Ray diffraction analysis
of pepsin. V. Conformation of the backbone of the enzyme.
Molecular Biology U.S.S.R. (Engl. Transl.) 12: 704-716.

Antonov, V. K., Ginodman, L. M., Kapitannikov, Y. V., Barshevskaya,
T. N., Gurova, A. G. & Rumsh, L. D.   Mechanism of pepsin
catalysis - general base catalysis by active-site carboxylate
ion. FEBS Lett. 88: 87-90.

Esumi, H., Sato, S., Sugimara, T. & Furihata, C.   Purification of
mouse pepsinogens by pepstatin-affinity chromatography. FEBS
Lett. 86: 33-36.

Kageyama, T. & Takahashi, K.   Monkey pepsinogens and pepsins. III.
Carbohydrate moiety of Japanese monkey pepsinogens and the amino
acid sequence around the site of its attachment to protein. J.
Biochem. 84: 771-778.

Kumegawa, M., Takuma, T., Hosoda, S., Kunii, S. & Kanada, Y.
Precocious induction of pepsinogen in the stomach of suckling
mice by hormones. Biochim. Biophys. Acta 543: 243-250.

Morris, B. J.   Activation of human inactive ("pro-") renin by
cathepsin D and pepsin. J. Clin. Endocrinol. Metab. 46: 153-
157.

Morris, B. J.   Properties of the activation by pepsin of inactive
renin in human amniotic fluid. Biochim. Biophys. Acta 527: 86-
97.

Roberts, N. B. & Taylor, W. H.   The preparation and purification
of individual human pepsins by using diethylaminoethyl-cellulose.
Biochem. J. 169: 607-615.

Tang, J., James, M. N. G., Hsu, I. N., Jenkins, J. A. & Blundell,
T. L.   Structural evidence for gene duplication in the evolution
of the acid proteases. Nature 271: 618-621.

Walker, V. & Taylor, W. H.   Ovalbumin digestion by human pepsins
1, 3 and 5. Biochem. J. 176: 429-432.

1979

Tang, J.   Evolution in the structure and function of carboxyl
proteases. Mol. Cell. Biochem. 26: 93-109.

Andreeva, N. S. & Gustchina, A. E.   On the supersecondary
structure of acid proteases. Biochem. Biophys. Res. Commun. 87:
32-42.

Irvine, G. B. & Elmore, D. T.   A radiochemical titrant for the
determination of the operational molarity of solutions of acid

proteinases. Biochem. J. 183: 389-394.

Roberts, N. B. & Taylor, W. H. Action of human pepsins 1, 2, 3
and 5 on the oxidized B-chain of insulin. Biochem. J. 179: 183-
190.

Ryle, A. P. & Auffret, C. A. The specificity of some pig and
human pepsins towards synthetic peptide substrates. Biochem. J.
179: 247-249.

Workman, R. J. & Burkitt, D. W. Pepsin inhibition by a high
specific activity radioiodinated derivative of pepstatin. Arch.
Biochem. Biophys. 194: 157-164.

1980

Foltmann, B. & Axelsen, N. H. Gastric proteinases and their
zymogens. Phylogenetic and developmental aspects. In: Enzyme
Regulation and Mechanism of Action (Mildner, P. & Ries, B. eds),
pp. 271-280, Pergamon, Oxford.

Silver, M. S. & James, L. T. Mechanistic features of pepsin-
catalyzed amino transfer reactions. J. Biol. Chem. 255: 555-
560.

Entry 3.02

PEPSIN B

Summary

EC Number: 3.4.23.2

Earlier names: Identical with porcine gelatinase and parapepsin I; related to human pepsin IV.

Distribution: Present in pig gastric juice, being formed from pepsinogen B produced by gastric mucosal cells. Probably secreted mainly in the antropyloric region of the stomach, with other minor forms of pepsin.

Source: Activation of pepsinogen B from gastric mucosa.

Action: Specificity more restricted than pepsin A; very active against Ac-Phe-Tyr(I_2), active against gelatin, but rather inactive against hemoglobin.

Requirements: Very acidic pH.

Substrate, usual: Ac-Phe-Tyr(I_2).

Substrate, special: Gelatin.

Inhibitors: -

Molecular properties: Generally similar to pepsin A, but lower anodal mobility in electrophoresis.

Comment: Correspondence between porcine pepsin B and pepsins in other species requires further study.

Bibliography

1932

Northrop, J. R.   The presence of a gelatin-liquefying enzyme in crude pepsin preparations.   J. Gen. Physiol. 15: 29-43.

1959

Ryle, A. P. & Porter, R. R.   Parapepsins: two proteolytic enzymes associated with porcine pepsin.   Biochem. J. 73: 75-86.

1964

Kushner, I., Rapp, W. & Burtin, P.   Electrophoretic and immunochemical demonstration of the existence of four human pepsinogens.   J. Clin. Invest. 43: 1983-1993.

1965

Ryle, A. P. Pepsinogen B: the zymogen of pepsin B. Biochem. J.
96: 6-16.

1970

Ryle, A. P. The porcine pepsins and pepsinogens. Methods
Enzymol. 19: 316-336.

1977

Foltmann, B. & Pedersen, V. B. Comparison of the primary
structures of acidic proteinases and of their zymogens. In:
Acid Proteases (Tang, J. ed.), pp. 3-22, Plenum Press, New York.

1980

Foltmann, B. & Axelsen, N. H. Gastric proteinases and their
zymogens. Phylogenetic and developmental aspects. In: Enzyme
Regulation and Mechanism of Action (Mildner, P. & Ries, B. eds),
pp. 271-280, Pergamon, Oxford.

Entry 3.03

PEPSIN C

Summary

EC Number: 3.4.23.3

Earlier names: Porcine pepsin C is identical with porcine parapepsin II, and related to human gastricsin, pyloruspepsin, PII and PIII.

Distribution: Zymogen is believed to be secreted by pyloric (rather than fundus) cells of the stomach. Found in several mammals, but apparently not Japanese monkey.

Source: Gastric mucosa (zymogen).

Action: Action on proteins and peptides is slightly more restricted than that of pepsin A. Ac-Phe-Tyr(I$_2$) is not cleaved at an appreciable rate.

Requirements: Optimum pH 3 with hemoglobin (higher than for pepsin A). More stable at pH 6.9 than other pepsins.

Substrate, usual: Hemoglobin, Z-Tyr-Ala.

Substrate, special: Ac-Tyr╪Leu-Val-His-NH$_2$, which is almost unaffected by other pepsins.

Inhibitors: Similar to pepsin A.

Molecular properties: Homologous with other pepsins. Estimates of mol.wt. range from 31,400 (human gastricsin) to 36,000 (porcine pepsin C). Contains no phosphate.

Comment: Correspondence of porcine pepsin C with pepsins in other species is not precise.

Bibliography

1958

Richmond, V., Tang, J., Wolf, S., Trucco, R. E. & Caputto, R. Chromatographic isolation of gastricsin, the proteolytic enzyme from gastric juice with pH optimum 3.2. Biochim. Biophys. Acta 29: 453-454.

1959

Ryle, A. P. & Porter, R. R. Parapepsins: two proteolytic enzymes associated with porcine pepsin. Biochem. J. 73: 75-86.

Tang, J., Wolf, S., Caputto, R. & Trucco, R. E.   Isolation and
    crystallisation of gastricsin from human gastric juice.   J.
    Biol. Chem. 234: 1174-1178.

1960

Ryle, A. P.   Parapepsinogen II: the zymogen of parapepsin II.
    Biochem. J. 75: 145-150.

1962

Taylor, W. H.   Proteinases of the stomach in health and disease.
    Physiol. Rev. 42: 519-553.

1964

Kushner, I., Rapp, W. & Burtin, P.   Electrophoretic and
    immunochemical demonstration of the existence of four human
    pepsinogens.   J. Clin. Invest. 43: 1983-1993.

1966

Chiang, L., Sanchez-Chiang, L., Wolf, S. & Tang, J.   The separate
    determination of human pepsin and gastricsin.   Proc. Soc. Exp.
    Biol. Med. 122: 700-704.

Hanley, W. B., Boyer, S. H. & Naughton, M. A.   Electrophoretic and
    functional heterogeneity of pepsinogen in several species.
    Nature 209: 996-1002.

Ryle, A. P. & Hamilton, M. P.   Pepsinogen C and pepsin C.
    Biochem. J. 101: 176-183.

1967

Mills, J. N. & Tang, J.   Molecular weight and amino acid
    composition of human gastricsin and pepsin.   J. Biol. Chem. 242:
    3093-3097.

1968

Ryle, A. P., Leclerc, J. R. & Falla, F.   The substrate specificity
    of pepsin C.   Biochem. J. 110: 4P only.

1969

Huang, W.-Y & Tang, J.   On the specificity of human gastricsin and
    pepsin.   J. Biol. Chem. 244: 1085-1091.

1970

Ryle, A. P.   The porcine pepsins and pepsinogens.   Methods
    Enzymol. 19: 316-336.

Tang, J.   Gastricsin and pepsin.   Methods Enzymol. 19: 406-421.

1971

Kay, J. & Ryle, A. P.   An active site peptide from pepsin C.
    Biochem. J. 123: 75-82.

1977

Foltmann, B. & Pedersen, V. B.   Comparison of the primary
    structures of acidic proteinases and of their zymogens.   In:
    Acid Proteases (Tang, J. ed.), pp. 3-22, Plenum Press, New York.

1978

Moriyama, A. & Takahashi, K.   The structure and function of acid
  proteases. VIII. Purification and characterization of cathepsins
  D from Japanese monkey lung. J. Biochem. 83: 441-451.

1979

Auffret, C. A. & Ryle, A. P.   The catalytic activity of pig pepsin
  C towards small synthetic substrates.   Biochem. J. 179: 239-246.
Ryle, A. P. & Auffret, C. A.   The specificity of some pig and
  human pepsins towards synthetic peptide substrates.   Biochem. J.
  179: 247-249.

1980

Foltmann, B. & Axelsen, N. H.   Gastric proteinases and their
  zymogens. Phylogenetic and developmental aspects.   In: Enzyme
  Regulation and Mechanism of Action (Mildner, P. & Ries, B. eds),
  pp. 271-280, Pergamon, Oxford.

Entry 3.04

SEMINAL PEPSIN

Summary

EC Number: (3.4.23.-)

Earlier names: Acidic protease of human seminal plasma.

Distribution: Zymogen detected in seminal plasma of various
mammals (but not bull), being formed in the seminal vesicles
(man) or coagulating gland (rat).

Source: Seminal plasma.

Action: After spontaneous activation at acid pH, clots milk,
hydolyzes N,N-dimethyl-casein and acid denatured hemoglobin.
(Albumin, ovalbumin and Z-Glu-Tyr are resistant to hydrolysis).

Requirements: pH optimum about 2.5.

Substrate, usual: Hemoglobin.

Inhibitors: p-Bromophenacylbromide, 1,2-epoxy-3-(p-nitrophenoxy)
propane, pepstatin.

Molecular properties: Zymogen, mol.wt. 42,000, is stable in the
range pH 6-9; activated below pH 6.0. Active enzyme, mol.wt.
35,000, is unstable above pH 6.5.

Comment: Detailed specificity of the human enzyme differs from
that of porcine pepsin A, but the zymogen is almost certainly
the human group II pepsinogen (that of pepsin C) detected
immunologically [1].

References
[1] Zöller et al. Biochim. Biophys. Acta 427: 708-718, 1976.

Bibliography

1952

Lundquist, F. & Seedorff, H. H. Pepsinogen in human seminal
fluid. Nature 170: 1115-1116.

1972

Samloff, I. M. & Liebman, W. M. Purification and immunochemical
characterisation of group II pepsinogens in human seminal fluid.
Clin. Exp. Immunol. 11: 405-414.

1974

Ruenwongsa, P. & Chulavatnatol, M. A new acidic protease in human seminal plasma. Biochem. Biophys. Res. Commun. 59: 44–50.

1975

Ruenwongsa, P. & Chulavatnatol, M. Acidic protease from human seminal plasma: purification and some properties of active enzyme and of proenzyme. J. Biol. Chem. 250: 7574–7578.

1976

Chulavatnatol, M. & Ruenwongsa, P. Activation of proenzyme of acidic protease from human seminal plasma. Biochim. Biophys. Acta 452: 525–532.

Zöller, M., Matzku, S. & Rapp, W. Purification of human gastric proteases by immunoadsorbents. Pepsinogen I group. Biochim. Biophys. Acta 427: 708–718.

1977

Ruenwongsa, P. & Chulavatnatol, M. Acid protease and its proenzyme from human seminal plasma. In: Acid Proteases (Tang, J. ed.), pp. 329–341, Plenum Press, New York.

Entry 3.05

UROPEPSIN

Summary

EC Number: (3.4.23.-)

Earlier names: -

Distribution: Detected in human urine. Presumably originates in kidney, although the plasma is also a possible source.

Source: Human urine.

Action: Degrades various proteins.

Requirements: pH 2-3.

Substrate, usual: Hemoglobin.

Inhibitors: As for pepsin A (#3.01).

Molecular properties: Not well characterized. Zymogen has immunological reactivity of a group I pepsinogen (the precursor of pepsin A) (in contrast to seminal pepsinogen). Enzyme alkali-labile.

Bibliography

1946

Farnsworth, E. B., Speer, E. & Alt, H. The quantitative determination of a pepsin-like substance in the urine of normal individuals and of patients with pernicious anemia. J. Lab. Clin. Med. 31: 1025-1028.

1948

Mirsky, I. A., Block, S., Osher, S. & Broh-Kahn, R. H. Uropepsin excretion by man. I. The source, properties and assay of uropepsin. J. Clin. Invest. 27: 825-833.

1961

Seijffers, M. J., Segal, M. L. & Miller, L. L. Separation of pepsinogen II and pepsinogen III from human urine. Am. J. Physiol. 206: 1106-1110.

1968

Raab, W. P. Enzymes and isoenzymes in urine. In: Enzymes in

Urine and Kidney (Dubach, U. C. ed.), pp. 17–83, Hans Huber,
Berne.

1969

Hirsch–Marie, H. & Conte, M.   Etude du pepsinogène urinaire et de
ses corrélations avec la sécrétion protéolytique gastrique.
Rev. Fr. Etud. Clin. Biol. 14: 977–983.

1976

Zöller, M., Matzku, S. & Rapp, W.   Purification of human gastric
proteases by immunoadsorbents. Pepsinogen I group.   Biochim.
Biophys. Acta 427: 708–718.

Entry 3.06

CHYMOSIN

Summary

EC Number: 3.4.23.4

Earlier names: Rennin, chymase (prior to 1939), (bovine) fetal or neonatal pepsin.

Distribution: Calf stomach. Fetal pepsins with generally similar properties occur in other species.

Source: Calf gastric mucosa (as prochymosin).

Action: Clots caseins by limited proteolysis: -Pro-His-Leu-Ser-Phe$_{105}$/Met$_{106}$-Ala-Ile- (-casein), -Phe$_{23}$/Phe$_{24}$- (α-casein), or several bonds (β-casein). Hydrolyzes Z-Glu-Tyr.

Requirements: Optimal pH 3-4 for general proteolytic activity; pH 5-6 for Z-Glu-Tyr hydrolysis, and pH 6-7 for milk clotting.

Substrate, usual: Casein (milk clotting assay), hemoglobin.

Substrate, special: Leu-Ser-Phe(NO$_2$)/Nle-Ala-Leu-OMe (also sensitive to pepsin, #3.01).

Inhibitors: Inhibitors of pepsins, including pepstatin.

Molecular properties: Prochymosin (mol.wt. 40,777, pI 5.0) is spontaneously converted to chymosin (mol.wt. 35,652, pI 4.6) and an activation peptide, at pH values below 5. Chymosin is one of the pepsin components of lowest electrophoretic mobility.

Comment: Evolved as one of the family of pepsins. Foltmann [1] speculates that the function of this very selective proteinase in animals with post-natal uptake of antibodies through the gut may be to clot casein with minimal destruction of immunoglobulin G.

References
[1] Foltmann & Axelsen In: Enzyme Regulation and Mechanism of Action (Mildner, P. & Ries, B. eds), pp. 271-280, Pergamon, Oxford, 1980.

Bibliography

1945
Berridge, N. J. The purification and crystallization of rennin. Biochem. J. 39: 179-186.

1953

De Baun, R. M., Connors, W. M. & Sullivan, R. A.   The preparation
   of crystalline rennin and a study of some of its properties.
   Arch. Biochem. Biophys. 43: 324-338.

1957

Fish, J. C.   Activity and specificity of rennin.   Nature 180: 345
   only.

1965

Cheeseman, G. C.   Denaturation of rennin: effect on activity and
   molecular configuration.   Nature 205: 1011-1012.
Delfour, A., Jollès, J., Alais, C. & Jollès, P.   Caseino-
   glycopeptides: characterization of a methionine residue and of
   the N-terminal sequence.   Biochem. Biophys. Res. Commun. 19:
   452-455.

1966

Foltmann, B.   A review of prorennin and rennin.   Compt. Rend. Lab.
   Carlsberg 35: 143-231.

1968

Hill, R. D.   The nature of the rennin-sensitive bond in casein and
   its possible relation to sensitive bonds in other proteins.
   Biochem. Biophys. Res. Commun. 33: 659-663.

1970

Foltmann, B.   Prochymosin and chymosin (prorennin and rennin).
   Methods Enzymol. 19: 421-435.

1971

Asato, N. & Rand, A. G., Jr.   Resolution of prorennin and rennin
   by polyacrylamide gel electrophoresis.   Anal. Biochem. 44: 32-
   41.

1972

Asato, M. & Rand, A. G.   Fractionation and isolation of the
   multiple forms of prorennin (prochymosin).   Biochem. J. 129:
   841-846.
Clement, G. E. & Cashell, G. S.   The rennin catalyzed hydrolysis
   of a tripeptide. Evidence for catalysis by two carboxyl groups.
   Biochem. Biophys. Res. Commun. 47: 328-332.
Raymond, M. N., Garnier, J., Bricas, E., Cilianu, S., Blasnic, M.,
   Chaix, A. & Lefrancier, P.   Studies on the specificity of
   chymosin (rennin). I. Kinetic parameters of the hydrolysis of
   synthetic oligopeptide substrates.   Biochimie 54: 145-154.

1973

Chang, W.-J. & Takahashi, K.   The structure and function of acid
   proteases. II. Inactivation of bovine rennin by acid protease-
   specific inhibitors.   J. Biochem. 74: 231-237.
Raymond, M. N., Mercier, J. C. & Bricas, E.   New oligopeptide

substrates of chymosin (rennin). In: Peptides 1972 (Hanson, H.
& Jukubke, H. D. eds), pp. 380-382, North-Holland Publishing
Co., Amsterdam.

1974

Chang, W.-J. & Takahashi, K. The structure and function of acid
proteases. III. Isolation and characterization of the active-site
peptides from bovine rennin. J. Biochem. 76: 467-474.
Hill, R. D., Lahav, E. & Givol, D. A rennin-sensitive bond in α_{s1}
B-casein. J. Dairy Res. 41: 147-153.

1975

Pedersen, V. B. & Foltmann, B. Amino-acid sequence of the peptide
segment liberated during activation of prochymosin (prorennin).
Eur. J. Biochem. 55: 95-103.

1976

Hirsch-Marie, H., Loisillier, F., Touboul, J. P. & Burtin, P.
Immunochemical study and cellular localization of human
pepsinogens during ontogenesis and in gastric cancers. Lab.
Invest. 34: 623-632.
Salesse, R. & Garnier, J. Synthetic peptides for chymosin and
pepsin assays: pH effect and pepsin-independent determination in
mixtures. J. Dairy Sci. 59: 1215-1221.
Visser, S., Van Rooijen, P. J., Schattenkerk, C. & Kerling, K. E.
T. Peptide substrates for chymosin (rennin). Kinetic studies
with peptides of different chain length including parts of the
sequence 101-112 of bovine κ-casein. Biochim. Biophys. Acta
438: 265-272.

1977

Foltmann, B. & Pedersen, V. B. Comparison of the primary
structures of acidic proteinases and of their zymogens. In:
Acid Proteases (Tang, J. ed.), pp. 3-22, Plenum Press, New York.
Foltmann, B., Pedersen, V. B., Jacobsen, H., Kauffman, D. &
Wybrandt, G. The complete amino acid sequence of prochymosin.
Proc. Natl. Acad. Sci. U.S.A. 74: 2321-2324.
Visser, S., Van Rooijen, P. J., Schattenkerk, C. & Kerling, K. E.
T. Peptide substrates for chymosin (rennin). Kinetic studies
with bovine κ-casein-(103-108)-hexapeptide analogues. Biochim.
Biophys. Acta 481: 171-176.

1978

Foltmann, B., Lønblad, P. & Axelsen, N. H. Demonstration of
chymosin (EC 3.4.23.4) in the stomach of newborn pig. Biochem.
J. 169: 425-427.

1979

Foltmann, B., Pedersen, V. B., Kauffman, D. & Wybrandt, G. The
primary structure of calf chymosin. J. Biol. Chem. 254: 8447-
8456.
Pedersen, V. B., Christensen, K. A. & Foltmann, B. Investigations

on the activation of bovine prochymosin.   Eur. J. Biochem. 94: 573-580.

1980

Foltmann, B. & Axelsen, N. H.   Gastric proteinases and their zymogens. Phylogenetic and developmental aspects.   In: Enzyme Regulation and Mechanism of Action (Mildner, P. & Ries, B. eds), pp. 271-280, Pergamon, Oxford.

Entry 3.07

GASTRIC CATHEPSIN

Summary

EC Number: (3.4.23.-)

Earlier names: Mucosal cathepsin.

Distribution: Detected in human gastric mucosa.

Source: Gastric mucosa.

Action: Degrades hemoglobin.

Requirements: pH 3-4.

Substrate, usual: Hemoglobin.

Inhibitors: -

Molecular properties: Mol.wt. 40,000-50,000. Lower electrophoretic mobility than pepsins. Active enzyme not denatured by neutral-alkaline pH.

Comment: Resembles cathepsin D and the fetal pepsins more than pepsin A, but not yet well characterized. Some early reports are attributable to misinterpretation of a bimodal pH-activity response of pepsin A; others to the activity of the minor forms of pepsin.

Enzymes from other sourses that seem to be intermediate in properties between cathepsin D and the pepsins include the human ascites proteinase (#3.14), cathepsin E (#3.15) and a monkey lung enzyme [1].

References
Moriyama et al. J. Biochem. 83: 441-451, 1978.

Bibliography

1974

Mangla, J. C., Guarasci, G. & Turner, M. D. Studies on human gastric mucosal cathepsin and its unique nature compared with other human pepsinogens. Biochem. Med. 10: 83-96.

1976

Rapp, W. & Lehmann, H. E. Isolation of components of pepsinogens I, II and mucosal cathepsin of human gastric mucosa by preparative polyacrylamide gel electrophoresis, and preparation of specific antisera. J. Clin. Chem. Clin. Biochem. 14: 569–576.

Zöller, M., Matzku, S. & Rapp, W. Purification of human gastric proteases by immunoadsorbents. Pepsinogen I group. Biochim. Biophys. Acta 427: 708–718.

1978

Roberts, N. B. & Taylor, W. H. The isolation and properties of a non-pepsin proteinase from human gastric mucosa. Biochem. J. 169: 617–624.

1980

Foltmann, B. & Axelsen, N. H. Gastric proteinases and their zymogens. Phylogenetic and developmental aspects. In: Enzyme Regulation and Mechanism of Action (Mildner, P. & Ries, B. eds), pp. 271–280, Pergamon, Oxford.

Entry 3.08

PLASMA ASPARTIC PROTEINASE

Summary

EC Number: (3.4.23.-)

Earlier names: -

Distribution: Human blood plasma.

Source: Human blood plasma.

Action: Solubilizes fibrin.

Requirements: Acidic conditions (about pH 3).

Substrate, usual: Fibrin.

Inhibitors: Pepstatin.

Molecular properties: pI may be near 5.0.

Comment: May be firmly associated with albumin. Suggested not to be pepsin(ogen) on the basis of lack of adsorption to alumina gel.

Bibliography

1977

Law, J. A. & Kemp, G. Partial purification of a fibrinolytic acid proteinase from human plasma. Biochem. Soc. Trans. 5: 1463-1465.

Entry 3.09

RENIN

Summary

EC Number: 3.4.99.19 (but being an aspartic proteinase, should become 3.4.23.?).

Earlier names: Angiotensin-forming enzyme, angiotensinogenase.

Distribution: Kidney (in non-lysosomal organelles) and plasma of many mammals.

Source: Kidney.

Action: Highly specific for limited proteolysis of angiotensinogen, releasing the decapeptide angiotensin I by cleaving a -Leu-Leu- bond. Little or no general proteolytic activity.

Requirements: Optimal pH varies with substrate, thus (for rat renin) optimum is pH 3.5 with the octapeptide substrate, and pH 6.0 with angiotensinogen.

Substrate, usual: Angiotensinogen.

Substrate, special: Z-Pro-Phe-His-Leu/Leu-Val-Tyr-Ser-2-NNap or Asp-Arg-Val-Tyr-Ile-His-Pro-Phe-His-Leu/Leu-Val-Tyr-Ser (the tetradecapeptide of which the free N-terminal decapeptide is angiotensin I).

Inhibitors: N_2Ac-Nle-OMe with Cu^{2+}, 1,2-epoxy-3-phenoxypropane, pepstatin (K_i 5 x 10^{-7} M). A phospholipid similar to phosphatidylserine. Analogues of the susceptible region of angiotensinogen.

Molecular properties: Glycoprotein; pI 4.5–5.7 (multiple forms). Mol.wt. 36,400 (sedimentation) to 42,500 (gel chromatography). Latent precursor forms of 60,000 and 140,000 mol.wt. exist in the kidney and plasma, and are activatable by an endogenous cysteine proteinase and pepsin.

Comment: Renin liberated from the kidney in response to low blood pressure in the organ acts on angiotensinogen in the plasma to generate angiotensin I, which has feeble pressor activity. The angiotensin-converting enzyme, peptidyl dipeptidase I, present on vascular endothelia especially in the lung, converts angiotensin I to the powerful pressor octapeptide, angiotensin II.
　　Some studies of "renin" have dealt with an enzyme(s) more active on hemoglobin, and more sensitive to pepstatin, than is

renin proper; these tissue angiotensinogenases ("isorenin", "pseudorenin") can often be identified as cathepsin D, but another form of renin (without a carbohydrate prosthetic group) occurs in the submandibular glands of rodents (see #3.10).

Bibliography

1958

Skeggs, L. T., Lentz, K. E., Kahn, J. R. & Schumway, N. P.   The synthesis of a tetradecapeptide renin substrate.   J. Exp. Med. 108: 283-297.

1962

Kemp, E. & Rubin, I.   Removal of angiotensinase from a preparation of renin.   Arch. Biochem. Biophys. 96: 56-59.

1964

Boucher, R., Veyrat, R., de Champlain, J. & Genest, J.   New procedures for measurement of human plasma angiotensin and renin activity levels.   Can. Med. Ass. J. 90: 194-201.

1966

Haas, E., Goldblatt, H., Gipson, E. C. & Lewis, L.   Extraction, purification and assay of human renin free of angiotensinase. Circ. Res. 19: 739-749.
Maier, G. D. & Morgan, W. S.   Purification of hog renin by column chromatography.   Biochim. Biophys. Acta 128: 193-195.
Roth, M. & Reinharz, A.   Un substrat pour le dosage chimique de la rénine.   Helv. Chim. Acta 49: 1903-1907.

1967

Skeggs, L. T., Lentz, K. E., Kahn, J. R. & Hochstrasser, H. Studies on the preparation and properties of renin.   Circ. Res. 11: 91-100.

1968

Ryan, J. W., McKenzie, J. K. & Lee, M. R.   A rapid simple method for the assay of renin in rabbit plasma.   Biochem. J. 108: 679-685.
Skeggs, L. T., Lentz, K. E., Kahn, J. R. & Hochstrasse, H. Kinetics of the reaction of renin with nine synthetic peptide substrates.   J. Exp. Med. 128: 13-34.

1969

Reinharz, A. & Roth, M.   Studies on renin with synthetic substrates.   Eur. J. Biochem. 7: 334-339.

1970

Lucas, C. P., Fukuchi, S., Conn, J. W., Berlinger, F. G., Waldhausl, W.-K., Cohen, E. L. & Rovner, D. R.   Purification of human renin.   J. Lab. Clin. Med. 76: 689-700.

Waldhäusl, W. K., Lucas, C. P., Conn, J. W., Lutz, J. H. & Cohen, E. L. Studies on the partial isolation of human renin. Biochim. Biophys. Acta 221: 536–548.

1971

Gross, F., Lazar, J. & Orth, H. Inhibition of the renin-angiotensinogen reaction by pepstatin. Science 175: 656 only.
Lee, H.-J. & Wilson, I. B. On the structure of renin substrate. Biochim. Biophys. Acta 243: 530–533.
Lumbers, E. R. Activation of renin in human ammiotic fluid by low pH. Enzymologia 40: 329–336.
Rakhit, S. On the structure of the renin inhibitor from hog kidney. Can. J. Biochem. 49: 1012–1014.

1972

Aoyagi, T., Morishima, H., Nishizawa, R., Kunimoto, S., Takeuchi, T. & Umezawa, H. Biological activity of pepstatins, pepstanone A and partial peptides on pepsin, cathepsin D and renin. J. Antibiot. (Tokyo) 25: 689–694.
Bath, N. M. & Gregerman, R. I. Labeled polymeric substrate for renin. Synthesis of N-acetyl-poly(L-glutamyl)-[^{125}I]-tridecapeptide and use for enzyme assay. Biochemistry 11: 2845–2853.
Gross, F., Lazar, J. & Orth, H. Inhibition of the renin-angiotensinogen reaction by pepstatin. Science 175: 656 only.
Miller, R. P., Poper, C. J., Wilson, C. W. & Devito, E. Renin inhibition by pepstatin. Biochem. Pharmacol. 21: 2941–2944.

1973

Corvol, P., Devaux, C. & Menard, J. Pepstatin, an inhibitor for renin purification by affinity chromatography. FEBS Lett. 34: 189–192.
Favre, L., Roussel-Deruyck, R. & Vallotton, M. B. Influence of pH on human renin activity with different substrates: role of substrate denaturation. Biochim. Biophys. Acta 302: 102–109.
Murakami, K., Inagami, T., Michelakis, A. M. & Cohen, S. An affinity column for renin. Biochem. Biophys. Res. Commun. 54: 482–487.
Parikh, I. & Cuatrecasas, P. Substrate analog competitive inhibitors of human renin. Biochem. Biophys. Res. Commun. 54: 1356–1361.
Poulson, K., Burton, J. & Haber, E. Competitive inhibitors of renin. Biochemistry 12: 3877–3882.

1974

Boyd, G. W. A protein-bound form of porcine renal renin. Circ. Res. 35: 426–438.
Inagami, T., Misono, K. & Michelakis, A. M. Definitive evidence for similarity in the active site of renin and acidic protease. Biochem. Biophys. Res. Commun. 56: 503–509.
McKown, M. M., Workman, R. J. & Gregerman, R. I. Pepstatin

inhibition of human renin. Kinetic studies and estimation of enzyme purity. J. Biol. Chem. 249: 7770-7774.

Michelakis, A. M., Yoshida, H., Menzie, J., Murakami, K. & Inagami, T. A radioimmunoassay for the direct measurement of renin in mice and its application to submaxillary gland and kidney studies. Endocrinology 94: 1101-1105.

Orth, H., Hackenthal, E., Lazar, J., Miksche, U. & Gross, F. Kinetics of the inhibitory effect of pepstatin on the reaction of hog renin with rat plasma substrate. Circ. Res. 35: 52-55.

Overturf, M., Leonard, M. & Kirkendall, W. M. Purification of human renin and inhibition of its activity by pepstatin. Biochem. Pharmacol. 23: 671-683.

Workman, R. J., McKown, M. M. & Gregerman, R. I. Renin: inhibition by proteins and peptides. Biochemistry 13: 3029-3035.

1975

Day, R. P. & Luetscher, J. A. Big renin: a possible prohormone in kidney and plasma of a patient with Wilm's tumor. J. Clin. Endocrinol. Metab. 38: 923-926.

Day, R. P., & Luetscher, J. A. Biochemical properties of big renin extracted from human plasma. J. Clin. Endocrinol. Metab. 40: 1085-1093.

Day, R. P., Luetscher, J. A. & Gonzales, C. M. Occurrence of big renin in human plasma, amniotic fluid and kidney extracts. J. Clin. Endocrinol. Metab. 40: 1078-1084.

Dew, M. E. & Heidrich, H.-G. The isolation of renin granules from rabbit kidney cortex by free flow electrophoresis. Hoppe-Seyler's Z. Physiol. Chem. 356: 621-623.

Gross D. M. & Barajas, L. The large-scale isolation of renin-containing granules from rabbit renal cortex by zonal centrifugation. J. Lab. Clin. Med. 85: 467-477.

Murakami, K. & Inagami, T. Isolation of pure and stable renin from hog kidney. Biochem. Biophys. Res. Commun. 62: 757-763.

1976

Day, R. P. & Reid, I. A. Renin activity in dog brain: enzymological similarity to cathepsin D. Endocrinology 99: 93-100.

Devaux, C., Ménard, J., Sicard, P. & Corvol, P. Partial characterization of hog renin purified by affinity chromatography. Eur. J. Biochem. 64: 621-627.

Guyene, T. T., Devaux, C., Menard, J. & Corvol, P. Inhibition of human plasma renin activity by pepstatin. J. Clin. Endocrinol. Metab. 43: 1301-1306.

Lauritzen, M., Damsgaard, J. J., Rubin, I. & Lauritzen, E. A comparison of the properties of renin isolated from pig and rat kidney. Biochem. J. 155: 317-323.

Poulsen, K., Haber, E. & Burton, J. On the specificity of human renin. Studies with peptide inhibitors. Biochim. Biophys. Acta 452: 533-537.

1977

Cooper, R. M., Murray, G. E. & Osmond, D. H.   Trypsin-induced
activation of renin precursor in plasma of normal and anephric
man. Circ. Res. 40: suppl. I, 171-179.

Corvol, P., Devaux, C., Ito, T., Sicard, P., Ducloux, J. & Menard,
J. Large scale purification of hog renin. Physicocochemical
characterization. Circ. Res. 41: 616-622.

Haber, E. & Slater, E. E.   Purification of renin. A review. Circ.
Res. 40: Suppl. I, 36-40.

Inagami, T. & Murakami, K.   Pure renin. Isolation from hog kidney
and characterization. J. Biol. Chem. 252: 2978-2983.

Inagami, T. & Murakami, K.   Purification of high molecular weight
forms of renin from hog kidney. Circ. Res. 41: suppl. II, 11-
16.

Inagami, T., Hirose, S., Murakami, K. & Matoba, T.   Native form of
renin in the kidney. J. Biol. Chem. 252: 7733-7737.

Inagami, T., Murakami, K., Misono, K., Workman, R. J., Cohen, S. &
Suketa, Y. Renin and precursors: purification, characterization,
and studies on active site. In: Acid Proteases (Tang, J. ed.),
pp. 225-247, Plenum Press, New York.

Leckie, B. J., McConnell, A. & Jordan, J.   Inactive renin - a
renin proenzyme? In: Acid Proteases (Tang, J. ed.), pp. 249-
269, Plenum Press, New York.

Malling, C. & Poulsen, K.   Direct measurement of high molecular
weight forms of renin in plasma. Biochim. Biophys. Acta 491:
542-550.

Sealey, J. E., Moon, C., Laragh, J. H. & Atlas, S. A.   Plasma
prorenin in normal, hypertensive, and anephric subjects and its
effect on renin measurements. Circ. Res. 40: suppl. I, 41-45.

1978

Atlas, S. A., Laragh, J. H. & Sealey, J. E.   Activation of
inactive plasma renin: evidence that both cryoactivation and
acid-activation work by liberating a neutral serine protease
from endogenous inhibitors.   Clin. Sci. Mol. Med. 55: 135s-138s.

Chou, H. J., Shaper, J. H. & Gregerman, R. I.   Separation of human
renal renin and pseudorenin by affinity chromatography on
hemoglobin-Sepharose 2B. Biochim. Biophys. Acta 524: 183-187.

Evin, G., Gardes, J., Kreft, C., Castro, B., Corvol, P. & Menard,
J. Soluble pepstatins: a new approach to blockade in vivo of
the renin-angiotensin system. Clin. Sci. Mol. Med. 55: 167s-
169s.

Galen, F. X., Devaux, C., Grogg, P., Menard, J. & Corvol, P.
Fluorimetric assay of renin. Biochim. Biophys. Acta. 523: 485-
493.

Hara, A., Matsunaga, M., Yamamoto, J., Morimoto, K., Nagai, H.,
Kanatsu, K., Pak, C. H., Ogino, K. & Kawai, C. Cryoactivation
of plasma renin. Clin. Sci. Mol. Med. 55: 139s-141s.

Hirose, S., Yokosawa, H. & Inagami, T.   Immunochemical
identification of renin in rat brain and distinction from acid

proteases. Nature 274: 392–393.
Matoba, T. Rat renin: purification and characterization.
Biochim. Biophys. Acta 526: 560–571.
Miller, J. A., Clappison, B. H. & Johnston, C. I. Kallikrein and
plasmin as activators of inactive renin. Lancet 1: 1375–1376.
Morris, B. J. Activation of human inactive ("pro-") renin by
cathepsin D and pepsin. J. Clin. Endocrinol. Metab. 46: 153–
157.
Morris, B. J. Properties of the activation by pepsin of inactive
renin in human amniotic fluid. Biochim. Biophys. Acta 527: 86–
97.
Nielsen, A. H., Malling, C. & Poulsen, K. Characteristics and
conversion of high molecular weight forms of renin in plasma and
their incomplete activation by the current acid treatment.
Biochim. Biophys. Acta 534: 246–257.
Osmond, D. H., & Loh, A. Y. Protease as endogenous activator of
inactive renin. Lancet 1: 102 only.
Sealey, J. E., Atlas, S. A., Laragh, J. H., Oza, N. B. & Ryan, J.
W. Human urinary kallikrein converts inactive to active renin
and is a possible physiological activator of renin. Nature 275:
144–145.
Slater, E. E. & Haber, E. A large form of renin from normal human
kidney. J. Clin. Endocrinol. Metab. 47: 105–109.
Slater, E. E., Cohn, R. C., Dzau, V. J. & Haber, E. Purification
of human renal renin. Clin. Sci. Mol. Med. 55: 117s–119s.
Wilson, C. M., Erdös, E. G., Wilson, J. D. & Taylor, B. A.
Location on chromosome 1 of Rnr, a gene that regulates renin in
the submaxillary gland of the mouse. Proc. Natl. Acad. Sci.
U.S.A. 75: 5623–5626.
Yokosawa, H., Inagami, T. & Haas, E. Purification of human renin.
Biochem. Biophys. Res. Commun. 83: 306–312.

1979

Dzau, V. J., Slater, E. E. & Haber, E. Complete purification of
dog renal renin. Biochemistry 18: 5224–5228.
Hirose, S., Workman, R. J. & Inagami, T. Specific antibody to hog
renal renin and its application to the direct radioimmunoassay
of renin in various organs. Circ. Res. 45: 275–281.
Inagami, T., Yokosawa, N., Takahashi, N. & Takii, Y. Partial
purification of prorenin and activation by kallikreins: a
possible link betwen renin and kallikrein systems. In: Kinins-
II. Systemic Proteases and Cellular Function (Fujii, S., Moriya,
H. & Suzuki, T. eds), pp. 415–428, Plenum Press, New York.
Leckie, B. J. An endogenous protease activating plasma inactive
renin. Clin. Sci. Mol. Med. 55: 133s–134s.
Overturf, M. L., Druilhet, R. E. & Fitz, A. The effects of
kallikrein, plasmin, and thrombin on hog kidney renin. J. Biol.
Chem. 254: 12078–12083.
Overturf, M. L., Druilhet, R. E. & Kirkendall, W. M. Renin:
multiple forms and prohormones. Life Sci. 24: 1913–1934.
Potter, D. M., Dunn, P. M. & McDonald, W. J. Evidence favoring

the existence of two high molecular weight precursor forms of
dog kidney renin. Endocrinology 105: 348-351.

Poulsen, K., Krøll, J., Nielsen, A. H., Jensenius, J. & Malling,
C. Renin binding proteins in plasma. Binding of renin to some
of the plasma protease inhibitors, to lipoproteins, and to a
non-trypsin binding unidentified plasma protein. Biochim.
Biophys, Acta. 577: 1-10.

Poulsen, K., Vuust, J., Lykkegaard, S., HøjNielsen, A. & Lund, T.
Renin is synthesized as a 50000 dalton single-chain polypeptide
in cell-free translation systems. FEBS Lett. 98: 135-138.

Yokosawa, N., Takahashi, N. & Inagami, T. Isolation of completely
inactive plasma prorenin and its activation by kallikreins: a
possible new link between renin and kallikrein. Biochim.
Biophys. Acta 569: 211-219.

Entry 3.10

MOUSE SUBMANDIBULAR RENIN

Summary

EC Number: 3.4.99.19 (but being an aspartic proteinase, should be 3.4.23.?).

Earlier names: -

Distribution: Submandibular salivary glands (equivalent to submaxillary glands) of rats and mice, especially adult males.

Source: Mouse submaxillary gland.

Action: Releases the decapeptide, angiotensin I, by cleavage of a -Leu-Leu- bond, from angiotensinogen.

Requirements: pH optimum is widely variable with mol.wt. and species of origin of the substrate: pH 5.4 (octapeptide), pH 6.5 (tetradecapeptide), or pH 8.3 (sheep plasma substrate).

Substrate, usual: Angiotensinogen; blocked octapeptide and tetradecapeptide synthetic substrates as for renin (#3.09).

Inhibitors: $N_2Ac-DL-Nle-OMe$ with Cu^{2+}, pepstatin.

Molecular properties: Mol.wt. in range 37,000 (sedimentation) to 43,000 (gel chromatography). Multiple forms, of which the major form A has pI 5.4. Amino acid composition similar to that of renal renin, but the submaxillary enzyme differs in containing no carbohydrate. A280,1% 10.0.

Comment: The submaxillary enzyme is a true extra-renal renin, whereas rat brain "isorenin" and hog spleen "pseudorenin" may be identical with cathepsin D. Antiserum to mouse submaxillary renin reacts with mouse renal renin.

Bibliography

1972

Cohen, S., Taylor, J. M., Murakami, K., Michelakis, A. M. & Inagami, T. Isolation and characterization of renin-like enzymes from mouse submaxillary glands. Biochemistry 11: 4286-4293.

1974

Inagami, T., Misono, K. & Michelakis, A. M. Definitive evidence

for similarity in the active site of renin and acidic protease.
Biochem. Biophys. Res. Commun. 56: 503-509.
Michelakis, A. M., Yoshida, H., Menzie, J., Murakami, K. &
Inagami, T. A radioimmunoassay for the direct measurement of
renin in mice and its application to submaxillary gland and
kidney studies. Endocrinology 94: 1101-1105.
Misono, K., Murakami, K., Michelakis, A., Inagami, T. & Lin, T.-Y.
Characterization of submaxillary renin and comparison with
cathepsin D. Fed. Proc. Fed. Am. Soc. Exp. Biol. 33: 1315 only.

1975

Suketa, Y. & Inagami, T. Active site directed inactivators of
mouse submaxillary renin. Biochemistry 14: 3188-3194.

1977

Inagami, T., Murakami, K., Misono, K., Workman, R. J., Cohen, S. &
Suketa, Y. Renin and precursors: purification, characterization,
and studies on active site. In: Acid Proteases (Tang, J. ed.),
pp. 225-247, Plenum Press, New York.

Summary

EC Number: 3.4.23.5

Earlier names: Cathepsin; isorenin, pseudorenin (see below).

Distribution: Enzymes with the major characteristics of cathepsin D seem to occur in all mammals, and a very wide range of other animals. Located in lysosomes, and red cell membranes.

Source: Tissues rich in lysosomes, e.g. spleen and liver.

Action: Digests proteins, and some peptides of at least five residues.

Requirements: pH optimum 2.8-5.0 depending on source and substrate.

Substrate, usual: Hemoglobin (much more sensitive than most other proteins for reasons that are not really clear).

Substrate, special: Ph-SO$_2$-Ph, Phe-Gly-His-Phe(NO$_2$)$\not$Phe-Val-Leu-OMe for spectrophotometric rate assays), Z-Arg-Gly-Phe$\not$Phe-Leu-2-NNapOMe (for histochemical staining, in conjunction with aminopeptidase M and diazo-coupling [1]).

Inhibitors: N$_2$Ac-Nle-OMe with Cu^{2+}. Pepstatin (K$_i$ 5 x 10^{-10} M). (Not Ascaris pepsin inhibitor).

Molecular properties: Mol.wt. about 42,000, depending on species. The molecule can occur as a single chain or as two chains of about 27,000 and 14,000 mol.wt. non-covalently linked; occurs in multiple forms, mainly in the range pI 5.5 - 6.5. Glycoprotein. Glycine N-terminus. A280,1% 10.0. May contain a free thiol group in rat, but not man.

Comment: Cathepsin D is probably identical with the isorenins of rat and dog brain, the pseudorenin of hog spleen and most of the non-renin, angiotensin-forming activities of various tissues.

References
[1] Smith & Van Frank In: Lysosomes in Biology and Pathology (Dingle J. T. & Dean, R. T. eds), vol. 4, pp. 193-249, North-Holland Publishing Co., Amsterdam, 1975.

Bibliography

1937

Anson, M. L. The estimation of cathepsin with hemoglobin and the partial purification of cathepsin. J. Gen. Physiol. 20: 565-574.

1938

Anson, M.L. The estimation of pepsin, trypsin, papain and cathepsin with hemoglobin. J. Gen. Physiol. 22: 79-89.

1940

Anson, M. L. The purification of cathepsin. J. Gen. Physiol. 23: 695-704.

1951

Adams, E. & Smith, E. L. Proteolytic activity of pituitary extracts. J. Biol. Chem. 191: 651-664.

1954

Greenbaum, A. L. & Greenwood, F. C. Some enzymic changes in the mammary gland of rats during pregnancy, lactation and mammary involution. Biochem. J. 56: 625-631.

1955

Dannenberg, A. M. & Smith, E. L. Action of proteinase I of bovine lung. Hydrolysis of the oxidised B chain of insulin; polymer formation from amino acid esters. J. Biol. Chem. 215: 55-66.

1957

Turk, V., Kregar, I. & Lebez, D. New observations of bovine spleen cathepsins D and E. Iugosl. Physiol. Pharacol. Acta 3: 448-500.

1960

Ellis, S. Pituitary proteinase I: purification and action on growth hormone and prolactin. J. Biol. Chem. 235: 1694-1699.

Lapresle, C. & Webb, T. Study of a proteolytic enzyme from rabbit spleen. Biochem. J. 76: 538-543.

Press, E. M., Porter, R. R. & Cebra, J. The isolation and properties of a proteolytic enzyme, cathepsin D, from bovine spleen. Biochem. J. 74: 501-51.

Todd, P. E. E. & Trikojus, V. M. Purification and properties of adrenal acid proteinase. Biochim. Biophys. Acta 45: 234-242.

1962

Lapresle, C. & Webb, T. The purification and properties of a proteolytic enzyme, rabbit cathepsin E, and further studies on rabbit cathepsin D. Biochem. J. 84: 455-462.

1963

McMaster, P. B. & Webb, T. Purification et proprietes de la

cathepsine D de la rate humaine. Ann. Inst. Pasteur 104: 90-101.

1964

Dopheide, T. A. A. & Todd, P. E. E. Hydrolysis of the A and B chains of oxidised insulin by adrenal acid proteinase. Biochim. Biophys. Acta 86: 130-135.

Firfarova, K. F., Morozkin, A. D. & Orekhovich, V. N. Isolation of proteinase from brain tissues. Biochemistry U.S.S.R. (Engl. Transl.) 29: 574-580.

1965

Balasubramaniam, K. & Deiss, W. P. Characteristics of thyroid lysosomal cathepsin. Biochim. Biophys. Acta 110: 564-575.

Woessner, J. F., Jr. Acid hydrolases of the rat uterus in relation to pregnancy, post-partum involution and collagen breakdown. Biochem. J. 97: 855-866.

1966

Cochrane, C. G. & Aiken, B. L. Polymorphonuclear leukocytes in immunological reactions. J. Exp. Med. 124: 733-752.

Dopheide, T. A. A. & Trikojus, V. M. Hydrolysis of the A & B chains of oxidised insulin by thyroid acid proteinase. Biochim. Biophys. Acta 118: 435-437.

Kress, L. F., Peanasky, R. J. & Klitgaard, H. M. Purification, properties and specificity of hog thyroid proteinase. Biochim. Biophys. Acta 113: 375-389.

Rangel, H. & Lapresle, C. Etude de la specificité sur la chaine B de l'insuline des cathepsines D et E de lapin. Biochim. Biophys. Acta 128: 372-379.

Schwabe, C. & Kalnitsky, G. A peptidohydrolase from mammalian fibroblasts (bovine dental pulp). Biochemistry, 5: 158-168.

1967

Barrett, A. J. Lysosomal acid proteinase of rabbit liver. Biochem. J. 104: 601-608.

Kregar, I., Turk, V. & Lebez, D. Purification and some properties of the cathepsin from the small intestinal mucosa of pig. Enzymologia 33: 80-88.

Turk, V., Kregar, I. & Lebez, D. Some new characteristics of bovine spleen cathepsins D and E. Acta Biol. Iugoslav. 3: 448-450.

Wood, C. & Psychoyos, A. Activité de certaines enzymes hydrolytiques dans l'endomètre et le myomètre au cours de la pseudogestation et de divers états de receptivité utérine chez la ratte. C. R. Acad. Sci. D (Paris) 265: 141-144.

1968

Atsushi, S. & Masao, F. Studies on proteolysis in stored muscle. II. Purification and properties of a proteolytic enzyme, cathepsin D, from rabbit muscle. Agr. Biol. Chem. (Tokyo) 32: 975-982.

Einstein, E. R., Csejtey, J. & Marks, N. Degradation of encephalitogen by purified brain acid proteinase. FEBS Lett. 1: 191–195.

Filipovic, I. & Buddecke, E. [Mutual activation and inracellular localization of hyaluronidase (EC. 3.2.1.35) and cathepsin D (EC 3.4.4.23) from bovine spleen.] Hoppe-Seyler's Z. Physiol. Chem. 349: 533–543.

Keilová, H. & Keil, B. On proteins. CXIII. Specificity of cleavage of heptapeptide substrate with bovine spleen cathepsin D. Coll. Czech. Chem. Commun. 33: 131–140.

Keilová, H., Bláha, K. & Keil, B. Effect of steric factors on digestibility of peptides containing aromatic amino acids by cathepsin D and pepsin. Eur. J. Biochem. 4: 442–447.

Lebez, D., Turk, V. & Kregar, I. The possibility of differentiation of cathepsins on the basis of their ability to degrade various protein substrates. Enzymologia 34: 344–348.

1969

Keilová, H., Markovič, O. & Keil, B. Characterisation and chemical modification of cathepsin D from bovine spleen. Coll. Czech. Chem. Commun. 34: 2154–2157.

Mzhel'skaya, T. I. & Orekhovich, V. N. Isolation of cathepsin D from bovine spleen. Biochemistry U.S.S.R. (Engl. Transl.) 34: 866–870.

Smith, G. D., Murray, M. A., Nichol, L. W. & Trikojus, V. M. Thyroid acid proteinase. Properties and inactivation by diazoacetyl-norleucine methyl ester. Biochim. Biophys. Acta 171: 288–298.

Stepanov, V. M., Orekhovich, V. N., Lobareva, L. S. & Mzhel'skaya, T. I. Action of pepsin inhibitor on beef spleen cathepsin D. Biochemistry U.S.S.R. (Engl. Transl.) 34: 170 only.

Weston, P. D. A specific antiserum to lysosomal cathepsin D. Immunology 17: 421–428.

Weston, P. D., Barrett, A. J. & Dingle, J. T. Specific inhibition of cartilage breakdown. Nature 222: 285–286.

Wood, J. C. Effect of an intrauterine thread on acid cathepsin activity in the endometrium of the pseudopregnant rat. Biochem. J. 114: 665–666.

1970

Barrett, A. J. Cathepsin D. Purification of isoenzymes from human and chicken liver. Biochem. J. 117: 601–607.

Callahan, P. X. & Ellis, S. Action of pituitary proteinase I on the insulin B_{ox} chain. Biochim. Biophys. Acta 221: 413–416.

De Lumen, B. O. & Tappel, A. L. Fluorescein-hemoglobin as a substrate for cathepsin D and other proteases. Anal. Biochem. 36: 22–29.

Hille, M. B., Barrett, A. J., Dingle, J. T. & Fell, H. B. Microassay for cathepsin D shows an unexpected effect of cycloheximide on limb-bone rudiments in organ culture. Exp. Cell Res. 61: 470–472.

Keilová, H.   Inhibition of cathepsin D by diazoacetylnorleucine
methyl ester.   FEBS Lett. 6: 312-314.

1971

Barrett, A. J.   Purification and properties of cathepsin D from
liver of chicken, rabbit and man.   In: Tissue Proteinases
(Barrettt, A. J. & Dingle, J. T. eds), pp. 109-133, North-Holland
Publishing Co., Amsterdam.

Dingle, J. T., Barrett, A. J. & Weston, P. D.   Cathepsin D:
immunoinhibition by specific antisera.   Biochem. J. 123: 1-13.

Firfarova, K. F. & Orekhovich, V. N.   On inactive precursor to
cathepsin D from chicken liver.   Biochem. Biophys. Res. Commun.
45: 911-916.

Reinharz, A. & Roth, M.   Studies on pituitary cathepsin D with
artificial substrates.   Enzyme 12: 458-466.

Roth, J. S. & Losty, T.   Assay of proteolytic enzyme activity
using a ^{14}C-labeled hemoglobin.   Anal. Biochem. 42: 214-221.

Woessner, J. F., Jr.   Cathepsin D: enzymic properties and role in
connective tissue breakdown.   In: Tissue Proteinases (Barrett,
A. J. & Dingle, J. T. eds), pp. 291-311, North-Holland
Publishing Co., Amsterdam.

Woessner, J. F., Jr. & Shamberger, R. J., Jr.   Purification and
properties of cathepsin D from bovine uterus.   J. Biol. Chem.
246: 1951-1960.

1972

Anderson, R. B., Gormsen, J. & Petersen, P. H.   Protease activity
in synovial tissue extracts.   Scand. J. Rheumatol. 1: 75-79.

Aoyagi, T., Morishima, H., Nishizawa, R., Kunimoto, S., Takeuchi,
T. & Umezawa, H.   Biological activity of pepstatins, pepstanone
A and partial peptides on pepsin, cathepsin D and renin.   J.
Antibiot. (Tokyo) 25: 689-694.

Barrett, A. J. & Dingle, J. T.   Inhibition of tissue acid
proteinases by pepstatin.   Biochem. J. 127: 439-441.

Dingle, J. T., Barrett, A. J., Poole, A. R. & Stovin, P.
Inhibition by pepstatin of human cartilage degradation.
Biochem. J. 127: 443-444.

Harris, E. D., Jr., Locke, S. & Lin, F. Y.   Limited degradation of
gelatin substrate by cathepsin D.   Connect. Tissue Res. 1: 103-
108.

Kazakova, O. V. & Orekhovich, V. N.   Study of functionally active
groups of D cathepsins.   Biochemistry U.S.S.R. (Engl. Transl.)
37: 859-861.

Kazakova, O. V., Orekhovich, V. N., Pourchot, L. & Schuck, J. M.
Effect of cathepsins D from normal and malignant tissues on
synthetic peptides.   J. Biol. Chem. 247: 4224-4228.

Keilová, H. & Tomásĕk, V.   Effect of pepsin inhibitor from Ascaris
lumbricoides on cathepsin D and E.   Biochim. Biophys. Acta 284:
461-464.

Mednis, A. & Remold, H. G.   A sensitive fluorometric assay for the
determination of cathepsin D.   Anal. Biochem. 49: 134-138.

Okitani, A., Suzuki, A., Yang, R. & Fujimaki, M. Effect of
cathepsin D treatment on ATPase activity of rabbit myofibril.
Agr. Biol. Chem. (Tokyo) 36: 2135-2141.
Pickup. J. C. & Hope, D. B. The limited proteolysis of bovine
neurophysins by cathepsin D. J. Neurochem. 19: 1049-1064.
Poole, A. R., Dingle, J. T. & Barrett, A. J. The
immunocytochemical demonstration of cathepsin D. J. Histochem.
Cytochem. 20: 261-265.
Sapolsky, A. I. & Woessner, J. F. Multiple forms of cathepsin D
from bovine uterus. J. Biol. Chem. 247: 2069-2076.
Schwabe, C. & Sweeney, S. C. The specificity of a bovine
fibroblast cathepsin D. I. Action on the S-sulfo derivatives of
the insulin A and B chains and of porcine glucagon. Biochim.
Biophys. Acta 284: 465-472.
Woessner, J. F., Jr. Pepstatin inhibits the digestion of
haemoglobin and protein-polysaccharide complex by cathepsin D.
Biochem. Biophys. Res. Commun. 47: 965-970.

1973

Bosmann, H. B. Protein catabolism. III. Proteolytic enzymes of
guinea pig cerebral cortex and synaptosomal localization. Int.
J. Pep. Prot. Res. 5: 135-147.
Dingle, J. T., Poole, A. R., Lazarus, G. S. & Barrett, A. J.
Immunoinhibition of intracellular protein digestion in
macrophages. J. Exp. Med. 137: 1124-1141.
Etherington, D. J. Collagenolytic-cathepsin and acid-proteinase
activities in the rat uterus during post partum involution.
Eur. J. Biochem. 32: 126-128.
Ferguson, J. B., Andrews, J. R., Voynick, I. M. & Fruton, J. S.
The specificity of cathepsin D. J. Biol. Chem. 248: 6701-6708.
Ghetie, V. & Mihaescu, S. The hydrolysis of rabbit immunoglobulin
G with purified cathepsins D and E. Immunochemistry 10: 251-
255.
Marks, N., Grynbaum, A. & Lajtha, A. Pentapeptide (pepstatin)
inhibition of brain acid proteinase. Science 181: 949-951.
McAdoo, M. H., Dannenberg, A. M., Hayes, C. J., James, S. P. &
Sanner, J. H. Inhibition of cathepsin D-type proteinase of
macrophages by pepstatin, a specific pepsin inhibitor, and other
substances. Infect. Immun. 7: 655-665.
Poole, A. R., Hembry, R. M. & Dingle, J. T. Extracellular
localization of cathepsin D in ossifying cartilage. Calcif.
Tissue Res. 12: 313-321.
Rojas-Espinosa, O., Dannenberg, A. M., Murphy, P. A., Straat, P.
A., Huang, P. C. & James, S. P. Purification and properties of
the cathepsin D type proteinase from beef and rabbit lung and
its identification in macrophages. Infect. Immun. 8: 1000-1008.
Sapolsky, A. I., Altman, R. D. & Howell, D. S. Cathepsin D
activity in normal and osteoarthritic human cartilage. Fed.
Proc. Fed. Am. Soc. Exp. Biol. 32: 1489-1493.
Sapolsky, A. I., Altman, R. D., Woessner, J. F. & Howell, D. S.
The action of cathepsin D in human articular cartilage on

proteoglycans. J. Clin. Invest. 52: 624-633.

Weston, P. D. & Poole, A. R. Antibodies to enzymes and their uses, with specific reference to cathepsin D and other lysosomal enzymes. In: Lysosomes in Biology and Pathology (Dingle, J. T. ed.), vol.3, pp. 425-464, North-Holland Publishing Co., Amsterdam.

Woessner, J. F., Jr. Cartilage cathepsin D and its action on matrix components. Fed. Proc. Fed. Am. Soc. Exp. Biol. 32: 1485-1488.

Woessner, J. F., Jr. Purification of cathepsin D from cartilage and uterus and its action on the protein-polysaccharide complex of cartilage. J. Biol. Chem. 248: 1634-1642.

1974

Brostoff, S. W., Reuter, W., Hichens, M. & Eylar, E. H. Specific cleavage of the Al protein from myelin by cathepsin D. J. Biol. Chem. 249: 559-567.

Huisman, W., Lanting, L., Doddema, H. J., Bouma, J. M. W. & Gruber, M. Role of individual cathepsins in lysosomal protein digestion as tested by specific inhibitors. Biochim. Biophys. Acta 370: 297-307.

Lazarus G. S. & Dingle, J. T. Cathepsin D of rabbit skin: an immunoenzymic study. J. Invest. Dermatol. 62: 61-66.

Marks, N., Benuck, M. & Hashim, G. Hydrolysis of myelin basic protein with brain acid proteinase. Biochem. Biophys. Res. Commun. 56: 68-74.

Poole, A. R., Hembry, R. M. & Dingle, J. T. Cathepsin D in cartilage: the immunohistochemical demonstration of extracellular enzyme in normal and pathological conditions. J. Cell Sci. 14: 139-161.

Rakitzis, E. T. Kinetics of irreversible enzyme inhibition by an unstable inhibitor. Biochem. J. 141: 601-603.

Reichelt, D., Jacobsohn, E. & Haschen, R. J. Purification and properties of a cathepsin D from human erythrocytes. Biochim. Biophys. Acta 341: 15-26.

Rojas-Espinosa, O., Dannenberg, A. M., Sternberger, L. A. & Tsuda, T. The role of cathepsin D in the pathogenesis of tuberculosis: a histo-chemical study employing unlabelled antibodies and the peroxidase-anti-peroxidase complex. Am. J. Pathol. 74: 1-10.

Sampath-Kumar, P. S. & Fruton, J. S. Studies on the extended active sites of acid proteinases. Proc. Natl. Acad. Sci. U.S.A. 71: 1070-1072.

Smith, R. & Turk, V. Cathepsin D: rapid isolation by affinity chromatography on haemoglobin-agarose resin. Eur. J. Biochem. 48: 245-254.

Wildenthal, K. & Mueller, E. A. Increased myocardial cathepsin D activity during regression of thyrotoxic cardiac hypertrophy. Nature 249: 478-479.

1975

Benuck, M., Marks, N. & Hashim, G. Metabolic instability of

myelin proteins. Breakdown of basic protein induced by brain cathepsin D. Eur. J. Biochem. 52: 615-621.

Dean, R. T. Direct evidence of the importance of lysosomes in the degradation of intracellular proteins. Nature 257: 414-416.

Greenbaum, L. M. Cathepsin D-generated, pharmacologically active peptides (leukokinins) and their role in ascites fluid accumulation. In: Proteases and Biological Control (Reich, E., Rifkin, D. B. & Shaw, E. eds), pp. 223-228, Cold Spring Harbor Laboratory, New York.

Hayasaka, S., Hara, S. & Mizuno, K. Degradation of rod outer segment proteins by cathepsin D. J. Biochem. 78: 1365-1367.

Heikkinen, J. E., Järvinen, M. & Jansen, C. T. Purification and biochemical characterization of rat skin cathepsin D. J. Invest. Dermatol. 65: 272-278.

Kazakova, O. V. & Orekhovich, V. N. Isolation of cathepsins of group D by affinity chromatography. Biochemistry U.S.S.R. (Engl. Trans.) 40: 826-828.

Kazakova, O. V. & Orekhovich, V. N. Some properties of cathepsins chemically fixed to carriers. Int. J. Pept. Protein Res. 7: 23-29.

Lin, T.-Y. & Williams, H. R. Inhibition of the local hemorrhagic Schwartzman reaction by an acid proteinase inhibitor, pepstatin. Experientia 31: 209-212.

Widmer, F. & Widmer, C. The effect of beef spleen cathepsin D on pig heart lactate dehydrogenase. Arch. Biochem. Biophys. 168: 252-258.

1976

Contractor, S. F. & Krakauer, K. Immunofluorescent localization of cathepsin D in trophoblastic cells in tissue culture. Beitr. Pathol. 158: 445-449.

Cunningham, M. & Tang, J. Purification and properties of cathepsin D from porcine spleen. J. Biol. Chem. 251: 4528-4536.

Day, R. P. & Reid, I. A. Renin activity in dog brain: enzymological similarity to cathepsin D. Endocrinology 99: 93-100.

Dean, R. T. The roles of cathepsins B1 and D in the digestion of cytoplasmic proteins in vitro by lysosomal extracts. Biochem. Biophys. Res. Commun. 68: 518-523.

Ducastaing, A., Azanza, J.-L., Raymond, J., Robin, J.-M. & Créac'h, P. La cathepsin D de rate de cheval. II. Etude de quelques propriétés enzymatiques. Biochimie 58: 783-791.

Ducastaing, A., Azanza, J.-L., Robin, J.-M., Raymond, J. & Créac'h, P. La cathepsin D de rate de cheval. I. Purification et étude de quelques proprietes physico-chimiques. Biochimie 58: 771-782.

Fräki, J. E. Human skin proteases. Separation and characterization of two acid proteases resembling cathepsin B1 and cathepsin D and of an inhibitor of cathepsin B1. Arch. Derm. Res. 255: 317-330.

Gubenšek, F., Barstow, I., Kregar, I. & Turk, V. Rapid isolation of cathepsin D by affinity chromatography on the immobilized synthetic inhibitor. FEBS Lett. 71: 42-44.

Iodice, A. A. The inhibition by pepstatin of cathepsin D and autolysis of dystrophic muscle. Life Sci. 19: 1351-1358.

Kazakova, O. V. & Orekhovich, V. N. Crystallization of cathepsin D. Biochem. Biophys. Res. Commun. 72: 747-752.

Keilová, H. & Tomášek, V. Contribution to studies on active site of cathepsin B_1 and D. In: Intracellular Protein Catabolism (Hanson, H. & Bohley, P. eds), pp. 237-251, J. A. Barth, Leipzig.

Knight, C. G. & Barrett, A. J. Interaction of human cathepsin D with the inhibitor pepstatin. Biochem. J. 155: 117-125.

Marks, N., Grynbaum, A. & Benuck, M. On the sequential cleavage of myelin basic protein by cathepsins A and D. J. Neurochem. 27: 765-768.

Poole, A. R., Hembry, R. M., Dingle, J. T., Pinder, I., Ring, E. F. J. & Cosh, J. Secretion and localization of cathepsin D in synovial tissues removed from rheumatoid and traumatized joints. An immunohistochemical study. Arthritis Rheum. 19: 1295-1307.

Rakitzis, E. T. & Malliopoulou, T. B. Inactivation of cathepsin D by dithiophosgene and by 2,2-dichloro-1,3-dithiacyclobutanone. Biochem. J. 153: 737-739.

Thorne, K. J. I., Oliver, R. C. & Barrett, A. J. Lysis and killing of bacteria by lysosomal proteinases. Infect. Immun. 14: 555-563.

Wiederanders, B., Ansorge, S., Bohley, P., Broghammer, U., Kirschke, H. & Langner, J. Intrazellulärer Proteinabbau. VI. Isolierung, Eigenschaften und biologische Bedeutung von Kathepsin D aus der Rattenleber. Acta Biol. Med. Ger. 35: 269-283.

1977

Barrett, A. J. Cathepsin D and other carboxyl proteinases. In: Proteinases in Mammalian Cells and Tissues (Barrett, A. J. ed.), pp. 209-248, North-Holland Publishing Co., Amsterdam.

Barrett, A. J. Human cathepsin D. In: Acid Proteases (Tang, J. ed.), pp. 291-300, Plenum Press, New York.

Bird, J. W. C., Schwartz, W. N. & Spanier, A. M. Degradation of myofibrillar proteins by cathepsins B and D. Acta Biol. Med. Ger. 36: 1587-1604.

Bowers, W. E., Beyer, C. F. & Yago, N. Cathepsin D of mouse leukemia L12010 cells: unusual intracellular localization and biochemical properties. Biochim. Biophys. Acta 497: 272-279.

Bowers, W. E., Panagides, J. & Yago, N. Unique biochemical and biological features of cathepsin D in rodent lymphoid tissues. In: Acid Proteases (Tang, J. ed.), pp. 301-312, Plenum Press, New York.

Fehr, K., Artmann, G., Velvart, M. & Böni, A. Cathepsin D agglutinators and neutral protease agglutinators in rheumatoid arthritis. Arthritis Rheum. 20: 1240-1248.

Hackenthal, E., Hackenthal, R. & Hilgenfeldt, U. Isorenin,

pseudorenin, cathepsin D and renin. A comparative enzymatic study of angiotensin-forming enzymes. Biochim. Biophys. Acta 522: 574-588.

Hackenthal, E., Hackenthal, R. & Hilgenfeldt, U. Purification and partial characterization of rat brain acid proteinase (isorenin). Biochim. Biophys. Acta 522: 561-573.

Ishikawa, I. & Cimasoni, G. Isolation of cathepsin D from human leucocytes. Biochim. Biophys. Acta 480: 228-240.

Johnson, R. L. & Poisner, A. M. Inhibition of pseudorenin by pepstatin. Biochem. Pharmacol. 26: 639-641.

Knook, D. L. The role of lysosomal enzymes in protein degradation in different types of rat liver cells. Acta Biol. Med. Ger. 36: 1747-1752.

Kregar, I., Urh, I., Smith, R., Umezawa, H. & Turk, V. Isolation of cathepsin D on pepstatin-agarose resin. In: Intracellular Protein Catabolism II (Turk, V. & Marks, N. eds.), pp. 250-254, Plenum Press, New York.

Lockwood, T. D. & Shier, W. T. Regulation of acid proteases during growth, quiescence and starvation in normal and transformed cells. Nature 267: 252-254.

Ramazzotto, L. J., Carlin, R. & Engstrom, R. Inhaled nitrous oxide and cathepsin D activity in the lung. J. Dent. Res. 56: 777-782.

Roughley, P. J. The degradation of cartilage proteolycans by tissue proteinases. Proteoglycan heterogeneity and the pathway of proteolytic degradation. Biochem. J. 167: 639-646.

Roughley, P. J. & Barrett, A. J. The degradation of cartilage proteoglycans by tissue proteinases. Proteoglycan structure and its susceptibility to proteolysis. Biochem. J. 167: 629-637.

Schwartz, W. N. & Bird, J. W. C. Degradation of myofibrillar proteins by cathepsins B and D. Biochem. J. 167: 811-820.

Turk, V., Kregar, I., Gubenšek, F., Lapanje, S., Urh, I. & Kovačič, M. Studies on bovine spleen cathepsin D. Acta Biol. Med. Ger. 36: 1531-1535.

Turk, V., Urh, I., Kregar, I., Babnik, J., Gubenšek, F. & Smith, R. Purification and some properties of native and immobilized cathepsin D. In: Intracellular Protein Catabolism II (Turk, V. & Marks, N. eds), pp. 240-249, Plenum Press, New York.

Woessner, J. F., Jr. Specificity and biological role of cathepsin D. In: Acid Proteases (Tang, J. ed.), pp. 313-327, Plenum Press, New York.

Woessner, J. F., Jr., Sapolsky, A. I., Morales, T. & Nagase, H. Proteinases that digest the connective tissue matrix. In: Intracellular Protein Catabolism II (Turk, V. & Marks, N. eds), pp. 215-223, Plenum Press, New York.

1978

Akopyan, T. N., Arutunyan, A. A., Lajtha, A. & Galoyan, A. A. Acid proteinase of hypothalamus. Purification, some properties, and action on somatostatin and substance P. Neurochem. Res. 3: 89-99.

Baggiolini, M., Bretz, U. & Dewald, B. Subcellular localization of granulocyte enzymes. In: Neutral Proteases of Human Polymorphonuclear Leukocytes (Havemann, K. & Janoff, A. eds), pp.3-17, Urban & Schwartzenberg, Baltimore.

Benuck, M., Grynbaum, A., Cooper, T. B. & Marks, N. Conversion of lipotropic peptides by purified cathepsin D of human pituitary: release of γ-endorphin by cleavage of the Leu[77]-Phe[78] bond. Neuroscience Lett. 10: 3-9.

Bird, J. W. C., Spanier, A. M. & Schwartz, W. N. Cathepsins B and D: proteolytic activity and ultrastructural localization in skeletal muscle. In: Protein Turnover and Lysosome Function (Segal, H. L. & Doyle, D. J. eds), pp. 589-604, Academic Press, New York.

Chou, H. J., Shaper, J. H. & Gregerman, R. I. Separation of human renal renin and pseudorenin by affinity chromatography on hemoglobin-Sepharose 2B. Biochim. Biophys. Acta 524: 183-187.

Dionyssiou-Asteriou, A. & Rakitzis, E. Inactivation of cathepsin D from human gastric mucosa and from stomach carcinoma by diazoacetyl-DL-norleucine methyl ester. Biochem. Pharmacol. 27: 827-829.

Dorer, F. E., Lentz, K. E., Kahn, J. R., Levine, M. & Skeggs, L. T. A comparison of the substrate specificities of cathepsin D and pseudorenin. J. Biol. Chem. 253: 3140-3142.

Greenbaum, L. M., Semente, G., Prakash, A. & Roffman, S. Leukokinins; their pharmacological properties and role in the pathology of fluid (ascites) accumulation. Agents Actions 8: 80-84.

Jasani, B., Jasani, M. K. & Talbot, M. D. Characterization of two acid proteinases found in rabbit skin homografts. Biochem. J. 169: 287-295.

Kazakova, O. V. & Orekhovich, V. N. Crystallization of cathepsin D. Biochem. Biophys. Res. Commun. 72: 747-752.

Kazakova, O. V., Orekhovich, V. N. & Vikha, G. V. Study of carbohydrate component of cathepsin D. Biochemistry U.S.S.R. (Engl. Transl.) 42: 1728-1732.

Lanoë, J. & Dunnigan, J. Improvements of the Anson assay for measuring proteolytic activities in acidic pH range. Anal. Biochem. 89: 461-471.

Moriyama, A. & Takahashi, K. The structure and function of acid proteases. VIII. Purification and characterization of cathepsins D from Japanese monkey lung. J. Biochem. 83: 441-451.

Morris, B. J. Activation of human inactive ("pro-") renin by cathepsin D and pepsin. J. Clin. Endocrinol. Metab. 46: 153-157.

Morris, B. J. & Reid, I. A. A "renin-like" enzymatic action of cathepsin D and the similarity in subcellular distributions of "renin-like" activity and cathepsin D in the midbrain of dogs. Endocrinology 103: 1289-1296.

Scott, P. G. & Pearson, C. H. Cathepsin D - cleavage of soluble collagen and crosslinked peptides. FEBS Lett. 88: 41-45.

Turk, V., Kregar, I., Gubensĕk, F. & Locnikar, P. Bovine spleen cathepsins D and S: purification, characterization and structural studies. In: Protein Turnover and Lysosome Function (Segal, H. L. & Doyle, D. J. eds), pp. 353-361, Academic Press, New York.

1979

Akopyan, T. N., Barchudaryan, N. A., Karabashyan, L. V., Arutunyan, A. A., Lajtha, A. & Galoyan, A. A. Hypothalamic cathepsin D: assay and isoenzyme composition. J. Neuroscience Res. 4: 365-370.

Tang, J. Evolution in the structure and function of carboxyl proteases. Mol. Cell. Biochem. 26: 93-109.

Dionyssiou-Asteriou, A. & Rakitzis, E. T. Activation of cathepsin D by glycine ethyl ester. Biochem. J. 177: 355-356.

Erickson, A. H. & Blobel, G. Early events in the biosynthesis of the lysosomal enzyme cathepsin D. J. Biol. Chem. 254: 11771-11774.

Gráf, L., Kenessey, A., Patthy, A., Grynbaum, A., Marks, N. & Lajtha, A. Cathepsin D generates γ-endorphin from β-endorphin. Arch. Biochem. Biophys. 193: 101-109.

Hasilik, A., Rome, L. H. & Neufeld, E. F. Processing of lysosomal enzymes in human skin fibroblasts. Fed. Proc. Fed. Am. Soc. Exp. Biol. 38: 467 only.

Huang, J. S., Huang, S. S. & Tang, J. Cathepsin D isozymes from porcine spleens. Large scale purification and polypeptide chain arrangements. J. Biol. Chem. 254: 11405-11417.

Irvine, G. B. & Elmore, D. T. A radiochemical titrant for the determination of the operational molarity of solutions of acid proteinases. Biochem. J. 183: 389-394.

Ito, A., Mori, Y. & Hirakawa, S. Purification and characterization of an acid proteinase from human uterine cervix. Chem. Pharm. Bull. 27: 969-973.

Lin, T.-Y. & Williams, H. R. Inhibition of cathepsin D by synthetic oligopeptides. J. Biol. Chem. 254: 11875-11883.

Murakami, T., Suzuki, Y. & Murachi, T. An acid protease in human erythrocytes and its localization in the inner membrane. Eur. J. Biochem. 96: 221-227.

Watabe, S., Terada, A., Ikeda, T., Kouyama, H., Taguchi, S. & Yago, N. Polyphosphate anions increase the activity of bovine spleen cathepsin D. Biochem. Biophys. Res. Commun. 89: 1161-1167.

Whitaker, J. N. & Seyer, J. M. The sequential limited degradation of bovine myelin basic protein by bovine brain cathepsin D. J. Biol. Chem. 254: 6956-6963.

Whitaker, J. N. & Seyer, J. M. Isolation and characterization of bovine brain cathepsin D. J. Neurochem. 32: 325-333.

Yamamoto, K., Katsuda, N., Himeno, M. & Kato, K. Cathepsin D of rat spleen. Affinity purification and properties of two types of cathepsin D. Eur. J. Biochem. 95: 459-467.

1980

Huang, J. S., Huang, S. S. & Tang, J. Structure and function of cathepsin D. In: Enzyme Regulation and Mechanism of Action (Mildner, P. & Ries, B. eds), pp. 289-306, Pergamon, Oxford.

Turk, V., Kregar, I., Gubensek, T., Popovic, P., Locnicar, P. & Lah, T. Carboxyl and thiol intracellular proteinases. In: Enzyme Regulation and Mechanism of Action (Mildner, P. & Ries, B. eds), pp. 317-330, Pergamon, Oxford.

Yamamoto, K., Kamata, O., Katsuda, N. & Kato, K. Immunochemical difference between cathepsin D and cathepsin E-like enzyme from rat spleen. J. Biochem. 87: 511-516.

Entry 3.12

THYROID ASPARTIC PROTEINASE

Summary

EC Number: 3.4.23.11

Earlier names: -

Distribution: Pig thyroid.

Source: Pig thyroid.

Action: Hydrolyzes a variety of proteins.

Requirements: pH optimum about 3.6.

Substrate, usual: Hemoglobin.

Inhibitors: N_2Ac-Nle-OMe.

Molecular properties: Mol.wt. 42,000, apparently composed of two subunits of 21,000 not covalently linked, but dissociating at low temperatures and pH values. Glycine N-terminus.

Comment: Further evidence is required to establish that this enzyme is not cathepsin D. The main difference is in the rather ready dissociation (cf. pig liver cathepsin D [1]).

References
[1] Cunningham & Tang   J. Biol. Chem. 251: 4528-4536, 1976.

Bibliography

1969

Smith, G. D., Murray, M. A., Nichol, L. W. & Triko jus, V. M. Thyroid acid proteinase. Properties and inactivation by diazoacetyl-norleucine methyl ester. Biochim. Biophys. Acta 171: 288-298.

Entry 3.13
LYMPHOCYTE ASPARTIC PROTEINASE

Summary

EC Number: (3.4.23.-)

Earlier names: Cathepsin D-type protease, cathepsins D_H and D_L.

Distribution: Rodent lymphocytes, partially soluble and partially in rapidly sedimenting organelles, in homogenates.

Source: Rapidly sedimenting organelles of mouse lymphocytes.

Action: Digests hemoglobin, with specific activity five-fold greater than cathepsin D.

Requirements: pH optimum 3.6.

Substrate, usual: Hemoglobin.

Inhibitors: Pepstatin.

Molecular properties: Two forms, molecular weight about 45,000 and 95,000, are interconvertible.

Comment: Not found in rabbit, or a variety of non-rodent species [1].

References
[1] Yago & Bowers   J. Biol. Chem. 250: 4749-4754, 1975.

Bibliography

1975

Yago, N. & Bowers, W. E.   Unique cathepsin D-type protease in rat thoracic duct lymphocytes and in rat lymphoid tissues.   J. Biol. Chem. 250: 4749-4754.

1977

Bowers, W. E., Beyer, C. G. & Yago, N.   Cathepsin D of mouse leukemia L12010 cells. Unusual intracellular localization and biochemical properties.   Biochim. Biophys. Acta 497: 272-279.

Bowers, W. E., Panagides, J. & Yago, N.   Unique biochemical and biological features of cathepsin D in rodent lymphoid tissues. In: Acid Proteases (Tang, J. ed.), pp. 301-312, Plenum Press, New York.

Entry 3.14

HUMAN ASCITES ASPARTIC PROTEINASE

Summary

EC Number: (3.4.23.-)

Earlier names: -

Distribution: Ascitic fluid of a patient with ovarian cancer, and similar activity in other ascites fluid samples.

Source: Human ovarian cancer ascitic fluid.

Action: Hydrolyzes hemoglobin and serum albumin at similar rates. Inactive against several synthetic substrates.

Requirements: Optimum pH 3.0.

Substrate, usual: Hemoglobin.

Inhibitors: Pepstatin.

Molecular properties: Mol.wt. 28,000 (gel chromatography); pI 4.1. Alkali-stable.

Comment: Possible relationships to the cathepsin D-like enzyme of Greenbaum et al. [1] and gastric cathepsin (#3.07) are of interest.

References
[1] Greenbaum et al. Agents Actions 8: 80-84, 1978.

Bibliography

1978

Esumi, H., Sato, S., Sugimura, T. & Okasaki, N. Purification and properties of an acid protease from human ascitic fluid. Biochim. Biophys. Acta 523: 191-197.

Entry 3.15

CATHEPSIN E

Summary

EC Number: (3.4.23.-)

Earlier names: -

Distribution: Rabbit bone marrow, polymorphonuclear leucocytes and macrophages. Human platelets.

Source: Rabbit bone marrow.

Action: Digests albumin, hemoglobin.

Requirements: pH optimum 2.5.

Substrate, usual: Albumin.

Inhibitors: $N_2Ac-Nle-OMe$ with Cu^{2+}, pepstatin. Ascaris pepsin inhibitor.

Molecular properties: Mol.wt. about 100,000 (may well be a dimer of identical subunits). pI very low.

Comment: There remains a good deal of uncertainty about the distribution and characteristics of this enzyme; in mol.wt. it resembles the rodent lymphocyte aspartic proteinase (#3.13).

Bibliography

1962

Lapresle, C. & Webb, T. The purification and properties of a proteolytic enzyme, rabbit cathepsin E, and further studies on rabbit cathepsin D. Biochem. J. 84: 455-462.

Štefanović, J., Webb, T. & Lapresle, C. Etude des cathepsines D et E dans des préparations de polynucléaires, de macrophages et de lymphocytes de lapin. Ann. Inst. Pasteur Paris 103: 276-284.

1966

Cochrane, C. G. & Aiken, B. S. Polymorphonuclear leukocytes in immunological reactions. J. Exp. Med. 124: 733-752.

Rangel, H. & Lapresle, C. Etude de la specificité sur la chaine B de l'insuline des cathepsines D et E de lapin. Biochim. Biophys. Acta 128: 372-379.

1967

Turk, V., Kregar, I. & Lebez, D. New observations of bovine
  spleen cathepsins D and E. Iugosl. Physiol. Pharmacol. Acta 3:
  448-500.
Turk, V., Kregar, I. & Lebez, D. Some new characteristics of
  bovine spleen cathepsins D and E. Acta Biol. Iugoslav. 3: 448-
  450.

1968

Turk, V., Kregar, I. & Lebez, D. Some properties of cathepsin E
  from bovine spleen. Enzymologia 34: 89-100.

1969

Turk, V., Kregar, I., Gubenšek, F. & Lebez, D. In vitro
  transition of beef spleen cathepsin E into cathepsin D.
  Enzymologia 36: 182-186.

1970

Keilová, H. & Lapresle, C. Inhibition of cathepsin E by
  diazoacetyl-norleucine methyl ester. FEBS Lett. 9: 348-350.

1971

Lapresle, C. Rabbit cathepsins D and E. In: Tissue Proteinases
  (Barrett, A. J. & Dingle, J. T. eds), pp. 135-155, North-Holland
  Publishing Co., Amsterdam.

1972

Barrett, A. J. & Dingle, J. T. Inhibition of tissue acid
  proteinases by pepstatin. Biochem. J. 127: 439-441.
Keilová, H. & Tomášek, V. Effect of pepsin inhibitor from Ascaris
  lumbricoides on cathepsin D and E. Biochim. Biophys. Acta 284:
  461-464.

1973

Ghetie, V. & Mihaescu, S. The hydrolysis of rabbit immunoglobulin
  G with purified cathepsin D and E. Immunochemistry 10: 251-255.

1976

Ehrlich, H. P. & Gordon, J. L. Proteinases in platelets. In:
  Platelets in Biology and Pathology (Gordon, J. L. ed.), pp. 353-
  372, North-Holland Publishing Co., Amsterdam.

1977

Barrett, A. J. Cathepsin D and other carboxyl proteinases. In:
  Proteinases in Mammalian Cells and Tissues (Barrett, A. J. ed.),
  pp. 209-248, North-Holland Publishing Co., Amsterdam.

1978

Jasani, B., Jasani, M. K. & Talbot, M. D. Characterization of two
  acid proteinases found in rabbit skin homografts. Biochem. J.
  169: 287-295.
Yamamoto, K., Katsuda, N. & Kato, K. Affinity purification and

properties of cathepsin E-like acid proteinase from rat spleen.
Eur. J. Biochem. 92: 499-508.

1980

Turk, V., Kregar, I., Gubensek, T., Popovic, P., Locnicar, P. &
Lah, T. Carboxyl and thiol intracellular proteinases. In:
Enzyme Regulation and Mechanism of Action (Mildner, P. & Ries,
B. eds), pp. 317-330, Pergamon, Oxford.
Yamamoto, K., Kamata, O., Katsuda, N. & Kato, K. Immunochemical
difference between cathepsin D and cathepsin E-like enzyme from
rat spleen. J. Biochem. 87: 511-516.

Section 4

METALLO-PROTEINASES

Entry 4.01

VERTEBRATE COLLAGENASE

Summary

EC Number: 3.4.23.7 (for all vertebrate collagenases).

Earlier names: Specific collagenase.

Distribution: Secreted by fibroblasts, macrophages and other types of cell, often in a latent form.

Source: Culture medium of fibroblasts, macrophages, synovial membrane, skin, etc.

Action: The triple helix of collagen (type I, II or III) is cleaved at one specific point, usually by hydrolysis of -Gly-Leu- or -Gly-Ile- bonds, about three quarters of the length of the molecule from the N-terminus. Gelatin (denatured collagen) is less susceptible, and the enzyme has very little general proteolytic activity. There is no action on the peptide substrates of bacterial collagenases.

Requirements: pH optimum about 7.5. Stimulated by Ca^{2+}. Latent forms activated by trypsin and certain other proteinases, and thiol-binding reagents, particularly 4-aminophenylmercuric acetate.

Substrate, usual: Collagen, normally radiolabelled, and used in solution or as reconstituted fibrils.

Inhibitors: EDTA, 1,10-phenanthroline. α_2-Macroglobulin [1], β_1-anticollagenase [2].

Molecular properties: Mol.wt. usually about 35,000 or 65,000. Oligomers can occur. pI about 8 (rabbit fibroblasts). Probably contains one atom of zinc/mole, and some less tightly bound calcium.

Comment: Apparently a family of enzymes with closely similar catalytic and immunological properties, but differences in mol.wt. The neutrophil collagenase is distinct, and is therefore listed separately (#4.02), although it has the same EC number.
    An EDTA-sensitive proteinase specific for type IV collagen has been described [3].

References
[1] Werb et al. Biochem. J. 139: 359-368, 1974.
[2] Woolley et al. Nature 261: 325-327, 1976.

[3] Liotta et al. Proc. Natl. Acad. Sci. U.S.A. 76: 2268-2272, 1979.

Bibliography

1965

Kaufman, E. J., Glimcher, M. J., Mechanic, G. L. & Goldhaber, P. Collagenolytic activity during active bone resorption in tissue culture. Proc. Soc. Exp. Biol. Med. 120: 632-637.

1966

Beutner, E. H., Triftshauser, C. & Hazen, S. P. Collagenase activity of gingival tissue from patients with periodontal diseases. Proc. Soc. Exp. Biol. Med. 121: 1082-1085.

1967

Evanson, J. M., Jeffrey, J. J. & Krane, S. M. Human collagenase: identification and characterisation of an enzyme from rheumatoid synovium in culture. Science 158: 499-502.
Fullmer, H. M. & Lazarus, G. Collagenase in human, goat and rat bone. Israel J. Med. Sci. 3: 758-761.
Jeffrey, J. J. & Gross, J. Isolation and characterisation of a mammalian collagenolytic enzyme. Fed. Proc. Fed. Am. Soc. Exp. Biol. 26: 670 only.
Riley, W. B., Jr. & Peacock, E. E., Jr. Identification, distribution, and significance of a collagenolytic enzyme in human tissues. Proc. Soc. Exp. Biol. Med. 124: 207-210.

1968

Anonymous. Collagenase and rheumatoid arthritis (editorial). N. Engl. J. Med. 279: 942-943.
Eisen, A. Z., Jeffrey, J. J. & Gross, J. Human skin collagenase. Isolation and mechanism of attack on the collagen molecule. Biochim. Biophys. Acta 151: 637-645.
Evanson, J. M., Jeffrey, J. J. & Krane, S. M. Studies on collagenase from rheumatoid synovium in tissue culture. J. Clin. Invest. 47: 2639-2651.
Lazarus, G. S., Decker, J. L., Oliver, C. H., Daniels, J. R., Multz, C. V. & Fullmer, H. M. Collagenolytic activity of synovium in rheumatoid arthritis. N. Engl. J. Med. 279: 914-918.

1969

Abramson, M. Collagenolytic activity in middle ear cholesteatoma. Ann. Otol. Rhinol. Laryngol. 78: 112-124.
Brown, S. I., Akiya, S. & Weller, C. A. Prevention of ulcers of the alkali-burned cornea: preliminary studies with collagenase inhibitors. Arch. Ophthalmol. 82: 95-97.
Brown, S. I., Weller, C. A. & Wassermann, H. E. Collagenolytic activity of alkali-burned corneas. Arch. Ophthalmol. 81: 370-373.

Eisen, A. Z. Human skin collagenase: localization and distribution in normal human skin. J. Invest. Dermatol. 52: 442–448.

Eisen, A. Z. Human skin collagenase: relationship to the pathogenesis of epidermolysis bullosa dystrophica. J. Invest. Dermatol. 52: 449–453.

Fullmer, H. M. & Lazarus, G. Collagenase in bone of man. J. Histochem. Cytochem. 17: 793–798.

Gnädinger, M. C., Itoi, M., Slansky, H. H. & Dohlman, C. H. The role of collagenase in the alkali-burned cornea. Am. J. Ophthalmol. 68: 478–483.

Harris, E. D., Jr., Cohen, G. L. & Krane, S. M. Synovial collagenase: its presence in culture from joint disease of diverse etiology. Arthritis Rheum. 12: 92–102.

Harris, E. D., Jr., DiBona, D. R. & Krane, S. M. Collagenases in human synovial fluid. J. Clin. Invest. 48: 2104–2113.

Itoi, M., Gnädinger, M. C., Slansky, H. H., Freeman, M. I. & Dohlman, C. H. Collagenase in the cornea. Exp. Eye Res. 8: 369–373.

Lazarus, G. S. & Fullmer, H. M. Collagenase production by human dermis. J. Invest. Dermatol. 52: 545–547.

Robertson, D. M. & Williams, D. C. In vitro evidence of neutral collagenase activity in an invasive mammalian tumor. Nature 221: 259–260.

Shimizu, M., Glimcher, M. J., Travis, D. & Goldhaber, P. Mouse bone collagenase: isolation, partial purification, and mechanism of action. Proc. Soc. Exp. Biol. Med. 130: 1175–1180.

Slansky, H. H., Gnädinger, M. C., Itoi, M. & Dohlman, C. H. Collagenase in corneal ulcerations. Arch. Ophthalmol. 82: 108–111.

1970

Bauer, E. A., Eisen, A. Z. & Jeffrey, J. J. Immunologic relationship of a purified human skin collagenase to other human and animal collagenases. Biochim. Biophys. Acta 206: 152–160.

Brown S. I., Weller, C. A. & Vidrich, A. M. Effect of corticosteroids on corneal collagenase of rabbits. Am. J. Ophthalmol. 70: 744–747.

Brown, S. I. & Weller, C. A. Collagenase inhibitors in prevention of ulcers of alkali-burned corneas. Arch. Ophthalmol. 83: 352–353.

Brown, S. I., Weller, C. A. & Akiya, S. Pathogenesis of ulcers of the alkali-burned cornea. Arch. Ophthalmol. 83: 205–208.

Brown. S. I. & Weller, C. A. Cell origin of collagenase in normal and wounded corneas. Arch. Ophthalmol. 83: 74–77.

Donoff, R. B. Wound healing: biochemical events and potential role of collagenase. J. Oral Surg. 28: 356–363.

Eisen, A. Z., Bauer, E. A. & Jeffrey, J. J. Animal and human collagenases. J. Invest. Dermatol. 55: 359–373.

Eisen, A. Z., Bloch, K. J. & Sakai, T. Inhibition of human skin collagenases by human serum. J. Lab. Clin. Med. 75: 258–263.

Gross, J. Animal collagenase. In: Chemistry and Molecular

Biology of the Intercellular Matrix (Balazs, E. A. ed.), vol. 3, pp. 1623-1636, Academic Press, New York.

Harris, E. D., Jr., Evanson, J. M., DiBona, D. R. & Krane, S. M. Collagenase and rheumatoid arthritis. Arthritis Rheum. 13: 83-94.

Hawley, P. R., Faulk, W. P., Hunt, T. K. & Dunphy, J. E. Collagenase activity in gastro-intestinal tract. Br. J. Surg. 57: 896-900.

Jeffrey, J. J. & Gross, J. Collagenase from rat uterus: isolation and partial characterization. Biochemistry 9: 268-273.

1971

Abramson, M. Collagenase in the mechanism of action of cholesteatoma. Ann. Otol. Rhinol. Laryngol. 80: 414 only.

Abramson, M. & Gross, J. Further studies on a collagenase in middle ear cholesteatoma. Ann. Otol. Rhinol. Laryngol. 80: 177-185.

Anderson, R. E., Kuns, M. D. & Dresden, M. H. Collagenase activity in the alkali-burned cornea. Ann. Ophthalmol. 3: 619-621.

Bauer, E. A., Eisen, A. Z. & Jeffrey, J. J. Studies on purified rheumatoid synovial collagenase in vitro and in vivo. J. Clin. Invest. 50: 2056-2064.

Bauer, E. A., Jeffrey, J. J. & Eisen, A. Z. Preparation of three vertebrate collagenases. Biochem. Biophys. Res. Commun. 44: 813-818.

Berman, M., Dohlman, C. H., Gnädinger, M. & Davison, P. Characterization of collagenolytic activity in the ulcerating cornea. Exp. Eye Res. 12: 255-257.

Brown, S. I. Collagenase and corneal ulcers. Invest. Ophthalmol. 10: 203-209.

Brown, S. I. & Hook, C. W. Isolation of stromal collagenase in corneal inflammation. Am. J. Ophthalmol, 72: 1139-1142.

Brown, S. I. Hook, C. W. & Weller, C. A. Treatment of corneal destruction with collagenase inhibitors. Trans. Am. Acad. Ophthalmol. Otolaryngol. 75: 1199-1207.

Donoff, R. B., McLennan, J. E. & Grillo, H. C. Preparation and properties of collagenases from epithelium and mesenchyme of healing mammalian wounds. Biochim. Biophys. Acta 227: 639-653.

Eisen, A. Z., Bauer, E. A. & Jeffrey, J. J. Human skin collagenase. The role of serum alpha-globulins in the control of activity in vivo and in vitro. Proc. Natl. Acad. Sci. U.S.A. 68: 248-251.

Evanson, J. M. Mammalian collagenases and their role in connective tissue breakdown. In: Tissue Proteinases (Barrett, A. J. & Dingle, J. T. eds), pp. 327-345.

Harper, E., Bloch, K. J. & Gross, J. The zymogen of tadpole collagenase. Biochemistry 10: 3035-3041.

Harris, E. D., Jr. & Krane, S. M. Effects of colchicine on collagenase in cultures of rheumatoid synovium. Arthritis Rheum. 14: 669-684.

Hook, C. W., Brown, S. I., Iwanij, W. & Nakanishi, I.
Characterization and inhibition of corneal collagenase. Invest.
Ophthalmol. 10: 496-503.

Jeffrey, J. J., Coffey, R. J. & Eisen, A. Z. Studies of uterine
collagenase in tissue culture. 1. Relationship of enzyme
production to collagen metabolism. Biochim. Biophys. Acta 252:
136-142.

Jeffrey, J. J., Coffey, R. J. & Eisen, A. Z. Studies on uterine
collagenase in tissue culture. II. Effect of steroid hormones on
enzyme production. Biochim. Biophys. Acta 252: 143-149.

Leibovich, S. J. & Weiss, J. B. Failure of human rheumatoid
synovial collagenase to degrade either normal or rheumatoid
arthritic polymeric collagen. Biochim. Biophys. Acta 251: 109-
121.

Pfister, R. R., McCulley, J. P., Friend, J. & Dohlman, C. H.
Collagenase activity of intact corneal epithelium in peripheral
alkali burns. Arch. Ophthalmol. 86: 308-313.

Ryan, J. N. & Woessner, J. F., Jr., Mammalian collagenase: direct
demonstration in homogenates of involuting rat uterus. Biochem.
Biophys. Res. Commun. 44: 144-149.

Seifter, S. & Harper, E. The collagenases. In: The Enzymes
(Boyer, P. D. ed.), 3rd edn., pp. 649-697, Academic Press, New
York.

Vaes, G. A latent collagenase released by bone and skin explants
in culture. Biochem. J. 123: 23P only.

                              1972

Abe, S. & Nagai, Y. Evidence for the presence of a latent form of
collagenase in human rheumatoid synovial fluid. J. Biochem. 71:
919-922.

Bauer, E. A., Eisen, A. Z. & Jeffrey, J. J. Radioimmunoassay of
human collagenase. 1. Specificity of the assay and quantitative
determination of in vivo and in vitro human skin collagenase.
J. Biol. Chem. 247: 6679-6685.

Bauer, E. A., Eisen, A. Z. & Jeffrey, J. J. Regulation of
vertebrate collagenase activity in vivo and in vitro. J.
Invest. Dermatol. 59: 50-55.

Brown, S. I. Collagenolytic enzymes in corneal pathology. Israel
J. Med. Sci. 8: 1537-1542.

François, J. & Cambie, E. Collagenase and the collagenase
inhibitors in torpid ulcers of the cornea. Ophthalm. Res. 3:
145-159.

Francois, J., Cambie, E., Feher, J. & van den Eeckhout, E.
Penicillamine in the treatment of alkali-burned corneas.
Ophthalmic Res. 4: 223-236.

Fullmer, H. M., Taylor, R. E. & Guthrie, R. W. Human gingival
collagenase: purification, molecular weight, and inhibitor
studies. J. Dent. Res. 51: 349-355.

Harper, E. & Gross, J. Collagenase, procollagenase and activator
relationships in tadpole tissue cultures. Biochem. Biophys.
Res. Commun. 48: 1147-1152.

Harris, E. D., Faulkner, C. S. & Wood, S. Collagenase in carcinoma cells. Biochem. Biophys. Res. Commun. 48: 1247-1253.

Harris, E. D., Jr. A collagenolytic system produced by primary cultures of rheumatoid nodule tissue. J. Clin. Invest. 51: 2973-2976.

Harris, E. D., Jr. & Farrell, M. E. Resistance to collagenase: a characteristic of collagen fibrils cross-linked by formaldehyde. Biochim. Biophys. Acta 278: 133-141.

Hook, C. W., Bull, F. G., Iwanij, V. & Brown, S. I. Purification of corneal collagenases. Invest. Ophthalmol. 11: 728-734.

Lazarus, G. S. Collagenase and connective tissue metabolism in epidermolysis bullosa. J. Invest. Dermatol. 58: 242-248.

Nagai, Y. & Hori, H. Entrapment of collagen in a polyacrylamide matrix and its application in the purification of animal collagenases. Biochim. Biophys. Acta 263: 564-573.

Nagai, Y. & Hori, H. Vertebrate collagenase: direct extraction from animal skin and human synovial membrane. J. Biochem. 72: 1147-1153.

Robertson, P. B. & Miller, E. J. Cartilage collagen: inability to serve as a substrate for collagenase active against skin and bone collagen. Biochim. Biophys. Acta 289: 247-250.

Robertson, P. B., Taylor, R. E. & Fullmer, H. M. A reproducible quantitative collagenase radiofibril assay. Clin. Chim. Acta 42: 43-45.

Ryan, J. N. & Woessner, J. F., Jr. Oestradiol inhibits collagen breakdown in the involuting rat uterus. Biochem. J. 127: 705-713.

Sakamoto, S., Goldhaber, P. & Glimcher, M. J. A new method for the assay of tissue collagenase. Proc. Soc. Exp. Biol. Med. 139: 1057-1059.

Sakamoto, S., Goldhaber, P. & Glimcher, M. J. Further studies on the nature of the components in serum which inhibit mouse bone collagenase. Calcif. Tissue Res. 10: 280-288.

Sakamoto, S., Goldhaber, P. & Glimcher, M. J. Maintenance of mouse bone collagenase activity in the presence of serum protein by addition of trypsin. Proc. Soc. Exp. Biol. Med. 139: 1038-1041.

Sakamoto, S., Goldhaber, P. & Glimcher, M. J. The further purification and characterization of mouse bone collagenase. Calcif. Tissue Res. 10: 142-151.

Tokoro, Y., Eisen, A. Z. & Jeffrey, J. J. Characterisation of a collagenase from rat skin. Biochim. Biophys. Acta 258: 289-302.

Vaes, G. Multiple steps in the activation of the inactive precursor of bone collagenase by trypsin. FEBS Lett. 28: 198-200.

Vaes, G. The release of collagenase as an inactive proenzyme by bone explants in culture. Biochem. J. 126: 275-289.

1973

Abe, S. & Nagai, Y. Evidence for the presence of a complex of collagenase with α_2-macroglobulin in human rheumatoid synovial fluid: a possible regulatory mechanism of collagenase activity in vivo. J. Biochem. 73: 897-900.

Abe, S., Shinmei, M. & Nagai, Y. Synovial collagenase and joint diseases: the significance of latent collagenase with special reference to rheumatoid arthritis. J. Biochem. 73: 1007-1011.

Berman, M. & Manabe, R. Corneal collagenases: evidence for zinc metallo-enzymes. Ann. Ophthalmol. 5: 1193-1209.

Berman, M. B., Kerza-Kwiatecki, A. P. & Dawson, P. F. Characterization of human corneal collagenase. Exp. Eye Res. 15: 367-373.

Bloomfield, S. E. & Brown, S. I. Treatment of corneal ulcers with collagenase inhibitors. Int. Ophthalmol. Clin. 13: 225-234.

Davison, P. F. & Berman, M. Corneal collagenase. Specific cleavage of types $(\alpha 1)_2 \alpha 2$ and $(\alpha 1)_3$ collagens. Connect. Tissue Res. 2: 57-64.

François, J., Cambie, E. & Feher, J. Collagenase inhibition by penicillamine. Ophthalmologica 166: 222-225.

Francois, J., Cambie, E., Feher, J. & van den Eeckhout, E. Collagenase inhibitors (penicillamine). Ann. Ophthalmol. 5: 391-408.

Fujiwara, K., Sakai, T., Oda, T. & Igarashi, S. The presence of collagenase in kupffer cells of the rat liver. Biochem. Biophys. Res. Commun. 54: 531-537.

Harper, E. & Gross, J. The activator of tadpole procollagenase. Fed. Proc. Fed. Am. Soc. Exp. Biol. 32: 2250 only.

Harper, E. & Toole, B. P. Collagenase and hyaluronidase stimulation by dibutyryl adenosine cyclic 3':5'-monophosphate. J. Biol. Chem. 248: 2625-2626.

Hashimoto, K., Yamanishi, Y., Maeyens, E., Dabbous, M. K. & Kanzaki, T. Collagenolytic activities of squamous-cell carcinoma of skin. Cancer Res. 33: 2790-2801.

Hook, R. M., Hook, C. W. & Brown, S. I. Fibroblast collagenase partial purification and characterization. Invest. Ophthalmol. 12: 771-776.

Lazarus, G. S. Studies on the degradation of collagen by collagenases. In: Lysosomes in Biology and Pathology (J. T. Dingle ed.), vol. 3, pp. 338-360, North-Holland Publishing Co., Amsterdam.

Leibovich, S. J. & Weiss, J. B. Elucidation of the exact sites of cleavage of tropocollagen by rheumatoid synovial collagenase: correlation of cleavage sites with fibril structure. Connect. Tissue Res. 2: 11-19.

McCroskery, P. A., Wood, S. & Harris, E. D. Gelatin: a poor substrate for a mammalian collagenase. Science 182: 70-71.

Nagai, Y. Vertebrate collagenase: further characterization and the significance of its latent form in vivo. Mol. Cell Biochem. 1: 137-145.

Sakamoto, S., Goldhaber, P. & Glimcher, M. J. Mouse bone collagenase: the effect of heparin on the amount of enzyme released in tissue culture and on the activity of the enzyme.

Calcif. Tissue Res. 12: 247-258.

Sakamoto, S., Sakamoto, M., Goldhaber, P. & Glimcher, M. J. Isolation of tissue collagenase from homogenates of embryonic chick bones. Biochem. Biophys. Res. Commun. 53: 1102-1108.

Sugar, A. & Wlatman, S. R. Corneal toxicity of collagenase inhibitors. Invest. Ophthalmol. 12: 779-782.

Woessner, J. F., Jr. Mammalian collagenases. Clin. Orthop. 96: 310-326.

Woessner, J. F., Jr. & Ryan, J. N. Collagenase activity in homogenates of the involuting rat uterus. Biochim. Biophys. Acta 309: 397-405.

Woolley, D. E., Glanville, R. W. & Evanson, J. M. Differences in the physical properties of collagenases isolated from rheumatoid synovium and human skin. Biochem. Biophys. Res. Commun. 51: 729-734.

Woolley, D. E., Glanville, R. W., Lindberg, K. A., Bailey, A. J. & Evanson, J. M. Action of human skin collagenase on cartilage collagen. FEBS Lett. 34: 267-269.

Yamanishi, Y., Maeyens, E., Dabbous, M. K. Ohyama, H. & Hashimoto, K. Collagenolytic activity in malignant melanoma: physicochemical studies. Cancer Res. 33: 2507-2512.

1974

Bloomfield, S. E. & Brown, S. I. Conjunctival collagenase: partial purification. Invest. Ophthalmol. 13: 547-550.

Chesney, C. M., Harper, E. & Colman, R. W. Human platelet collagenase. J. Clin. Invest. 53: 1647-1654.

Colman, R. W., Chesney, C. M. & Harper, E. Collagenase in human platelets. Thromb. Diath. Haemorrh. Suppl. 60: 117-126.

Eeckhout, Y., Otto, K. & Vaes, G. Properties and nature of a lysosomal activator of the proenzyme of bone collagenase. Hoppe-Seyler's Z. Physiol. Chem. 355: 21 only.

Eisen, A. Z., Bauer, E. A., Stricklin, G. P. & Jeffrey, J. J. Vertebrate collagenase. Methods Enzymol. 34: 420-424.

Eisen, A. Z., Nepute, J., Stricklin, G. P., Bauer, E. A. & Jeffrey, J. J. Solid-phase radioimmunoassay of rat skin collagenase. Comparative studies and in vivo measurement of the enzyme. Biochim. Biophys. Acta 348: 442-454.

Fujiwara, K., Sakai, T., Oda, T. & Igarishi, S. Demonstration of collagenase activity in rat liver homogenate. Biochem. Biophys. Res. Commun. 60: 166-171.

Gross, J., Harper, E., Harris, E. D., McCroskery, P. A., Highberger, J. H., Corbett, C. & Kang, A. H. Animal collagenases: specificity of action and structures of the substrate cleavage site. Biochem. Biophys. Res. Commun. 61: 605-612.

Harper, E. & Seifter, S. Studies on the mechanism of action of collagenase: inhibition by cysteine and other chelating agents. Israel J. Chem. 12: 515-528.

Harris, E. D., Jr. & Krane, S. M. Collagenases. N. Engl. J. Med. 291: 557-563, 605-609, 652-661 (three parts).

Harris, E. D., Jr. & McCroskery, P. A. Influence of temperature
and fibril stability on degradation of cartilage collagen by
rheumatoid synovial collagenase. N. Engl. J. Med. 290: 1-6.

Koob, T. J. & Jeffrey, J. J. Hormonal regulation of collagen
degradation in uterus: inhibition of collagenase expression by
progesterone and cyclic AMP. Biochim. Biophys. Acta 354: 61-70.

Koob, T. J., Jeffrey, J. J. & Eisen, A. Z. Regulation of human
skin collagenase activity by hydrocortisone and dexamethasone in
organ culture. Biochem. Biophys. Res. Commun. 61: 1083-1088.

Krane, S. M. Joint erosion in rheumatoid arthritis. Arthritis
Rheum. 17: 306-312.

Michaelson, J. D., Schmidt, J. D., Dresden, M. H. & Duncan, W. C.
Vitamin E treatment of epidermolysis bullosa. Changes in tissue
collagenase levels. Arch Dermatol. 109: 67-69.

Pardo, A. & Pérez-Tamayo, R. The collagenase of carrageenin
granuloma. Connect. Tissue Res. 2: 243-252.

Reddick, M. E., Bauer, E. A. & Eisen, A. Z. Immunocytochemical
localization of collagenase in human skin and fibroblasts in
monolayer culture. J. Invest. Dermatol. 62: 361-366.

Robertson, P. B., Cobb, C. M., Taylor, R. E. & Fullmer, H. M.
Activation of latent collagenase by microbial plaque. J.
Periodont. Res. 9: 81-83.

Ryan, J. N. & Woessner, J. F., Jr. Oestradiol inhibition of
collagenase role in uterine involution. Nature 248: 526-528.

Sakamoto, S., Sakamoto, M., Goldhaber, P. & Glimcher, M. J. The
inhibition of mouse bone collagenase by lysozyme. Calcif.
Tissue Res. 14: 291-299.

Shinkai, H., Fujiwara, N., Matsubayashi, S. & Sano, S. Effect of
thiocyanate on human skin collagenase. J. Dermatol. 1: 145-151.

Simpson, J. W. & Taylor, A. C. Regulation of gingival collagenase:
a possible role for a mast cell factor. Proc. Soc. Exp. Biol.
Med. 145: 42-47.

Usuku, G., Honya, M., Oyamada, I. & Ogata, T. Studies on animal
collagenase. Acta Pathol. Jap. 24: 1-19.

Wahl, L. M., Wahl, S. M., Mergenhagen, S. E. & Martin, G. R.
Collagenase production by endotoxin-activated macrophages.
Proc. Natl. Acad. Sci. U.S.A. 71: 3598-3601.

Werb, Z. & Burleigh, M. C. A specific collagenase from rabbit
fibroblasts in monolayer culture. Biochem. J. 137: 373-385.

Werb, Z. & Reynolds, J. J. Stimulation by endocytosis of
secretion of collagenase and neutral proteinase from rabbit
synovial fibroblasts. J. Exp. Med. 140: 1482-1497.

Werb, Z., Burleigh, M. C., Barrett, A. J. & Starkey, P. M. The
interaction of α_2-macroglobulin with proteinases. Binding and
inhibition of mammalian collagenases and other metal proteinases.
Biochem. J. 139: 359-368.

1975

Abramson, M., Schilling, R. W., Huang, C.-C. & Salome, R. G.
Collagenase activity in epidermoid carcinoma of the oral cavity
and larynx. Ann. Otol. Rhinol. Laryngol. 84: 158-163.

Bauer, E. A., Stricklin, G. P., Jeffrey, J. J. & Eisen, A. Z.
Collagenase production by human skin fibroblasts. Biochem.
Biophys. Res. Commun. 64: 232-240.

Berman, M. Collagenase inhibitors: rationale for their use in
treating corneal ulceration. In: International Ophthalmology
Clinics, vol. 15, no. 4, Ocular Viral Diseases (Pavan-Longston,
D. ed.), pp. 49-66, Little Brown, Boston.

Berman, M. & Dohlman, C. Collagenase inhibitors. Arch.
Ophthalmol. (Paris) 35: 95-108.

Berman, M., Gordon, J., Garcia, L. A. & Gage, J. Corneal
ulceration and the serum antiproteases. II. Complexes of corneal
collagenases and α-macroglobulins. Exp. Eye Res. 20: 231-244.

Birkedal-Hansen, H., Cobb, C. M., Taylor, R. E. & Fullmer, H. M.
Activation of latent bovine gingival collagenase. Arch. Oral
Biol. 20: 681-685.

Birkedal-Hansen, H., Cobb, C. M., Taylor, R. E. & Fullmer, H. M.
Bovine gingival collagenase demonstration and initial
characterization. J. Oral Pathol. 3: 232-238.

Birkedal-Hansen, H., Cobb, C. M., Taylor, R. E. & Fullmer, H. M.
Serum inhibition of gingival collagenase. J. Oral Pathol. 3:
284-290.

Birkedal-Hansen, H., Cobb, C. M., Taylor, R. E. & Fullmer, H. M.
Trypsin activation of latent collagenase from several mammalian
sources. Scand. J. Dent. Res. 83: 302-305.

Brady, A. H. Collagenase in scleroderma. J. Clin. Invest. 56:
1175-1180.

Brown, S. I. Clinical aspects, experimental proof of proteolytic
activities in corneal pathology (without therapy). Arch.
Ophthalmol. (Paris) 35: 91-94.

Brown, S. I. Mooren's ulcer. Histopathology and proteolytic
enzymes of adjacent conjunctiva. Br. J. Ophthalmol. 59: 670-
674.

Halme, J. & Woessner, J. F., Jr. Effect of progesterone on
collagenolytic activity in the involuting rat uterus. J.
Endocrinol. 66: 357-362.

Harris, E. D., Jr., Faulkner, C. S., II. & Brown, F. E.
Collagenolytic systems in rheumatoid arthritis. Clin. Orthop.
110: 303-316.

Harris, E. D., Jr., Reynolds, J. J. & Werb, Z. Cytochalasin B
increases collagenase production by cells in vitro. Nature 257:
243-244.

Huang, C.-C. & Abramson, M. Purification and characterisation of
collagenase from guinea pig skin. Biochim. Biophys. Acta 384:
484-492.

Janoff, A. At least three human neutrophil lysosomal proteases
are capable of degrading joint connective tissues. Ann. N. Y.
Acad. Sci. 256: 402-408.

Jeffrey, J. J., Koob, T. J. & Eisen, A. Z. Hormonal regulation of
mammalian collagenases. In: Dynamics of Connective Tissue
Macromolecules (Burleigh, P. M. C. & Poole, A. R. eds), pp. 147-

158, North-Holland Publishing Co., Amsterdam.

Krane, S. M. Collagenase production by human synovial tissues. Ann. N. Y. Acad. Sci. 256: 289-303.

LeLous, M., Bazin, S. & Delaunay, A. [Activity of specific collagenases in evolving granulation tissues.] C. R. Acad. Sci. D (Paris) 281: 961-963.

McCroskery, P. A., Richards, J. F. & Harris, E. D., Jr. Purification and characterization of a collagenase extracted from rabbit tumours. Biochem. J. 152: 131-142.

Montfort, I. & Pérez-Tamayo, R. Distribution of collagenase in rat uterus during postpartum involution. Connect. Tissue Res. 3: 245-252.

Montfort, I. & Pérez-Tamayo, R. The distribution of collagenase in normal rat tissues. J. Histochem. Cytochem. 23: 910-920.

Nagai, Y., Hori, H., Kawamoto, T. & Komiya, M. A regulation mechanism of collagenase activity in vitro and in vivo. In: Dynamics of Connective Tissue Macromolecules (Burleigh, P. M. C. & Poole, A. R. eds), pp. 171-187, North-Holland Publishing Co., Amsterdam.

Nagy, C. F., Coffey, J. W. & Salvador, R. A. Factors influencing the yield and apparent molecular weight of human collagenase in the medium from organ cultures of skin. Fed. Proc. Fed. Am. Soc. Exp. Biol. 34: 697 only.

Pirie, A., Werb, Z. & Burleigh, M. C. Collagenase and other proteinases in the cornea of the retinol-deficient rat. Br. J. Nutr. 34: 297-309.

Sakamoto, S., Sakamoto, M., Goldhaber, P. & Glimcher, M. J. Collagenase and bone resorption: isolation of collagenase from culture medium containing serum after stimulation of bone resorption by addition of parathyroid hormone extract. Biochem. Biophys. Res. Commun. 63: 172-178.

Sakamoto, S., Sakamoto, M., Goldhaber, P. & Glimcher, M. J. Studies on the interaction between heparin and mouse bone collagenase. Biochim. Biophys. Acta 385: 41-50.

Vaes, G. & Eeckhout, Y. Procollagenase and its activation. In: Dynamics of Connective Tissue Macromolecules (Burleigh, M. C. & Poole, A. R. eds), pp. 129-146, North-Holland Publishing Co., Amsterdam.

Vaes, G. & Eeckhout, Y. The precursor of bone collagenase and its activation. Protides Biol. Fluids Proc. Colloqu. 22: 391-397.

Wahl, L. M., Wahl, S. M., Mergenhagen, S. E. & Martin, G. R. Collagenase production by lymphokine-activated macrophages. Science 187: 261-263.

Werb, Z. & Gordon, S. Secretion of a specific collagenase by stimulated macrophages. J. Exp. Med. 142: 346-360.

Werb, Z. & Reynolds, J. J. Immunochemical studies with a specific antiserum to rabbit fibroblast collagenase. Biochem. J. 151: 655-663.

Werb, Z. & Reynolds, J. J. Purification and properties of a specific collagenase from rabbit synovial fibroblasts. Biochem.

J. 151: 645-653.

Werb, Z. & Reynolds, J. J. Rabbit collagenase. Immunological identity of the enzymes released from cells and tissues in normal and pathological conditions. Biochem. J. 151: 665-669.

Wirl, G. Extraction of collagenase from the 6000xg sediment of uterine and skin tissues of mice. A comparative study. Hoppe-Seyler's Z. Physiol. Chem. 356: 1289-1295.

Woolley, D. E., Glanville, R. W., Crossley, M. J. & Evanson, J. M. Purification of rheumatoid synovial collagenase and its action on soluble and insoluble collagen. Eur. J. Biochem. 54: 611-622.

Woolley, D. E., Lindberg, K. A., Glanville, R. W. & Evanson, J. M. Action of rheumatoid synovial collagenase on cartilage collagen. Ann. Rheum. Dis. 34: suppl., 71-72.

Woolley, D. E., Lindberg, K. A., Glanville, R. W. & Evanson, J. M. Action of rheumatoid synovial collagenase on cartilage collagen. Different susceptibilities of cartilage and tendon collagen to collagenase attack. Eur. J. Biochem. 50: 437-444.

Woolley, D. E., Roberts, D. R. & Evanson, J. M. Inhibition of human collagenase activity by a small molecular weight serum protein. Biochem. Biophys. Res. Commun. 66: 747-754.

1976

Baker, J. R., Baker, P. A., Bassett, E. G. & Myers, D. B. Collagenolytic activity in peritoneal eosinophils from rats. J. Physiol. (Lond.) 263: 242P-243P.

Berman, M. B., Cavanagh, H. D. & Gage, J. Regulation of collagenase activity in the ulcerating cornea by cyclic-AMP. Exp. Eye Res. 22: 209-218.

Birdedal-Hansen, H., Taylor, R. E. & Fullmer, H. M. Rabbit alveolar macrophage collagenase: evidence of polymeric forms. Biochim. Biophys. Acta 420: 428-432.

Birkedal-Hansen, H., Cobb, C. M., Taylor, R. E. & Fullmer, H. M. Activation of fibroblast procollagenase by mast cell proteases. Biochim. Biophys. Acta 438: 273-286.

Birkedal-Hansen, H., Cobb, C. M., Taylor, R. E. & Fullmer, H. M. Fibroblastic origin of bovine gingival collagenase. Arch. Oral Biol. 21: 297-305.

Birkedal-Hansen, H., Cobb, C. M., Taylor, R. E. & Fullmer, H. M. In vivo and in vitro stimulation of collagenase production by rabbit alveolar macrophages. Arch. Oral Biol. 21: 21-25.

Birkedal-Hansen, H., Cobb, C. M., Taylor, R. E. & Fullmer, H. M. Procollagenase from bovine gingiva. Biochim. Biophys. Acta 429: 229-238.

Birkedal-Hansen, H., Cobb, C. M., Taylor, R. E. & Fullmer, H. M. Synthesis and release of procollagenase by cultured fibroblasts. J. Biol. Chem. 251: 3162-3168.

Boxer, P. A. & Leibovich, S. J. Production of collagenase by mouse peritoneal macrophages in vitro. Characterization of sites of cleavage of tropocollagen. Biochim. Biophys. Acta 444: 626-632.

Cruess, R. L. & Hong, K. C.   Effect of estrogen on the
  collagenolytic activity of rat bone.   Calcif. Tissue Res. 20:
  317–320.
Dayer, J.-M., Krane, S. M., Russell, R. G. G. & Robinson, D. R.
  Production of collagenase and prostaglandins by isolated
  adherent rheumatoid synovial cells.   Proc. Natl. Acad. Sci.
  U.S.A. 73: 945–949.
Dowsett, M., Eastman, A. R., Easty, D. M., Powles, T. J. &
  Neville, A. M.   Prostaglandin mediation of collagenase-induced
  bone resorption.   Nature 263: 72–74.
Evanson, J. M.   Preparation of collagenase from mammalian tissue.
  In: Methodology of Connective Tissue Research (Hall, D. A. ed.),
  pp. 175–180, Joynson-Bruvvers, Oxford.
Gordon, S.   Macrophage neutral proteinases and chronic
  inflammation.   Ann. N. Y. Acad. Sci. 278: 176–189.
Gross, J.   Aspects of animal collagenases.   In: Biochemistry of
  Collagen (Ramachandran, G. N. & Reddi, A. H. eds), pp 275–317,
  Plenum Press, New York.
Heard, S. B.   Collagenase studies in bones of guinea pigs.   J.
  Oral Pathol. 5: 17–32.
Horwitz, A. L. & Crystal, R. G.   Collagenase from rabbit pulmonary
  alveolar macrophages.   Biochem. Biophys. Res. Commun. 69: 296–
  303.
Horwitz, A. L., Kelman, J. A. & Crystal, R. G.   Activation of
  alveolar macrophage collagenase by a neutral protease secreted
  by the same cell.   Nature 264: 772–774.
Kuettner, K. E., Hiti, J., Eisenstein, R. & Harper, E.   Collagenase
  inhibition by cationic proteins derived from cartilage and
  aorta.   Biochem. Biophys. Res. Commun. 72: 40–46.
Lichtenstein, J. R., Bauer, E. A. & Uitto, J.   Cleavage of human
  fibroblast type 1 procollagen by mammalian collagenase:
  demonstration of amino- and carboxy-terminal extension peptides.
  Biochem. Biophys. Res. Commun. 73: 665–672.
Miller, E. J., Harris, E. D., Jr., Chung, E., Finch, J. E., Jr.,
  McCroskery, P. A. & Butler, W. T.   Cleavage of type II and III
  collagens with mammalian collagenase: site of cleavage and
  primary structure at the NH_2-terminal portion of the smaller
  fragment released from both collagens.   Biochemistry 15: 787–
  792.
Nagai, Y., Masui, Y. & Sakakibara, S.   Substrate specificity of
  vertebrate collagenase.   Biochim. Biophys. Acta 445: 521–524.
Puzas, J. E. & Brand, J. S.   Collagenolytic activity from isolated
  bone cells.   Biochim. Biophys. Acta 429: 964–974.
Reed, F. B.   Collagenase assays.   In: Methodology of Connective
  Tissue Research (Hall, D. A. ed.), pp. 275–282, Joynson-Bruvvers,
  Oxford.
Robertson, P. B. & Simpson, J.   Collagenase: current concepts and
  relevance to periodontal disease.   J. Periodontol. 47: 29–33.
Seltzer, J. L., Welgus, H. G., Jeffrey, J. J. & Eisen, A. Z.   The
  function of Ca^{2+} in the action of mammalian collagenases.   Arch.

Biochem. Biophys. 173: 355-361.

Terato, K., Nagai, Y., Kawanishi, K. & Yamamoto, S. A rapid assay method of collagenase activity using ^{14}C-labelled soluble collagen as substrate. Biochim. Biophys. Acta 445: 753-762.

Weeks, J. G., Halme, J. & Woessner, J. F., Jr. Extraction of collagenase from the involuting rat uterus. Biochim. Biophys. Acta 445: 205-214.

Weiss, J. B. Enzymatic degradation of collagen. In: International Review of Connective Tissue Research (Hall, D. A. & Jackson, D. S. eds), vol. 7, pp. 101-157, Academic Press, New York.

Woolley, D. E., Crossley, M. J. & Evanson, J. M. Antibody to rheumatoid synovial collagenase. Its characterization, specificity and immunological cross-reactivity. Eur. J. Biochem. 69: 421-428.

Woolley, D. E., Roberts, D. R. & Evanson, J. M. Small molecular weight β_1 serum protein which specifically inhibits human collagenases. Nature 261: 325-327.

Woolley, D. E., Tucker, J. S., Green, G. & Evanson, J. M. A neutral collagenase from human gastric mucosa. Biochem. J. 153: 119-126.

1977

Abramson, M. & Huang, C.-C. Localization of collagenase in human middle ear cholesteatoma. Laryngoscope 87: 771-791.

Anonymous. Enzyme secretion in the pathogenesis of rheumatoid arthritis (editorial). Lancet 1: 1190-1191.

Bauer, E. A. Cell culture density as a modulator of collagenase expression in normal human fibroblast cultures. Exp. Cell Res. 107: 269-276.

Bauer, E. A. Recessive dystophic epidermolysis bullosa: evidence for an altered collagenase in fibroblast cultures. Proc. Natl. Acad. Sci. U.S.A. 74: 4646-4650.

Bauer, E. A., Gedde-Dahl, T. & Eisen, A. Z. The role of human skin collagenase in epidermolysis bullosa. J. Invest. Dermatol. 68: 119-124.

Bauer, E. A., Gordon, J. M., Reddick, M. E. & Eisen, A. Z. Quantitation and immunocytochemical localization of human skin collagenase in basal cell carcinoma. J. Invest. Dermatol. 69: 363-367.

Berman, M., Leary, R. & Gage, J. Latent collagenase in the ulcerating rabbit cornea. Exp. Eye Res. 25: 435-445.

Brown, R. A., Hukins, D. W., Weiss, J. B. & Twose, T. M. Do mammalian collagenase and DNA restriction endonucleases share a similar mechanism for cleavage site recognition? Biochem. Biophys. Res. Commun. 74: 1102-1108.

Burleigh, M. C., Werb, Z. & Reynolds, J. J. Evidence that species specificity and rate of collagen degradation are properties of collagen, not collagenase. Biochim. Biophys. Acta 494: 198-208.

Cartwright, E., Murphy, G., Sellers, A. & Reynolds, J. J. Collagenase activity from cultured non-gravid rabbit uterus. Biochem. Soc. Trans. 5: 229-231.

Cathcart, E. S. Rheumatoid arthritis – a "collagenase disease" (editorial). N. Engl. J. Med. 296: 1058-1059.

Dabbous, M. K., Roberts, A. N. & Brinkley, B. Collagenase and neutral protease activities in cultures of rabbit VX-2 carcinoma. Cancer Res. 37: 3537-3544.

Dabbous, M. K., Sobhy, C., Roberts, A. N. & Brinkley, B. Changes in the collagenolytic activity released by primary VX-2 carcinoma cultures as a function of tumor growth. Mol. Cell. Biochem. 16: 37-42.

Dayer, J.-M., Russell, R. G. G. & Krane, S. M. Collagenase production by rheumatoid synovial cells: stimulation by a human lymphocyte factor. Science 195: 181-183.

Eekhout, Y. & Vaes, G. Further studies on the activation of procollagenase, the latent precursor of bone collagenase. Effects of lysosomal cathepsin B, plasmin and kallikrein, and spontaneous activation. Biochem. J. 166: 21-31.

Ehrlich, M. G., Mankin, H. J., Jones, H., Wright, R., Crispen, C. & Vigliani, G. Collagenase and collagenase inhibitors in osteoarthritic and normal human cartilage. J. Clin. Invest. 59: 226-233.

Fiedler-Nagy, C., Coffey, J. W. & Salvador, R. A. Factors influencing the apparent molecular weight of collagenase produced by human-skin explants. Eur. J. Biochem. 76: 291-297.

Francois, J. Collagenase and collagenase inhibitors. Trans. Am. Ophthalmol. Soc. 75: 285-315.

Gillet, C., Eeckhout, Y. & Vaes, G. Purification of procollagenase and collagenase by affinity chromotography on Sepharose-collagen. FEBS Lett. 74: 126-128.

Harris, E. D., Jr. & Cartwright, E. C. Mammalian collagenases. In: Proteinases in Mammalian Cells and Tissues (Barrett, A. J. ed.), pp. 249-283, North-Holland Publishing Co., Amsterdam.

Kuettner, K. E., Soble, L., Croxen, R. L., Marczynska, B., Hiti, J. & Harper, E. Tumor cell collagenase and its inhibition by a cartilage-derived protease inhibitor. Science 196: 653-654.

Masui, Y., Takemoto, T., Sakakibara, S., Hori, H. & Nagai, Y. Synthetic substrates for vertebrate collagenase. Biochem. Med. 17: 215-221.

Menzel, J. & Steffen, C. Detection of collagenase activity in rheumatoid arthritis synovial fluids using soluble carbon-14-labeled collagen. Z. Rheumatol. 36: 364-377.

Murphy, G., Cartwright, E. C., Sellers, A. & Reynolds, J. J. The detection and chracterisation of collagenase inhibitors from rabbit tissues in culture. Biochim. Biophys. Acta 483: 493-498.

Newsome, D. A. & Gross, J. Prevention by medroxyprogesterone of perforation in the alkali-burned rabbit cornea: inhibition of collagenolytic activity. Invest. Ophthalmol. Vis. Sci. 16: 21-31.

Ohyama, H. & Hashimoto, K. Collagenase of human skin basal cell epithelioma. J. Biochem. 82: 175-183.

Rose, G. G. & Robertson, P. B. Collagenolysis by human gingival fibroblast cell lines. J. Dent. Res. 56: 416-424.

Sedowofia, K. A. & Weiss, J. B. Active and latent collagenases from uncultured rheumatoid synovial membranes. Ann. Rheum. Dis. 36: 285 only.

Sellers, A. & Reynolds, J. J. Identification and partial characterization of an inhibitor of collagenase from rabbit bone. Biochem. J. 167: 353-360.

Sellers, A., Cartwright, E., Murphy, G. & Reynolds, J. J. An inhibitor of mammalian collagenase from foetal rabbit bone in culture. Biochem. Soc. Trans. 5: 227-228.

Sellers, A., Cartwright, E., Murphy, G. & Reynolds, J. J. Evidence that latent collagenases are enzyme-inhibitor complexes. Biochem. J. 163: 303-307.

Seltzer, J. L., Jeffrey, J. J. & Eisen, A. Z. Evidence for mammalian collagenases as zinc ion metalloenzymes. Biochim. Biophys. Acta 485: 179-187.

Shinkai, H. & Nagai, Y. A latent collagenase from embryonic human skin explants. J. Biochem. 81: 1261-1268.

Shinkai, H., Kawamoto, T., Hori, H. & Nagai, Y. A complex of collagenase with low molecular weight inhibitors in the culture medium of embryonic chick skin explants. J. Biochem. 81: 261-263.

Sorgente, N., Brownell, A. G. & Slavkin, H. C. Basal lamina degradation: the identification of mammalian-like collagenase activity in mesenchymal-derived matrix vesicles. Biochem. Biophys. Res. Commun. 74: 448-454.

Steven, F. S. & Itzhaki, S. Evidence for a latent form of collagenase extracted from rabbit tumour cells. Biochim. Biophys. Acta 496: 241-246.

Stricklin, G. P., Bauer, E. A., Jeffrey, J. J. & Eisen, A. Z. Human skin collagenase: isolation of precursor and active forms from both fibroblast and organ cultures. Biochemistry 16: 1607-1615.

Talbot, M. W. & Strausser, H. R. Increase in serum IgE levels of ovalbumin-sensitized cats and the detection of elastase and collagenase activities in secretions of sensitized feline alveolar macrophages challenged in vitro. Int. Arch. Allergy Appl. Immunol. 54: 198-204.

Wahl, L. M. Hormonal regulation of macrophage collagenase activity. Biochem. Biophys. Res. Commun. 74: 838-845.

Wahl, L. M., Olsen, C. E., Sandberg, A. L. & Mergenhagen, S. E. Prostaglandin regulation of macrophage collagenase production. Proc. Natl. Acad. Sci. U.S.A. 74: 4955-4958.

Werb, Z., Mainardi, C. L., Vater, C. A. & Harris, E. D., Jr. Endogenous activation of latent collagenase by rheumatoid synovial cells. N. Engl. J. Med. 296: 1017-1023.

Wirl, G. Croton oil induced collagenolytic activity in the mouse skin. In: Intracellular Protein Catabolism (Turk, V. & Marks, N. eds), pp. 154-162, Plenum Press, New York.

Wirl, G. Extractable collagenase and carcinogenesis of the mouse skin. Connect. Tissue Res. 5: 171-178.

Woessner, J. F., Jr. A latent form of collagenase in the involuting rat uterus and its activation by a serine proteinase. Biochem. J. 161: 535-542.

Woolley, D. E. & Evanson, J. M. Collagenase and its natural inhibitors in relation to the rheumatoid joint. Connect. Tissue Res. 5: 31-35.

Woolley, D. E. & Evanson, J. M. Effect of cartilage proteoglycans on human collagenase activities. Biochim. Biophys. Acta 497: 144-150.

Woolley, D. E., Crossley, M. J. & Evanson, J. M. Collagenase at sites of cartilage erosion in the rheumatoid joint. Arthritis Rheum. 20: 1231-1239.

Yankeelov, J. A., Jr., Wacker, W. B., & Schweri, M. M. Radial diffusion assay of tissue collagenase and its application in evaluation of collagenase inhibitors. Biochim. Biophys. Acta 482: 159-172.

1978

Bauer, E. A. & Eisen, A. Z. Recessive dystrophic epidermolysis bullosa. Evidence for increased collagenase as a genetic characteristic in cell culture. J. Exp. Med. 148: 1378-1387.

Biswas, C., Moran, W. P., Bloch, K. J. & Gross, J. Collagenolytic activity of rabbit V_2-carcinoma growing at multiple sites. Biochem. Biophys. Res. Commun. 80: 33-38.

Brinckerhoff, C. E. & Harris, E. D., Jr. Collagenase production by cultures containing multinucleated cells derived from synovial fibroblasts. Arthritis Rheum. 21: 745-753.

Ehrlich, M. G., Houle, P. A., Vigliani, G. & Mankin, H. J. Correlation between articular cartilage collagenase activity and osteoarthritis. Arthritis Rheum. 21: 761-766.

Gantz, B. J., Abramson, M. & Huang, C. C. Localization of collagenase in chronically inflamed guinea-pig temporal bone. Otolaryngology 86: 236-248.

Harris, E. D., Jr. Role of collagenases in joint distruction. In: Joints and Synovial Fluid (Sokoloff, L. ed.), vol. 1, pp. 243-272, Academic Press, New York.

Harris, E. D., Jr., Vater, C. A., Mainardi, C. L. & Werb, Z. Cellular control of collagen breakdown in rheumatoid arthritis. Agents Actions 8: 36-42.

Hiti-Harper, J., Wohl, H. & Harper, E. Platelet factor 4: an inhibitor of collagenase. Science 199: 991-992.

Horton, J. E., Wezeman, F. H. & Kuettner, K. E. Inhibition of bone resorption in vitro by a cartilage-derived anticollagenase factor. Science 199: 1342-1345.

Hu, C.-L., Crombie, G. & Franzblau, C. New assay for collagenolytic activity. Anal. Biochem. 88: 638-643.

Johnson-Muller, B. & Gross, J. Regulation of corneal production: epithelial-stromal cell interactions. Proc. Natl. Acad. Sci. U.S.A. 75: 4417-4421.

Menzel, E. J. & Smolen, J. S. [Degradation of Clq, the first subcomponent of the complement sequence, by synovial collagenase from patients with rheumatoid arthritis.] Wien. Klin. Wochschr. 90: 727–729.

Morales, T. I., Woessner, J. F., Jr, Howell, D. S., Marsh, J. M. & LeMaire, W. J. A microassay for the direct demonstration of collagenolytic activity in Graafian follicles of the rat. Biochim. Biophys. Acta 524: 428–434.

Nagai, Y., Shinkai, H. & Ninomiya, Y. The release of collagenase inhibitors from procollagen additional peptides by pepsin treatment. Proc. Jpn. Acad. B 54: 140–144.

Nolan, J. C., Ridge, S., Oronsky, A. L., Slakey, L. L. & Kerwar, S. S. Synthesis of a collagenase inhibitor by smooth muscle cells in culture. Biochem. Biophys. Res. Commun. 83: 1183–1190.

Pérez-Tamayo, R. Pathology of collagen degradation: review. Am. J. Pathol. 92: 509–566.

Peltonen, L. Collagenase in synovial fluid. Scand. J. Rheumatol. 7: 49–54.

Pettigrew, D. W., Ho, G. H., Sodek, J., Brunette, D. M. & Wang, H. M. Effect of oxygen tension and indomethacin on production of collagenase and neutral proteinase enzymes and their latent forms by porcine gingival explants in culture. Arch. Oral Biol. 23: 767–777.

Roughley, P. J., Murphy, G. & Barrett, A. J. Proteinase inhibitors of bovine nasal cartilage. Biochem. J. 169: 721–724.

Sakamoto, S., Sakamoto, M., Goldhaber, P. & Glimcher, M. J. Mouse bone collagenase. Purification of the enzyme by heparin-substituted Sepharose 4B affinity chromatography and preparation of specific antibody to the enzyme. Arch. Biochem. Biophys. 188: 438–449.

Sakamoto, S., Sakamoto, M., Matsumoto, A., Goldhaber, P. & Glimcher, M. J. Latent collagenase from the culture medium of embryonic chick bones. FEBS Lett. 88: 53–58.

Seltzer, J. L., Eisen, A. Z., Jeffrey, J. J. & Feder, J. A component of normal human serum which enhances the activity of vertebrate collagenases. Biochem. Biophys. Res. Commun. 80: 637–645.

Stricklin, G. P., Eisen, A. Z., Bauer, E. A. & Jeffrey, J. J. Human skin fibroblast collagenase: chemical properties of precursor and active forms. Biochemistry 17: 2331–2337.

Tane, N., Hashimoto, K., Kanzaki, T. & Ohyama, H. Collagenolytic activities of cultured human malignant melanoma cells. J. Biochem. 84: 1171–1176.

Uitto, V.-J. & Raeste, A. M. Activation of latent collagenase of human leukocytes and gingival fluid bacterial plaque. J. Dent. Res. 57: 844–851.

Uitto, V.-J., Turto, H. & Saxén, L. Extraction of collagenase from human gingiva. J. Periodont. Res. 13: 207–214.

Vaes, G., Eeckhout, Y., Lenaers-Claeys, G., Francois-Gillet, C. & Druetz, J.-E. The simultaneous release by bone explants in culture and the parallel activation of procollagenase and of a

latent neutral proteinase that degrades cartilage proteoglycans
and denatured collagen. Biochem. J. 172: 261–274.

Vater, C. A., Mainardi, C. L. & Harris, E. D., Jr. Activation in
vitro of rheumatoid synovial collagenase from cell cultures. J.
Clin. Invest. 62: 987–992.

Vater, C. A., Mainardi, C. L. & Harris, E. D., Jr. Binding of
latent rheumatoid synovial collagenase to collagen fibrils.
Biochim. Biophys. Acta 539: 238–247.

Werb, Z. Biochemical actions of glucocorticoids on macrophages in
culture. Specific inhibition of elastase, collagenase, and
plasminogen activator secretion and effects on other metabolic
functions. J. Exp. Med. 147: 1695–1712.

Werb, Z. & Aggeler, J. Proteases induce secretion of collagenase
and plasminogen activator by fibroblasts. Proc. Natl. Acad.
Sci. U.S.A. 75: 1839–1843.

Woolley, D. E., Akroyd, C., Evanson, J. M., Soames, J. V. &
Davies, R. M. Characterization and serum inhibition of neutral
collagenase from cultured dog gingival tissue. Biochim.
Biophys. Acta 522: 205–217.

Woolley, D. E., Glanville, R. W., Roberts, D. R. & Evanson, J. M.
Purification, characterization and inhibition of human skin
collagenase. Biochem. J. 169: 265–276.

Woolley, D. E., Harris, E. D., Jr., Mainardi, C. L. & Brinckerhoff,
C. E. Collagenase immunolocalization in cultures of rheumatoid
synovial cells. Science 200: 773–775.

1979

Bauer, E. A., Uitto, J., Walters, R. C. & Eisen, A. Z. Enhanced
collagenase production by fibroblasts derived from human basel
cell carcinomas. Cancer Res. 39: 4594–4599.

Biswas, C. & Dayer, J.-M. Stimulation of collagenase production
by collagen in mammalian cell cultures. Cell 18: 1035–1041.

Cawston, T. E. & Barrett, A. J. A rapid and reproducible assay
for collagenase using [1-^{14}C]acetylated collagen. Anal.
Biochem. 99: 340–345.

Cawston, T. E. & Tyler, J. A. Purification of pig synovial
collagenase to high specific activity. Biochem. J. 183: 647–
656.

Huang, C.-C., Wu, C.-H. & Abramson, M. Collagenase activity in
cultures of rat prostate carcinoma. Biochim. Biophys. Acta 570:
149–156.

Ishikawa, T. & Nimni, M. E. A modified collagenase assay method
based on the use of p-dioxane. Anal. Biochem. 92: 136–143.

Kishi, J.-I., Iijima, K.-I, & Hayakawa, T. Dental pulp
collagenase: initial demonstration and characterization.
Biochem. Biophys. Res. Commun. 86: 27–31.

Lenaers-Claeys, G. & Vaes, G. Collagenase, procollagenase and
bone resorption. Effects of heparin, parathyroid hormone and
calcitonin. Biochim. Biophys. Acta 584: 375–388.

Liotta, L. A., Abe, S., Robey, P. G. & Martin, G. R. Preferential
digestion of basement membrane collagen by an enzyme derived

from a metastatic murine tumor. Proc. Natl. Acad. Sci. USA 76:
2268-2272.

Sellers, A., Murphy, G., Meikle, M. C. & Reynolds, J. J. Rabbit
bone collagenase inhibitor blocks the activity of other neutral
metalloproteases. Biochem. Biophys. Res. Commun. 87: 581-587.

Valle, K.-J. & Bauer, E. A. Biosynthesis of collagenase by human
skin fibroblasts in monolayer culture. J. Biol. Chem. 254:
10115-10122.

Vater, C. A., Mainardi, C. L. & Harris, E. D., Jr. Inhibitor of
human collagenase from cultures of human tendon. J. Biol. Chem.
254: 3045-3053.

Welgus, H. G., Stricklin, G. P., Eisen, A. Z., Bauer, E. A.,
Cooney, R. V. & Jeffrey, J. J. A specific inhibitor of
vertebrate collagenase produced by human skin fibroblasts. J.
Biol. Chem. 254: 1938-1943.

Woessner, J. F., Jr. Separation of collagenase and a metal-
dependent endopepidase of rat uterus that hydrolyzes a
heptapeptide related to collagen. Biochim. Biophys. Acta 571:
313-320.

                              1980

Liotta, L. A., Tryggvason, K., Garbisa, S., Hart, I., Foltz, C. M.
& Shafie, S. Metastatic potential correlates with enzymatic
degradation of basement membrane collagen. Nature 284: 67-68.

Entry 4.02

NEUTROPHIL COLLAGENASE

Summary

EC Number: 3.4.24.7 (for all vertebrate collagenases).

Earlier names: Granulocyte collagenase.

Distribution: Specific (primary) granules of human neutrophilic polymorphonuclear leukocytes, initially in latent form. Also reported from granules of rabbit polymorphonuclear leukocytes [1], but this has been questioned [2].

Source: Granules of human blood leukocytes.

Action: Catalyzes a specific cleavage of all three chains in the helical region of the tropocollagen molecule at the same point as most other specific collagenases (see #4.01). Little action on insoluble collagen.

Requirements: pH optimum about 8.0-8.5. Stimulated by Ca^{2+} (5mM). Thiol-blocking reagents (e.g. 4-aminophenylmercuric acetate) and trypsin activate the latent form.

Substrate, usual: Collagen in solution or as reconstituted fibrils, usually radiolabelled.

Inhibitors: EDTA, 1,10-phenanthroline and other chelating reagents, including thiol compounds. α_2-Macroglobulin (slowly). Not β_1-anticollagenase.

Molecular properties: Mol.wt. 61,000.

Comment: The enzyme is distinguished from some other collagenases by its higher molecular weight. Activity against collagen may be more restricted (e.g. less active against type III) [3]. Also antigenically distinct from tissue collagenases [4].

   There has been confusion between this enzyme and a serine proteinase from the azurophil granules that was described by Ohlsson and co-workers (see #1.38).

References
   [1] Robertson et al. Science 177: 64-65, 1972.
   [2] Werb & Reynolds Biochem. J. 151: 665-669, 1975.
   [3] Horwitz et al. Proc. Natl. Acad. Sci. U.S.A. 74: 897-901, 1977.
   [4] Woolley et al. Eur. J. Biochem. 69: 421-428, 1976.

Bibliography

1968

Lazarus, G. S., Brown, R. S., Daniels, J. R. & Fullmer, H. M. Human granulocyte collagenase. Science 159: 1483-1485.

Lazarus, G. S., Daniels, J. R., Brown, R. S., Bladon, H. A. & Fullmer, H. M. Degradation of collagen by a human granulocyte collagenolytic system. J. Clin. Invest. 47: 2622-2629.

1969

Daniels, J. R., Lian, J. & Lazarus, G. S. Granulocyte collagenase preparation and kinetic studies. Clin. Res. 17: 154 only.

1972

Kruze, D. & Wojtecka, E. Activation of leucocyte collagenase proenzyme by rheumatoid synovial fluid. Biochim. Biophys. Acta 285: 436-446.

Lazarus, G. S., Daniels, J. R., Lian, J. & Burleigh, M. C. Role of granulocyte collagenase in collagen degradation. Am. J. Pathol. 68: 565-576.

Robertson, P. B., Ryel, R. B., Fullman, H. M., Taylor, R. E. & Shyu, K. W. Polymorphonuclear leukocyte collagenase. Localization in granules subjected to zonal centrifugation. J. Oral Pathol. 1: 103-107.

Robertson, P. B., Ryel, R. B., Taylor, R. E., Shyu, K. W. & Fullmer, H. M. Collagenase: localisation in polymorphonuclear leukocyte granules in the rabbit. Science 177: 64-65.

1973

Lazarus, G. S. Studies on the degradation of collagen by collagenases. In: Lysosomes in Biology and Pathology (J. T. Dingle ed.), vol. 3, pp. 338-360, North-Holland Publishing Co., Amsterdam.

Oronsky, A. L., Perper, R. J. & Schroder, H. C. Phagocytic release and activation of human leukocyte procollagenase. Nature 246: 417-419.

1974

Merryman, P. F. Isolation and purification of collagenase from leukemic granulocytes. Diss. Abst. Int. B 35: 1546 only.

Sopata, I., Wojtecka-Łukasik, E. & Dancewicz, A. M. Solubilization of collagen fibrils by human leucocyte collagenase activated by rheumatoid synovial fluid. Acta Biochim. Pol. 21: 283-289.

Werb, Z., Burleigh, M. C., Barrett, A. J. & Starkey, P. M. The interaction of α_2-macroglobulin with proteinases. Binding and inhibition of mammalian collagenases and other metal proteinases. Biochem. J. 139: 359-368.

Wojtecka-Lukasik, E. & Dancewicz, A. M. Inhibition of human leucocyte collagenase by some drugs used in the therapy of rheumatic diseases. Biochem. Pharmacol. 23: 2077-2081.

1975

Oronsky, A. L. & Perper, R. J.   Connective tissue-degrading
enzymes of human leukocytes.   Ann. N. Y. Acad. Sci. 256: 233-
253.
Wize, J., Abgarowicz, T., Wojtecka-Łukasik, E., Księżny, S. &
Dancewicz, A. M.   Activation of human leucocyte procollagenase
by rheumatoid synovial tissue culture medium.   Ann. Rheum. Dis.
34: 520-523.
Wize, J., Sopata, I., Wojtecka-Łukasik, E., Księżny, S. &
Dancewicz, A. M.   Isolation, purification and properties of a
factor from rheumtoid synovial fluid activating the latent forms
of collagenolytic enzymes.   Acta Biochim. Pol. 22: 239-250.

1976

Bassett, E. G., Baker, J. R., Baker, P. A. & Myers, D. B.
Comparison of collagenase activity in eosinophil and neutrophil
fractions from rat peritoneal exudates.   Aust. J. Exp. Biol.
Med. Sci. 54: 459-465.
Williams, H. R., Lin, T.-Y. & Perper, R. J.   Activation of latent
collagenase from human leukocytes.   Arthritis Rheum. 19: 829
only.
Woolley, D. E., Crossley, M. J. & Evanson, J. M.   Antibody to
rheumatoid synovial collagenase.   Its characterization,
specificity and immunological cross-reactivity.   Eur. J.
Biochem. 69: 421-428.

1977

Horwitz, A. L., Hance, A. J. & Crystal, R. G.   Granulocyte
collagenase - selective digestion of type-1 relative to type-3
collagen.   Proc. Natl. Acad. Sci. U.S.A. 74: 897-901.
Murphy, G., Reynolds, J. J., Bretz, U. & Baggiolini, M.
Collagenase is a component of the specific granules of human
neutrophil leucocytes.   Biochem. J. 162: 195-197.
Turto, H., Lindy, S., Uitto, V.-J., Wigelius, O. & Uitto, J.
Human leukocyte collagenase: characterization of enzyme kinetics
by a new method.   Anal. Biochem. 83: 557-569.

1978

Baggiolini, M., Bretz, U. & Dewald, B.   Subcellular localization
of granulocyte enzymes.   In: Neutral Proteases of Human
Polymorphonuclear Leukocytes (Havemann, K. & Janoff, A. eds),
pp. 3-17, Urban & Schwartzenberg, Baltimore.
Dancewicz, A. M., Wize, J., Sopata, I., Wojtecka-Łukasik, E. &
Księżny, S.   Specific and nonspecific activation of latent
collagenolytic proteases of human polymorphonuclear leukocytes.
In: Neutral Proteases of Human Polymorphonuclear Leukocytes
(Havemann, K. & Janoff, A. eds), pp. 373-383, Urban &
Schwartzenberg, Baltimore.
Fletcher, D. S., Williams, H. R. & Lin, T.-Y.   Effects of human
polymorphonuclear leukocyte collagenase on sub-component C1q of
the first component of human complement.   Biochim. Biophys. Acta

540: 270-277.

Kaczanowska, J., Sopata, I., Wojtecka-Lukasik, E. & Ksiezny, S. [Activation of latent forms of collagenolytic enzymes in human leukocytes by neutral protease of these cells.] Reumatologia 16: 101-102.

Kobayashi, S. & Nagai, Y. Human leucocyte neutral proteases, with special reference to collagen metabolism. J. Biochem. 84: 559-567.

Kruze, D., Salgam, P., Cohen, G., Fehr, K. & Böni, A. Purification and some properties of collagenase proenzyme activator from rheumatoid synovial fluid. Z. Rheumatol. 37: 355-365.

Uitto, V.-J. & Raeste, A. M. Activation of latent collagenase of human leukocytes and gingival fluid bacterial plaque. J. Dent. Res. 57: 844-851.

Entry 4.03

NEUTROPHIL GELATINASE

Summary

EC Number: (3.4.24.-)

Earlier names: -

Distribution: Found in latent form in C-particles of human neutrophil leukocytes.

Source: Human leukocyte granules.

Action: Degrades gelatin. No action on collagen or Pz-Pro-Leu-Gly-Pro-D-Arg.

Requirements: Optimal pH 8.0-8.5. Activated by trypsin and 4-aminophenylmercuric acetate from the latent form.

Substrate, usual: Gelatin.

Inhibitors: EDTA, 1,10-phenanthroline. α_2-Macroglobulin (slow reaction).

Molecular properties: Mol.wt. in the region of 150,000.

Comment: This enzyme is distinct from the tissue gelatinase, and is quite unrelated to the "gelatinase", pepsin B (#3.02).

Bibliography

1972

Harris, E. D., Jr. & Krane, S. M.   An endopeptidase from rheumatoid synovial tissue culture.   Biochim. Biophys. Acta 258: 566-576.

1974

Sopata, I. & Dancewicz, A. M.   Presence of a gelatin-specific proteinase and its latent form in human leucocytes.   Biochim. Biophys. Acta 370: 510-523.

1975

Galdston, M. & Carter, C.   Elastase and neutral, metal-dependent enzyme activity of human leukocyte lysosomal granules.   Am. Rev. Respir. Dis. 111: 873-877.

1978

Baggiolini, M., Bretz, U. & Dewald, B.   Subcellular localization of granulocyte enzymes. In: Neutral Proteases of Human Polymorphonuclear Leukocytes (Havemann, K. & Janoff, A. eds), pp. 3-17, Urban & Schwartzenberg, Baltimore.

Dancewicz, A. M., Wize, J., Sopata, I., Wojtecka-Łukasik, E. & Księżny, S. Specific and nonspecific activation of latent collagenolytic proteases of human polymorphonuclear leukocytes. In: Neutral Proteases of Human Polymorphonuclear Leukocytes (Havemann, K. & Janoff, A. eds), pp. 373-383, Urban & Schwartzenberg, Baltimore.

Fletcher, D. S., Williams, H. R. & Lin, T.-Y. Effects of human polymorphonuclear leukocyte collagenase on sub-component Clq of the first component of human complement. Biochim. Biophys. Acta 540: 270-277.

Kobayashi, S. & Nagai, Y. Human leucocyte neutral proteases, with special reference to collagen metabolism. J. Biochem. 84: 559-567.

1979

Sopata, I. & Wize, J. A latent gelatin specific proteinase of human leucocytes and its activation. Biochim. Biophys. Acta 571: 305-312.

Entry 4.04

PROCOLLAGEN N-PEPTIDASE

Summary

EC Number: (3.4.24.-)

Earlier names: Procollagen peptidase.

Distribution: Detected in bone, tendon, etc. Found extracellularly in tissues and tissue culture.

Source: Bovine tendon.

Action: Cleaves N-terminal extension peptides from procollagen in the formation of tropocollagen, which then assembles into fibrils. No general proteolytic activity has been detected.

Requirements: Optimum about pH 7.4. Ca^{2+} (less active with Mg^{2+} or Mn^{2+}).

Substrate, usual: Procollagen (type I).

Inhibitors: EDTA, 1,10-phenanthroline. (Not Dip-F or soybean trypsin inhibitor).

Molecular properties: Apparently glycoprotein. Mol.wt. 260,000 (chick), but lower estimates also exist.

Comment: Heritable deficiency of activity is associated with dermatosparaxis of cattle and human Type VII Ehlers-Danlos syndrome. There are indications that the activity on type I (and probably II) procollagen is due to a different enzyme from that on type III.

Bibliography

1971

Lapière, C., Lenaers, A. & Kohn, L. D. Procollagen peptides: an enzyme excising the coordination peptides of procollagen. Proc. Natl. Acad. Sci. U.S.A. 68: 3054-3058.
Lenaers, A., Ansay, M., Nusgens, B. V. & Lapiere, C. M. Collagen made of extended α-chains, procollagen, in genetically-defective dermatosparaxic calves. Eur. J. Biochem. 23: 533-543.

1972

Bornstein, P., Ehrlich, H. P. & Wyke, A. W. Procollagen: conversion of the precursor to collagen by a neutral protease.

Science 175: 544-546.

1974

Kohn, L. D., Isersky, C., Zupnik, J., Lenaers, A., Lee, G., & Lapière, C. M. Calf tendon procollagen peptidase: its purification and endopeptidase mode of action. Proc. Natl. Acad. Sci. U.S.A. 71: 40-44.
Lapière, C. M. & Pierard, G. Skin procollagen peptidase in normal and pathological conditions. J. Invest. Dermatol. 62: 582-586.

1975

Bornstein, P., Davidson, J. M. & Monson, J. M. The biosynthetic conversion of procollagen to collagen by limited proteolysis. In: Proteases in Biological Control (Reich, E., Rifkin, D. B. & Shaw, E. eds), pp. 579-590, Cold Spring Harbor Laboratory, New York.
Goldberg, B., Taubman, M. B. & Radin, A. Procollagen peptidase: its mode of action on the native substrate. Cell 4: 45-50.

1976

Uitto, J. & Lichtenstein, J. R. Removal of amino-terminal and carboxy-terminal extension peptides from procollagen during synthesis of chick embryo tendon collagen. Biochem. Biophys. Res. Commun. 71: 60-67.

1978

Hörlein, D., Fietzek, P. P. & Kühn, K. Pro-Gln: the procollagen cleavage site in the $\alpha_1(I)$ chain of dermatosparactic calf skin collagen. FEBS Lett. 89: 279-282.
Tuderman, L., Kivirikko, K. I. & Prockop, D. J. Partial purification and characterization of a neutral protease which cleaves the N-terminal propeptides from procollagen. Biochemistry 17: 2948-2954.

1979

Leung, M. K. K., Fessler, L. I., Greenberg, D. B., & Fessler, J. H. Separate amino and carboxyl procollagen peptidases in chick embryo tendon. J. Biol. Chem. 254: 224-232.
Morris, N. P., Fessler, L. I. & Fessler, J. H. Procollagen propeptide release by procollagen peptidases and bacterial collagenase. J. Biol. Chem. 254: 11024-11032.
Nusgens, B. & Lapiere, C. M. A simplified procedure for measuring amino-procollagen peptidase type I. Anal. Biochem. 95: 406-412.

Entry 4.05

PROCOLLAGEN C-PEPTIDASE

Summary

EC Number: (3.4.24.-)

Earlier names: -

Distribution: Presumed to be widely distributed in connective tissues.

Source: Tendon, cultured tendon cells.

Action: Cleaves the disulfide cross-linked C-terminal extension peptides of procollagen in the maturation to tropocollagen.

Requirements: Neutral pH.

Substrate, usual: Procollagen (usually radiolabelled type I).

Inhibitors: EDTA, 1,10-phenanthroline.

Molecular properties: -

Bibliography

1976

Uitto, J. & Lichtenstein, J. R.   Removal of amino-terminal and carboxy-terminal extension peptides from procollagen during synthesis of chick embryo tendon collagen.   Biochem. Biophys. Res. Commun. 71: 60-67.

1977

Davidson, J. M., McEneany, L. S. G. & Bornstein, P.   Intermediates in the conversion of procollagen to collagen. Evidence for stepwise limited proteolysis of the COOH-terminal peptide extensions.   Eur. J. Biochem. 81: 349-355.

Duksin, D. & Bornstein, P.   Impaired conversion of procollagen to collagen by fibroblasts and bone treated with tunicamycin, an inhibitor of protein glycosylation.   J. Biol. Chem. 252: 955-962.

1978

Kessler, E. & Goldberg, B.   A method for assaying the activity of the endopeptidase which excises the nonhelical carboxyterminal extensions from type 1 procollagen.   Anal. Biochem. 86: 463-469.

1979

Leung, M. K. K., Fessler, L. I., Greenberg, D. B. & Fessler, J. H. Separate amino and carboxyl procollagen peptidases in chick embryo tendon. J. Biol. Chem. 254: 224-232.

Morris, N. P., Fessler, L. I. & Fessler, J. H. Procollagen propeptide release by procollagen peptidases and bacterial collagenase. J. Biol. Chem. 254: 11024-11032.

KIDNEY MICROVILLUS PROTEINASE

Summary

EC Number: 3.4.24.11

Earlier names: -

Distribution: Kidney of rabbit, rat and pig; also pig intestinal mucosa.

Source: Microvillus preparations from rabbit kidney cortex or rat kidney.

Action: Substrates include insulin B chain, glucagon and some other peptides, but not insulin, or proteins. The rabbit enzyme cleaves Z-Gly-Phe-NH$_2$ and furylacryloyl-Gly-Leu-NH$_2$, like the bacterial metallo-proteinase, thermolysin.

Requirements: pH optimum 6.0-6.5, depending on the buffer ion and ionic strength.

Substrate, usual: ^{125}I-labelled insulin.

Inhibitors: Strongly inhibited by EDTA and 1,10-phenanthroline; also phosphoramidon (K_i 2 x 10^{-6} M).

Molecular properties: Rabbit: mol.wt. 93,000, single chain glycoprotein with one atom of zinc/mole. Mol.wt. may be higher in the rat.

Comment: The properties reported for the rat and rabbit enzymes are similar in many respects, although the differences in mol.wt. and action on synthetic substrates point to the enzymes being distinct.

Bibliography

1974

Kerr, M. A. & Kenny, A. J. The molecular weight and properties of a neutral metallo-endopeptidase from rabbit kidney brush border. Biochem. J. 137: 489-495.

Kerr, M. A. & Kenny, A. J. The purification and specificity of a neutral endopeptidase from rabbit kidney brush border. Biochem. J. 137: 477-488.

1977

Kenny, A. J. Proteinases associated with cell membranes. In:
  Proteinases in Mammalian Cells and Tissues (Barrett, A. J. ed.),
  pp. 393-444, North-Holland Publishing Co., Amsterdam.
Varandani, P. T. & Shroyer, L. A. A rat kidney neutral peptidase
  that degrades B chain of insulin, glucagon and ACTH: purification
  by affinity chromatography and some properties. Arch. Biochem.
  Biophys. 181: 82-93.

1979

Duckworth, W. C., Stentz, F. B., Heinemann, M. & Kitabchi, A. E.
  Initial site of insulin cleavage by insulin protease. Proc.
  Natl. Acad. Sci. U.S.A. 76: 635-639.
Phelps, B. H., Varandani, P. T. & Stroyer, L. A. Insulin B chain-
  degrading neutral peptidase activity in the rat. Tissue
  distribution and the effects of starvation and streptozotocin-
  induced diabetes. Arch. Biochem. Biophys. 195: 1-11.

Entry 4.07

ACROLYSIN

Summary

EC Number: (3.4.24.-)

Earlier names: -

Distribution: Rabbit spermatozoan acrosome.

Source: Rabbit spermatozoa.

Action: Activates proacrosin. Cleaves the insulin A chain on the amino side of hydrophobic residues - thus resembling thermolysin (also an activator of proacrosin).

Requirements: -

Substrate, usual: Proacrosin.

Inhibitors: EDTA, 1,10-phenanthroline, phosphoramidon.

Molecular properties: -

Comment: Acrolysin has not yet been purified or characterized in any detail.

Bibliography

1976

McRorie, R. A., Turner, R. B., Bradford, M. M. & Williams, W. L. Acrolysin, the aminoproteinase catalyzing the initial conversion of proacrosin to acrosin in mammalian fertilization. Biochem. Biophys. Res. Commun. 71: 492-498.

Entry 4.08

MACROPHAGE ELASTASE

Summary

EC Number: (3.4.24.-)

Earlier names: -

Distribution: Culture medium of mouse peritoneal and alveolar
macrophages; human monocytes; monkey alveolar macrophages.

Source: Not purified.

Action: Solubilizes elastin, and degrades other proteins; has
little or no action on the usual synthetic substrates of
pancreatic elastase (#1.05) and leukocyte elastase (#1.34).

Requirements: Neutral pH. Ca^{2+} stabilizes and activates. SDS
enhances elastolytic activity.

Substrate, usual: Elastin.

Inhibitors: EDTA. Probably α_2-macroglobulin. Not Dip-F, peptidyl-
chloromethane inhibitors of pancreatic elastase, soybean trypsin
inhibitor or α_1-proteinase inhibitor.

Molecular properties: Mol.wt. about 25,000.

Comment: May be identical with the macrophage metallo-proteinase
(#4.09).
    Some of the uncertainty that has existed about the properties
of macrophage elastase may originate from the capacity of
macrophages to acquire proteinases from neutrophil leukocytes
(as well as mast cells) by phagocytosis [1].

References
Campbell et al. J. Clin. Invest. 64: 824-833, 1979.

Bibliography

1971

Janoff, A., Rosenberg, R. & Galdston, M. Elastase-like,
esteroprotease activity in human and rabbit alveolar macrophage
granules. Proc. Soc. Exp. Biol. Med. 136: 1054-1058.

1972

Rosenberg, R., Sandhaus, R. & Janoff, A. Further studies of human
alveolar macrophage alanine p-nitrophenyl esterase. Evidence of

resistance to endogenous antiproteases. Am. Rev. Respir. Dis.
106: 114-115.

1975

Werb, Z. & Gordon, S. Elastase secretion by stimulated
macrophages. Characterization and regulation. J. Exp. Med. 142:
361-377.

1976

Levine, E. A., Senior, R. M. & Butler, J. V. The elastase
activity of alveolar macrophages: measurements using synthetic
substrates and elastin. Am. Rev. Respir. Dis. 113: 25-30.

1977

Rifkin, D. B. & Crowe, R. M. A sensitive assay for elastase
employing radioactive elastin coupled to Sepharose. Anal.
Biochem. 79: 268-275.
Rodriguez, R. J., White, R. R., Senior, R. M. & Levine, E. A.
Elastase release from human alveolar macrophages: comparison
between smokers and non-smokers. Science 198: 313-314.
Talbot, M. W. & Strausser, H. R. Increase in serum IgE levels of
ovalbumin-sensitized cats and the detection of elastase and
collagenase activities in secretions of sensitized feline
alveolar macrophages challenged in vitro. Int. Arch. Allergy
Appl. Immunol. 54: 198-204.
White, R. R. & Kuhn, C. The effect of tetracycline on elastase
secretion by mouse peritoneal exudative and alveolar macrophages.
Am. Rev. Respir. Dis. 115: 387 only.
White, R., Lin, H.-S. & Kuhn, C., III. Elastase secretion by
peritoneal exudative and alveolar macrophages. J. Exp. Med.
146: 802-808.

1978

De Cremoux, H., Hornebeck, W., Jaurand, M.-C., Bignon, J. &
Robert, L. Partial characterisation of an elastase-like enzyme
secreted by human and monkey alveolar macrophages. J. Pathol.
125: 171-177.
Werb, Z. Biochemical actions of glucocorticoids on macrophages in
culture. Specific inhibition of elastase, collagenase, and
plasminogen activator secretion and effects on other metabolic
functions. J. Exp. Med. 147: 1695-1712.

1979

Banda, M. J. & Werb, Z. Purification and characterization of
macrophage elastase. Fed. Proc. Fed. Am. Soc. Exp. Biol. 38:
1339 only.

MACROPHAGE METALLO-PROTEINASE

Summary

EC Number: (3.4.24.-)

Earlier names: -

Distribution: Rabbit bone marrow macrophages in culture.

Source: As above.

Action: Degrades cartilage proteoglycan, Azocoll and casein.

Requirements: Optimal pH about 7.0.

Substrate, usual: ^{35}S-Labelled proteoglycan in cartilage pieces.

Inhibitors: EDTA (0.1 mM), 1,10-phenanthroline (1 mM), cysteine, serum. (Not Dip-F, soybean trypsin inhibitor, 4-hydroxymercuribenzoate or aprotinin.)

Molecular properties: Mol.wt. 17,000 (by gel chromatography). No latent form detected.

Comment: Not yet purified or characterized in detail, and may be identical with the macrophage elastase (#4.08).

Bibliography

1978

Cammer, W., Bloom, B. R., Norton, W. T. & Gordon, S. Degradation of basic protein in myelin by neutral proteases secreted by stimulated macrophages - possible mechanism of inflammatory demyelination. Proc. Natl. Acad. Sci. U.S.A. 75: 1554-1558.

Hauser, P. & Vaes, G. Degradation of cartilage proteolycans by a neutral proteinase secreted by rabbit bone-marrow macrophages in culture. Biochem. J. 172: 275-284.

Entry 4.10

TISSUE METALLO-PROTEINASE

Summary

EC Number: (3.4.24.-)

Earlier names: -

Distribution: Produced by connective tissues (bone, cartilage, etc.), in culture.

Source: Organ culture medium of bones of neonatal mouse, rabbit etc.

Action: Catalyzes limited degradation of cartilage proteoglycan and casein.

Requirements: Optimum pH 7-8. 1mM-CaCl$_2$.

Substrate, usual: Casein, proteoglycan.

Substrate, special: ^{3}H-Labelled casein, ^{35}S-labelled proteoglycan in polyacrylamide beads.

Inhibitors: EDTA, 1,10-phenanthroline, cysteine, dithiothreitol. α_2-Macroglobulin.

Molecular properties: Apparent mol.wt. 51,000 for latent form, decreasing to 30,000 on activation by aminophenylmercuric acetate or trypsin (these values being determined by gel chromatography in the presence of 1 M NaCl).

Comment: This is one of at least two metallo-proteinases produced by tissues in culture (and presumably in vivo); each occurs in a latent form with similar requirements for activation, but the latent enzymes are separable by gel chromatography.
    A separate enzyme has been obtained from bovine nasal cartilage [1]; this is activated by Co^{2+} and Mn^{2+}, and inhibited by EDTA, it has an apparent mol.wt. of 17,000 in gel chromatography, but passes through dialysis membranes.

References
[1] Nagase & Woessner Fed. Proc. Fed. Am. Soc. Exp. Biol. 36: 749 only, 1977.

Bibliography

1976

Sapolsky, A. I., Keiser, H., Howell, D. S. & Woessner, J. F.

Metalloproteases of human articular cartilage that digest
cartilage proteoglycan at neutral and acid pH. J. Clin. Invest.
58: 1030-1041.

1977

Nagase, H. & Woessner, J. F., Jr. A metalloprotease from bovine
nasal cartilage that digests cartilage proteoglycan. Fed. Proc.
Fed. Am. Soc. Exp. Biol. 36: 749 only.

1978

Sellers, A., Reynolds, J. J. & Meikle, M. C. Neutral metallo-
proteinases of rabbit bone. Separation in latent forms of
distinct enzymes that when activated degrade collagen, gelatin,
and proteoglycans. Biochem. J. 171: 493-496.
Vaes, G., Eeckhout, Y., Lenaers-Claeys, G., Francois-Gillet, C. &
Druetz, J.-E. The simultaneous release by bone explants in
culture and the parallel activation of procollagenase and of a
latent neutral proteinase that degrades cartilage proteoglycans
and denatured collagen. Biochem. J. 172: 261-274.

1979

Sellers, A., Murphy, G., Meikle, M. C. & Reynolds, J. J. Rabbit
bone collagenase inhibitor blocks the activity of other neutral
metalloproteases. Biochem. Biophys. Res. Commun. 87: 581-587.

Entry 4.11
TISSUE GELATINASE

Summary

EC Number: (3.4.24.-)

Earlier names: -

Distribution: Produced as latent enzyme by connective tissues (bone, cartilage, etc.) in culture.

Source: Organ culture media of suitable tissues.

Action: Degrades gelatin and Azocoll.

Requirements: Optimal pH 7-8. Ca^{2+}.

Substrate, usual: Radiolabelled gelatin, Azocoll.

Inhibitors: EDTA, 1,10-phenanthroline. α_2-Macroglobulin.

Molecular properties: The latent enzyme has mol.wt. 78,000 (by gel chromatography), but this decreases to 58,000 after activation by aminophenylmercuric acetate or trypsin.

Bibliography

1978

Sellers, A., Reynolds, J. J. & Meikle, M. C. Neutral metalloproteinases of rabbit bone. Separation in latent forms of distinct enzymes that when activated degrade collagen, gelatin, and proteoglycans. Biochem. J. 171: 493-496.

Vaes, G., Eeckhout, Y., Lenaers-Claeys, G., Francois-Gillet, C. & Druetz, J.-E. The simultaneous release by bone explants in culture and the parallel activation of procollagenase and of a latent neutral proteinase that degrades cartilage proteoglycans and denatured collagen. Biochem. J. 172: 261-274.

1979

Sellers, A., Murphy, G., Meikle, M. C. & Reynolds, J. J. Rabbit bone collagenase inhibitor blocks the activity of other neutral metalloproteases. Biochem. Biophys. Res. Commun. 87: 581-587.

Section 5

UNCLASSIFIED PROTEINASES

Entry 5.01

COLLAGENASE-LIKE PEPTIDASE

Summary

EC Number: 3.4.99.31

Earlier names: "Tissue endopeptidase degrading collagenase synthetic substrate", Pz-peptidase (because the enzyme is commonly assayed with a Pz-blocked peptide substrate - see below), collagen peptidase. Abbreviation: CL-peptidase.

Distribution: Activity is widely distributed in mammalian tissues.

Source: Monkey kidney, rat granuloma, bovine dental follicle, rat liver mitochondria.

Action: Cleaves Pz-Pro-Leu/Gly-Pro-D-Arg, and probably other peptides of 5-30 residues, especially those with collagen-like sequences. There is controversy as to whether denatured collagen and high mol.wt. fragments are also degraded (i.e. whether this is also a gelatinase).

Requirements: Optimal pH 8.0-8.5 (bovine dental follicle form A), 6.5-7.2 (form B) or 7.5-8.5 (chick). Stabilized by dithiothreitol (kidney). Often activated by Ca^{2+}, Sr^{2+} or Mg^{2+}.

Substrate, usual: Pz-Pro-Leu-Gly-Pro-D-Arg.

Inhibitors: These vary with the source of the enzyme: EDTA (some forms are then reactivated by Ca^{2+}, Zn^{2+} or Mn^{2+} [1]), 4-chloromercuribenzoate, heavy metal cations. Cysteine (spermatozoa). Pms-F, 1,10-phenanthroline and Zn^{2+} (rabbit serum, [2]). (Not inhibited by serum.)

Molecular properties: Probably several enzymes: mol.wt. 220,000 and 20,000 (bovine dental follicle), 95,000 (rat liver mitochondria) or 56,000 (monkey kidney). Rabbit serum: 124,000 mol.wt., pI 5.1 [2]. Single polypeptide chains.

Comment: These enzymes are "collagenase-like" in that their synthetic substrates were originally designed for clostridial collagenase, but the name is not really satisfactory.

References
[1] Lukac & Koren J. Reprod. Fertil. 49: 95-99, 1977.
[2] Nagelschmidt et al. Biochim. Biophys. Acta 571: 105-111, 1979.

Bibliography

1963

Wünsch, E. & Heidrich, H. G. Zur quantitativen Bestimmung der Kollagenase. Hoppe-Seyler's Z. Physiol. Chem. 333: 149-151.

1968

Strauch, L., Vencelj, H. & Hannig, K. Kollagenase in Zellen höher entwickelter Tiere. Hoppe-Seyler's Z. Physiol. Chem. 349: 171-178.

1969

Heidrich, H. G., Prokopova, D. & Hannig, K. The use of synthetic substrates for the determination of mammalian collagenases. Is collagenolytic activity present in mitochondria? Hoppe-Seyler's Z. Physiol. Chem. 350: 1430-1436.

1971

Wünsch, E., Schönsteiner-Altmann, G. & Jaeger, E. [Chromophore substrates. III. An improved method for the preparation of the coloured collagenase substrate 4-phenylazobenzyloxycarbonyl-L-prolyl-L-leucyl-glycyl-L-prolyl-D-arginine.] Hoppe-Seyler's Z. Physiol. Chem. 352: 1424-1428.

1972

Aswanikumar, S. & Radhakrishnan, A. N. Purification and properties of a peptidase acting on a synthetic collagenase substrate from experimental granuloma tissue in the rat. Biochim. Biophys. Acta 276: 241-250.

Harris, E. D., Jr. & Krane, S. M. An endopeptidase from rheumatoid synovial tissue culture. Biochim. Biophys. Acta 258: 566-576.

1973

Heidrich, H.-G., Kronschnabl, O. & Hannig, K. Eine Endopeptidase aus der Matrix von Rattenlebermitochondrien. Isolierung, Reinigung und Charakterisierung des Enzymes. Hoppe-Seyler's Z. Physiol. Chem. 354: 1399-1404.

Koren, E. & Milković, S. 'Collagenase-like' peptidase in human rat and bull spermatozoa. J. Reprod. Fertil. 32: 349-356.

1974

Koren, E., Lukač, J. & Milković, S. The effect of collagenase-like peptidase from rat testis and clostridial collagenase A on the rat seminal vesicle secretion and its coagulation. J. Reprod. Fertil. 36: 161-167.

1975

Aswanikumar, S. & Radhakrishnan, A. N. Purification and properties of a peptidase acting on a synthetic substrate for collagenase from monkey kidney. Biochim. Biophys. Acta 384: 194-202.

Koren, E., Schön, E. & Lukač, J. The coagulation of insoluble and basic proteins from rat seminal vesicle secretion with vesiculase: influence of collagenase–like peptidase from rat testis. J. Reprod. Fertil. 42: 491–495.

1976

Hino, M. & Nagatsu, T. Separation of two Pz–peptidases from bovine dental follicle. Biochim. Biophys. Acta 429: 555–563.

1977

Ito, A., Naganeo, K., Mori, Y., Hirakawa, S. & Hayashi, M. PZ–peptidase activity in human uterine cervix in pregnancy at term. Clin. Chim. Acta 78: 267–270.

Lukač, J. & Koren, E. The metalloenzymic nature of collagenase–like peptidase of the rat testis. J. Reprod. Fertil. 49: 95–99.

Morales, T. I. & Woessner, J. F., Jr. PZ–peptidase from chick embryos. Purification, properties, and action on collagen peptides. J. Biol. Chem. 252: 4855–4860.

1979

Kojima, K., Kinoshita, H., Kato, T., Nagatsu, T., Takada, K. & Sakakibara, S. A new and highly sensitive fluorescence assay for collagenase–like peptidase activity. Anal. Biochem. 100: 43–50.

Nagelschmidt, M., Unger, T. & Struck, H. Purification and properties of a collagen peptidase (PZ–peptidase) from rabbit serum. Biochim. Biophys. Acta 571: 105–111.

LENS NEUTRAL PROTEINASE

Summary

EC Number: 3.4.24.5 (but see Comment).

Earlier names: -

Distribution: Lens of the eye in cattle, man, sheep, pig, rabbit and rat.

Source: Bovine and human lens.

Action: Digests α-crystallin; no general proteolytic activity, or known synthetic substrates.

Requirements: Optimal pH 7.5. Assayed at 55°C; stimulated by Ca^{2+} (10mM) and Mg^{2+}.

Substrate, usual: α-Crystallin.

Inhibitors: EDTA, 4-chloromercuribenzoate, Zn^{2+}, MalNEt. (Scarcely inhibited by 1,10-phenanthroline, dithiothreitol or Dip-F.)

Molecular properties: Apparently of very high molecular weight.

Comment: Resembles calcium-dependent cysteine proteinases (#2.07 and #2.08) in some properties, and is therefore difficult to allocate to a catalytic group, at present.

Bibliography

1962

Van Heyningen, R. & Waley, S. G.   Search for a neutral proteinase in bovine lens.   Exp. Eye Res. 1: 336-342.

Waley, S. G. & van Heyningen, R.   Neutral proteinases in the lens. Biochem. J. 83: 274-283.

Waley, S. G. & van Heyningen, R.   Purification and properties of a neutral proteinase in the lens.   Exp. Eye Res. 1: 343-349.

1963

Van Heyningen, R. & Waley, S. G.   Neutral proteinases in the lens. 2. Partial purification and properties.   Biochem. J. 86: 92-101.

1975

Blow, A. M. J., van Heyningen, R. & Barrett, A. J.   Metal-dependent proteinase of the lens. Assay, purification and properties of the bovine enzyme. Biochem. J. 145: 591–599.

1976

Trayhurn, P. & van Heyningen, R. Neutral proteinase activity in the human lens. Exp. Eye Res. 22: 251–257.

1977

Blow, A. M. J. Proteolysis in the lens. In: Proteinases in Mammalian Cells and Tissues (Barrett, A. J. ed.), pp. 501–514, North-Holland Publishing Co., Amsterdam.

Entry 5.03

CATHEPSIN F

Summary

EC Number: (3.4.99.-)

Earlier names: –

Distribution: Cartilage (rabbit ear, human articular).

Source: Rabbit ear cartilage.

Action: Degrades cartilage proteoglycan. No known synthetic substrates.

Requirements: pH optimum 4.5.

Substrate, usual: ^{35}S-Labelled cartilage proteoglycan.

Inhibitors: α_2-Macroglobulin, whole egg white. Resists inhibitors of major classes of proteinase (Dip-F, Pms-F, pepstatin, leupeptin, EDTA, iodoacetate, etc.).

Molecular properties: Mol.wt. 50,000–70,000.

Comment: Instability of the partially purified enzyme prevented further characterization.

Bibliography

1975

Blow, A. M. J. The detection and characterisation of cathepsin F, a cartilage enzyme that degrades proteoglycan. Ital. J. Biochem. 24: 13–14.

1977

Dingle, J. T., Barrett, A. J., Blow, A. M. J. & Martin, P. E. N. Proteoglycan-degrading enzymes. A radiochemical assay method, and the detection of a new enzyme, cathepsin F. Biochem. J. 167: 775–785.

Entry 5.04

TONIN

Summary

EC Number: (3.4.99.-)

Earlier names: -

Distribution: Rat submaxillary gland.

Source: Rat submaxillary gland.

Action: Cleaves a -Phe-His- bond in the synthetic tetradecapeptide renin substrate, to yield angiotensin II directly.

Requirements: pH optimum 6.0-7.0 (no effect of Cl⁻).

Substrate, usual: Tetradecapeptide renin substrate.

Inhibitors: Serum. (Not EDTA, Dip-F or PCA-Glu-Lys-Trp-Ala-Pro.)

Molecular properties: Mol.wt. 31,400 by gel chromatography.

Comment: Little is known of this enzyme: possibly a cysteine proteinase?

Bibliography

1974

Boucher, R., Asselin, J. & Genest, J. A new enzyme leading to the direct formation of angiotensin II. Circ. Res. suppl. I: 203-209.

Entry 5.05

BRAIN NEUTRAL PROTEINASE

Summary

EC Number: (3.4.99.-)

Earlier names: (See Comment.) Brain neutral proteinases have been
called kininase A [1], endopeptidase A [2], kininase B [1],
endopeptidase B [2] and cathepsin M [3], among other names. Some
of these names may yet prove entirely appropriate when the
enzymes have been more fully studied.

Distribution: Brain.

Source: Proteinases have been purified from brain of rat, rabbit,
cattle and other species.

Action: Casein, hemoglobin, myelin basic protein and other
proteins are degraded, but much work has been done on the
degradation of biologically active peptides, including
bradykinin, angiotensin, oxytocin, substance P, etc. and there
is evidince that some of the brain proteinases are "endo-
oligopeptidases" with specificity for these lower mol.wt.
substrates [2].

Requirements: Neutral pH. The "kininases" have pH optima of 7.5,
and require thiol activators. Other activities require Ca^{2+},
too.

Substrate, usual: Casein and bradykinin have been much used.

Substrate, special: Radiolabelled oligopeptides.

Inhibitors: Obviously these vary greatly between enzymes: EDTA,
thiol-blocking reagents and Dip-F have each been found effective
against some activities.

Molecular properties: Again, one cannot generalize, and the data
are sketchy. Kininase A has mol.wt. 71,000, pI 5.2, and kininase
B mol.wt. 68,000 and pI 4.9 [1]. Cathepsin M is excluded from
Sephadex G-200.

Comment: This entry is exceptional in that it certainly covers
more than one enzyme, but none of the individual enzymes has
been purified and characterized sufficiently thoroughly to merit
individual description, in our view. Neutral proteinase
activities detected in brain probably include those of prolyl
endopeptidase (#1.55) (similar to kininase B), and calcium
dependent cysteine proteinase (#2.08).

References
[1] Oliviera et al. Biochemistry 15: 1967-1974, 1976.
[2] Camargo et al. J. Biol. Chem. 254: 5304-5307, 1979.
[3] Marks In: Intracellular Protein Catabolism II (Turk & Marks, eds), pp. 85-102, Plenum Press, New York, 1977.

Bibliography

1963

Marks, N. & Lajtha, A. Protein breakdown in brain. Subcellular distribution and properties of neutral and acid proteinases. Biochem. J. 89: 438-447.

1964

Guroff, H. A neutral calcium-activated proteinase from the soluble fraction of rat brain. J. Biol. Chem. 239: 149-155.

1965

Marks, N. & Lajtha, A. Separation of acid and neutral proteinases of brain. Biochem. J. 97: 74-83.

1968

Riekkinen, P. J. & Rinne, U. K. A new neutral proteinase from the rat brain. Brain Res. 9: 126-135.

1970

Oja, S. S. & Oja, H. Cerebral proteinases in the growing rat. J. Neurochem. 17: 901-912.
Riekkinen, P. J., Clausen, J. & Arstila, A. V. Further studies on neutral proteinase activity of CNS myelin. Brain Res. 19: 213-220.

1971

Marks, N. & Lajtha, A. Protein and polypeptide breakdown. In: Handbook of Neurochemistry (Lajtha, A. ed.), vol. 5A, pp. 49-123, Plenum Press, New York.

1972

Camargo, A. C. M., Ramalho-Pinto, F. J. & Greene, L. J. Brain peptidases: conversion and inactivation of kinin hormones. J. Neurochem. 19: 37-49.

1973

Bosmann, H. B. Protein catabolism. III. Proteolytic enzymes of guinea pig cerebral cortex and synapotosomal localization. Int. J. Pept. Protein Res. 5: 135-147.
Camargo, A. C. M., Shapanka, R. & Greene, L. J. Preparation, assay and partial characterization of a neutral endopeptidase from rabbit brain. Biochemistry 12: 1838-1844.

1976

Marks, N. Biodegradation of hormonally active peptides in the central nervous system. In: Subcellular Mechanisms in Reproductive Neuroendocrinology (Naftolin, F., Ryan, K. J. & Davies, J. eds), pp. 129-147, Elsevier Scientific Publishing Co., Amsterdam.

Oliveira, E. B., Martins, A. R. & Camargo, A. C. M. Isolation of brain endopeptidases: influence of size and sequence of substrates structurally related to bradykinin. Biochemistry 15: 1967-1974.

1977

Marks, N. Conversion and inactivation of neuropeptides. In: Peptides in Neurobiology (Gainer, H. ed.), pp. 221-258, Plenum Press, New York.

Marks, N. Specificity of breakdown based on the inactivation of active proteins and peptides by brain proteolytic enzymes. In: Intracellular Protein Catabolism II (Turk, V. & Marks, N. eds), pp. 85-102, Plenum Press, New York.

1979

Camargo, A. C. M., Caldo, H. & Reis, M. L. Susceptibility of a peptide derived from bradykinin to hydrolysis by brain endo-oligopeptidases and pancreatic proteinases. J. Biol. Chem. 254: 5304-5307.

Orlowski, M., Wilk, E., Pearce, S. & Wilk, S. Purification and properties of a prolyl endopeptidas from rabbit brain. J. Neurochem. 33: 461-469.

Wilk, S. & Orlowski, M. Degradation of bradykinin by isolated neutral endopeptidases of brain and pituitary. Biochem. Biophys. Res. Commun. 90: 1-6.

Entry 5.06
LEUCOKININOGENASE

Summary

EC Number: (3.4.99.-)

Earlier names: Leucokininase.

Distribution: Rabbit polymorphonuclear leukocyte cell membrane.

Source: As above.

Action: Releases tuftsin, Thr-Lys-Pro-Arg, from the C-terminus of a fragment of the H chain of IgG (which is generated by a spleen enzyme).

Requirements: Acts at neutral pH.

Substrate, usual: Leucokininogen.

Inhibitors: -

Molecular properties: -

Comment: There has been very little characterization of this enzyme.

Bibliography

1970

Najjar, V. A. & Nishioka, K. "Tuftsin": a natural phagocytosis stimulating peptide. Nature 228: 672-673.

1972

Nishioka, K., Constantopoulos, A., Satoh, P. S. & Najjar, V. A. The characteristics, isolation and synthesis of the phagocytosis stimulating peptide tuftsin. Biochem. Biophys. Res. Commun. 47: 172-179.

1974

Najjar, V. A. The physiological role of γ-globulin. Adv. Enzymol. 41: 129-178.

INDEX

The preferred names of the enzymes are given in slightly bolder type than the alternative names.